Microbiomes of the Built Environment

A RESEARCH AGENDA FOR INDOOR MICROBIOLOGY, HUMAN HEALTH, AND BUILDINGS

Committee on Microbiomes of the Built Environment: From Research to Application

Board on Life Sciences

Board on Environmental Studies and Toxicology

Board on Infrastructure and the Constructed Environment

Division on Earth and Life Studies

Health and Medicine Division

Division on Engineering and Physical Sciences

National Academy of Engineering

A Consensus Study Report of

The National Academies of
SCIENCES · ENGINEERING · MEDICINE

THE NATIONAL ACADEMIES PRESS
Washington, DC
www.nap.edu

THE NATIONAL ACADEMIES PRESS 500 Fifth Street, NW Washington, DC 20001

This activity was supported by Grant No. 2014-13628 from the Alfred P. Sloan Foundation, a grant from the Gordon and Betty Moore Foundation, Grant No. NNX16AC85G from the National Aeronautics and Space Administration, Contract No. HHSN263201200074I with the National Institutes of Health, Contract No. EP-C-14-005/0007 with the U.S. Environmental Protection Agency, and with additional support from the National Academy of Sciences Cecil and Ida Green Fund. Any opinions, findings, conclusions, or recommendations expressed in this publication do not necessarily reflect the views of any organization or agency that provided support for the project.

International Standard Book Number-13: 978-0-309-44980-9
International Standard Book Number-10: 0-309-44980-4
Digital Object Identifier: https://doi.org/10.17226/23647
Library of Congress Control Number: 2017952589

Additional copies of this publication are available for sale from the National Academies Press, 500 Fifth Street, NW, Keck 360, Washington, DC 20001; (800) 624-6242 or (202) 334-3313; http://www.nap.edu.

Copyright 2017 by the National Academy of Sciences. All rights reserved.

Printed in the United States of America

Suggested citation: National Academies of Sciences, Engineering, and Medicine. 2017. *Microbiomes of the Built Environment: A Research Agenda for Indoor Microbiology, Human Health, and Buildings*. Washington, DC: The National Academies Press. doi: https://doi.org/10.17226/23647.

The National Academies of
SCIENCES • ENGINEERING • MEDICINE

The **National Academy of Sciences** was established in 1863 by an Act of Congress, signed by President Lincoln, as a private, nongovernmental institution to advise the nation on issues related to science and technology. Members are elected by their peers for outstanding contributions to research. Dr. Marcia McNutt is president.

The **National Academy of Engineering** was established in 1964 under the charter of the National Academy of Sciences to bring the practices of engineering to advising the nation. Members are elected by their peers for extraordinary contributions to engineering. Dr. C. D. Mote, Jr., is president.

The **National Academy of Medicine** (formerly the Institute of Medicine) was established in 1970 under the charter of the National Academy of Sciences to advise the nation on medical and health issues. Members are elected by their peers for distinguished contributions to medicine and health. Dr. Victor J. Dzau is president.

The three Academies work together as the **National Academies of Sciences, Engineering, and Medicine** to provide independent, objective analysis and advice to the nation and conduct other activities to solve complex problems and inform public policy decisions. The National Academies also encourage education and research, recognize outstanding contributions to knowledge, and increase public understanding in matters of science, engineering, and medicine.

Learn more about the National Academies of Sciences, Engineering, and Medicine at **www.nationalacademies.org**.

The National Academies of
SCIENCES • ENGINEERING • MEDICINE

Consensus Study Reports published by the National Academies of Sciences, Engineering, and Medicine document the evidence-based consensus on the study's statement of task by an authoring committee of experts. Reports typically include findings, conclusions, and recommendations based on information gathered by the committee and the committee's deliberations. Each report has been subjected to a rigorous and independent peer-review process and it represents the position of the National Academies on the statement of task.

Proceedings published by the National Academies of Sciences, Engineering, and Medicine chronicle the presentations and discussions at a workshop, symposium, or other event convened by the National Academies. The statements and opinions contained in proceedings are those of the participants and are not endorsed by other participants, the planning committee, or the National Academies.

For information about other products and activities of the National Academies, please visit www.nationalacademies.org/about/whatwedo.

COMMITTEE ON MICROBIOMES OF THE BUILT ENVIRONMENT: FROM RESEARCH TO APPLICATION

Committee Members

JOAN WENNSTROM BENNETT (*Chair*), Rutgers University
JONATHAN ALLEN, Lawrence Livermore National Laboratory
JEAN COX-GANSER, National Institute for Occupational Safety and Health
JACK GILBERT, University of Chicago
DIANE GOLD, Brigham and Women's Hospital, Harvard Medical School, and Harvard T.H. Chan School of Public Health
JESSICA GREEN, University of Oregon
CHARLES HAAS, Drexel University
MARK HERNANDEZ, University of Colorado Boulder
ROBERT HOLT, University of Florida
RONALD LATANISION, Exponent, Inc.
HAL LEVIN, Building Ecology Research Group
VIVIAN LOFTNESS, Carnegie Mellon University
KAREN NELSON, J. Craig Venter Institute
JORDAN PECCIA, Yale University
ANDREW PERSILY, National Institute of Standards and Technology
JIZHONG ZHOU, University of Oklahoma

Project Staff

KATHERINE BOWMAN, Study Director and Senior Program Officer, Board on Life Sciences
ELIZABETH BOYLE, Program Officer, Board on Environmental Studies and Toxicology
DAVID A. BUTLER, Scholar, Health and Medicine Division
ANDREA HODGSON, Postdoctoral Fellow, Board on Life Sciences
JENNA OGILVIE, Research Associate, Board on Life Sciences
CAMERON OSKVIG, Director, Board on Infrastructure and the Constructed Environment
PROCTOR REID, Director, National Academy of Engineering Program Office
FRANCES SHARPLES, Director, Board on Life Sciences

Consultants

RONA BRIERE, Editor
HELAINE RESNICK, Editor

Acknowledgments

This Consensus Study Report was reviewed in draft form by individuals chosen for their diverse perspectives and technical expertise. The purpose of this independent review is to provide candid and critical comments that will assist the National Academies of Sciences, Engineering, and Medicine in making each published report as sound as possible and to ensure that it meets the institutional standards for quality, objectivity, evidence, and responsiveness to the study charge. The review comments and draft manuscript remain confidential to protect the integrity of the deliberative process.

We thank the following individuals for their review of this report:

William P. Bahnfleth, The Pennsylvania State University
Rita R. Colwell, University of Maryland, College Park
Richard Corsi, The University of Texas at Austin
Pieter C. Dorrestein, University of California, San Diego
Peter B. Hutt, Covington & Burling LLP
Susan Lynch, University of California, San Francisco
Janet Macher, California Department of Public Health (retired)
Mihai Pop, University of Maryland, College Park
Joan B. Rose, Michigan State University
Sarah Slaughter, Built Environment Coalition
Martin Täubel, National Institute for Health and Welfare, Finland
Mary E. Wilson, Harvard T.H. Chan School of Public Health

Although the reviewers listed above provided many constructive comments and suggestions, they were not asked to endorse the conclusions or recommendations of this report, nor did they see the final draft before its release. The review of this report was overseen by **Michael R. Ladisch**, Purdue University, and **William W. Nazaroff**, University of California, Berkeley. They were responsible for making certain that an independent examination of this report was carried out in accordance with the standards of the National Academies and that all review comments were carefully considered. Responsibility for the final content rests entirely with the authoring committee and the National Academies.

Preface

Ours is a microbial world. Although we cannot see microbes with the naked eye, we all live with microbial consortia. The microbes that are indigenous to our bodies are an essential component of our biology. Moreover, the indoor environments in which we live also harbor a complicated constellation of microbial types. The levels of microbial diversity, and the sheer numbers of organisms, are incongruous with our visual experience, but current microbiome research is changing the way we look not only at ourselves but also at the built environments we have created. DNA sequencing technologies provide a new view of the ubiquity and diversity of microbes in our lives. In looking back on centuries of human experience with buildings, we can see that people have developed many systems that support human comfort and convenience. The vision articulated in this report is that microbiome research can guide improvements to future buildings to enhance human healthfulness.

Do we know enough to rationally manage the microbial communities around us in built environments? The answer is "no." However, there are provocative hints that in the future, coherent management of the indoor microbiome can help prevent the spread of disease and contribute to human longevity, health, and well-being.

To produce this Consensus Study Report, the National Academies of Sciences, Engineering, and Medicine brought together a group of experts to discuss the microbial communities inside our built environments and their potential effects on human health. The committee sought to understand indoor microbiome research, a discipline that is dedicated to studying buildings, the microbial communities found inside of buildings, and the complex interactions that impact human health and well-being. Of necessity, this report touches on a number of extremely dissimilar areas of research and therefore required a committee with diverse expertise. I am grateful to the informed and insightful group of professionals who so generously shared their time and knowledge during the process of writing this report. Their collective expertise was reflective of the range of subject matter covered during our deliberations. The report also was informed by a number of excellent speakers and other participants who came to our open sessions. We thank all of these contributors for sharing their perspectives and research with us. Their contributions were invaluable in further developing our ideas and filling gaps in our expertise. In addition, we thank the report reviewers who provided insightful and instrumental feedback.

On behalf of the committee, I extend our greatest appreciation to the staff of the National Academies who worked with us throughout the process of creating this report. Without their time and guidance, this report would not have been possible. Finally, we thank the sponsors of the study for their financial support and for their astute vision of what this report could accomplish.

Joan Wennstrom Bennett, *Chair*
Committee on Microbiomes of the Built Environment:
From Research to Application

Contents

ACRONYMS AND ABBREVIATIONS — xvii

SUMMARY — 1

1 INTRODUCTION — 15
Study Charge, 16
Study Focus and Scope, 16
Emerging Tools That Facilitate Analysis, 19
Studying the Intersection of Microbial Communities,
    Built Environments, and Human Occupants, 19
Prior Efforts on Which This Report Builds, 26
Organization of the Report, 28
References, 29

2 MICROORGANISMS IN BUILT ENVIRONMENTS:
IMPACTS ON HUMAN HEALTH — 31
Influence of Building Microbiomes on Human Health:
    Ecologic and Biologic Plausibility, 32
Transmission of Infection in Indoor Environments, 37
Damp Indoor Environments, Indoor Microbial Exposures,
    and Respiratory or Allergic Disease Outcomes, 43
Nonairway and Nonallergy Effects, 50
Beneficial Effects of Microbes, 56
Summary Observations and Knowledge Gaps, 61
References, 63

3 THE BUILT ENVIRONMENT AND MICROBIAL COMMUNITIES   91
Introduction to Microbial Reservoirs in Commercial and Residential Buildings, 92
The Diversity of Buildings and Its Impact on Their Microbiomes, 95
Indoor Air Sources and Reservoirs of Microbes, 100
Indoor Water Sources and Reservoirs of Microbes, 108
Building Surfaces and Reservoirs of Microbes, 117
Impacts of Microbes on Degradation of Building Materials and Energy Usage, 128
Building Codes and Standards That May Affect the Microbiome, 129
The Influence of Climate and Climate Change on the Built Environment and Microbial Communities, 131
Summary Observations and Knowledge Gaps, 134
References, 138

4 TOOLS FOR CHARACTERIZING MICROBIOME–BUILT ENVIRONMENT INTERACTIONS   147
The Built Environment as a Complex Experimental Environment, 149
Characterizing Buildings, 150
Characterizing Indoor Microbial Communities, 157
Linking Analysis of Microbial Communities to Building Characteristics and Human Health Impacts, 166
Needs for Future Progress, 172
Moving from Research Toward Practice, 176
Summary Observations and Knowledge Gaps, 179
References, 181

5 INTERVENTIONS IN THE BUILT ENVIRONMENT   189
Physical Interventions to Reduce Exposure to Hazardous Microbes, 190
Chemical Interventions to Reduce Exposure to Hazardous Microbes, 200
Interventions to Encourage Exposure to Beneficial Microbes, 202
A Framework for Assessing Built Environment Interventions, 205
Summary Observations and Knowledge Gaps, 211
References, 213

6 MOVING FORWARD: A VISION FOR THE FUTURE AND RESEARCH AGENDA 219
A Vision for the Future of the Field: Microbiome-Informed Built Environments, 219
A Research Agenda for Achieving the Vision, 223
References, 235

APPENDIXES

A	An Assessment of Molecular Characterization Tools	237
B	Study Methods	277
C	Committee Member Biographies	283
D	Glossary	291

Boxes, Figures, and Tables

BOXES

S-1 Key Terms Used in This Report, 3
S-2 Knowledge Gaps Identified in This Report, 9
S-3 A Research Agenda for Moving to Practical Application, 12

1-1 Statement of Task, 17
1-2 Modeling Built Environment Ecology, 24

2-1 Transmission of Infectious and Noninfectious Organisms in Hospital Neonates, 35
2-2 Flint, Michigan: An Exemplar of the Role of the Municipal Water Supply in Triggering a Legionnaires' Disease Outbreak, 42
2-3 Asthma, Early-Life Exposures, and Farm-Type Environments, 57

3-1 Concealed Spaces, 93
3-2 Challenges for the Management of Microbes: The International Space Station, 99

4-1 Design and Use of Longitudinal Building Studies to Understand Microbiome–Built Environment–Human Interactions, 152

FIGURES

1-1   The complex interactions among human occupants, built environments, and associated microbial communities, 20
1-2   Transport and life cycle of indoor microbes, 22

2-1   Modes of transmission of microorganisms from the airborne environment, 40

3-1   The influence of water chemistry and flow on the microbiome of bulk water pipes, 111

5-1   Physical processes govern the assembly of indoor microbial communities, 207

TABLES

2-1   Mode of Transmission for Selected Pathogens Implicated in Infections Due to Inhalation or Fomite Interactions, 39
2-2   Associations Between Health Outcomes and Exposure to Damp Indoor Environments, 44
Annex Table 2-1: Selected Studies on Building/Home-Based Exposure Reduction and Asthma Outcomes in Children (2000–2017), 79
Annex Table 2-2: Beneficial Associations of Indoor Microbiota with Asthma or Allergy Outcomes in Selected Studies Using Metagenomics, Molecular Biologic, or Culture Methods to Measure Indoor Environmental Microbiota, 86

3-1   Buildings and Surfaces Where Viruses Have Been Detected or Survived, 119
3-2   Home High-Touch Surfaces and Bacterial Reservoirs, 121
3-3   Hospital High-Touch Surfaces and Bacterial Reservoirs, 121
3-4   Environmental, Location, and Surface Parameters That May Influence Microbial Populations and Communities, 122
3-5   Sustainable, Green, and Healthy Codes, Standards, Guidelines, and Certifications That Address Microbiome-Related Issues, 131
3-6   WELL Building Standard Features That Address Microbiome-Related Issues, 132

4-1   Selected Building Simulation Tools, 156

A-1   Overview of Molecular Characterization Tools, 240

Acronyms and Abbreviations

ACGIH	American Conference of Governmental Industrial Hygienists
AHAM	Association of Home Appliance Manufacturers
AIHA	American Industrial Hygiene Association
ANSI	American National Standards Institute
ASHRAE	American Society of Heating, Refrigerating and Air-Conditioning Engineers
ASM	American Society for Microbiology
ASTM	American Society for Testing and Materials
a_w	water activity
BASE	Building Assessment Survey and Evaluation
BOMA	Building Owners and Managers Association
CADR	clean air delivery rate
CBECS	Commercial Buildings Energy Consumption Survey
CDC	U.S. Centers for Disease Control and Prevention
CHAMPS	Combined Heat, Air, Moisture, and Pollutant Simulation
CHILD	Canadian Healthy Infant Longitudinal Development
CHW	community health worker
CMPBS	Center for Maximum Potential Building Systems
CNS	central nervous system
DNA	deoxyribonucleic acid
DOAS	dedicated outdoor air system

DOD	U.S. Department of Defense
DOE	U.S. Department of Energy
EBI	European Bioinformatics Institute
ECHO	Environmental Influences on Child Health Outcomes
ECRHS	European Community Respiratory Health Survey
EPA	U.S. Environmental Protection Agency
eQUEST	Quick Energy Simulation Tool
ERH	equilibrium relative humidity
ETS	environmental tobacco smoke
FIFRA	Federal Insecticide, Fungicide, and Rodenticide Act
GSA	General Services Administration
HCWH	Health Care Without Harm
HEPA	high-efficiency particulate air (filter)
HHS	U.S. Department of Health and Human Services
HUD	U.S. Department of Housing and Urban Development
HVAC	heating, ventilation, and air conditioning
IAPMO	International Association of Plumbing and Mechanical Officials
IAQA	Indoor Air Quality Association
IBPSA	International Building Performance Simulation Association
ICAS	Inner City Asthma Study
ICC	International Code Council
ICS	inhaled corticosteroid
ICU	intensive care unit
IDA ICE	IDA Indoor Climate and Energy simulation tool
IES	Illuminating Engineering Society; also Integrated Environmental Solutions
IgCC	International Green Construction Code
IICRC	Institute for Inspection, Cleaning and Restoration Certification
ILFI	International Living Future Institute
IPM	integrated pest management
IR	infrared
ISAAC	International Study of Asthma and Allergies in Childhood
ISIAQ	International Society of Indoor Air Quality and Climate
ISS	International Space Station
ITS	internally transcribed spacer
IV	intravenous

ACRONYMS AND ABBREVIATIONS

IWBI	International WELL Building Institute
LEED	Leadership in Energy and Environmental Design
LPS	lipopolysaccharide
MBARC-26	Mock Bacteria and ARchaea Community
MCAN	Merck Childhood Asthma Network
MERCCURI	Microbial Ecology Research Combining Citizen and University Researchers on ISS
MERS	Middle East respiratory syndrome
MERV	Minimum Efficiency Reporting Value
MIxS-BE	Minimum Information about any (X) Sequence-extension for the Built Environment
MoBE	Microbiomes of the Built Environment
mRNA	messenger ribonucleic acid
MRSA	methicillin-resistant *Staphylococcus aureus*
MSSA	methicillin-sensitive *Staphylococcus aureus*
MVOC	microbial volatile organic compound
NASA	National Aeronautics and Space Administration
NCBI	U.S. National Center for Biotechnology Information
NCS	National Children's Study
NGS	next-generation sequencing
NHANES	National Health and Nutrition Examination Survey
NHAPS	National Human Activity Pattern Survey
NHLBI	National Heart, Lung, and Blood Institute
NIAID	National Institute of Allergy and Infectious Diseases
NICU	neonatal intensive care unit
NIEHS	National Institute of Environmental Health Sciences
NIH	National Institutes of Health
NIOSH	National Institute for Occupational Safety and Health
NIST	National Institute of Standards and Technology
NSF	National Science Foundation
OFEE	Office of the Federal Environmental Executive
OSHA	Occupational Safety and Health Administration
OTU	operational taxonomic unit
PAMP	pathogen-associated molecular pattern
PCR	polymerase chain reaction
PM	particulate matter
QOL	quality of life

qPCR	quantitative PCR
RCT	randomized controlled trial
RECS	Residential Energy Consumption Survey
RH	relative humidity
RNA	ribonucleic acid
rRNA	ribosomal ribonucleic acid
SARS	severe acute respiratory syndrome
SSCP	single-strand conformation polymorphism analysis
SVOC	semivolatile organic compound
TAB	testing, adjusting, and balancing
TB	tuberculosis
tRFLP	terminal restriction fragment length polymorphism
TRNSYS	Transient System Simulation Tool
USGBC	U.S. Green Building Council
UV	ultraviolet
UVGI	ultraviolet germicidal irradiation
VA	U.S. Department of Veterans Affairs
VOC	volatile organic compound
VRE	vancomycin-resistant enterococci
WHO	World Health Organization
WUFI	Wärme Und Feuchte Instationär

Summary[1]

People's desire to understand the environments in which they live is a natural one. People spend most of their time in spaces and structures designed, built, and managed by humans, and it is estimated that people in developed countries now spend 90 percent of their lives indoors. As people move from homes to workplaces, traveling in cars and on transit systems, microorganisms are continually with and around them. These microorganisms reside outdoors in soil and water and coexist indoors where people live and work. They are found in and on pets, plants, and rodents; in water; in dirt tracked indoors on shoes; and in the air that enters buildings. Microorganisms also live on human skin and in systems such as the digestive tract, and the human-associated microbes that are shed, along with the human behaviors that affect their transport and removal, make significant contributions to the diversity of the indoor microbiome. What microorganisms are people exposed to in these indoor settings? What factors control their abundance, diversity, persistence, and other community characteristics? What effects could these organisms have on the health of human occupants and on such other factors as degradation of building materials?

The characteristics of "healthy" indoor environments cannot yet be defined, nor do microbial, clinical, and building researchers yet understand how to modify features of indoor environments—such as building ventilation systems and the chemistry of building materials—in ways that would have predictable impacts on microbial communities to promote health and

---

[1]This Summary does not include references. Citations for the findings presented in this Summary appear in subsequent chapters of the report.

prevent disease. The factors that affect the environments within buildings, the ways in which building characteristics influence the composition and function of indoor microbial communities, and the ways in which these microbial communities relate to human health and well-being are extraordinarily complex and can be explored only as a dynamic, interconnected ecosystem by engaging the fields of microbial biology and ecology, chemistry, building science,[2] and human physiology.

This Consensus Study Report reviews both what is known about the intersection of these disciplines and how new tools may facilitate advances in understanding the ecosystem of built environments, indoor microbiomes, and effects on human health and well-being. The report provides a vision of a future in which indoor microbial communities are better understood, and built environments can be designed and operated to improve human health. To advance this vision, the report offers a research agenda to generate the information needed so that stakeholders with an interest in understanding the impacts of built environments will be able to make more informed decisions.[3] The key terms used in the report are defined in Box S-1.

EFFECTS OF INDOOR MICROORGANISMS ON HUMAN HEALTH

More is understood about transmission of infectious microorganisms than about noninfectious health impacts. Concern about diseases spreading from person to person inside buildings and in enclosed spaces is long-standing, and the increasing prevalence of hospital-associated infections further motivates the desire to understand how humans are exposed to disease-causing microorganisms and how the microbial agents associated with infection and disease move through, live, evolve, and die within a building. It is well established that humans can become sick after being exposed to infectious microorganisms indoors (e.g., live virus on a doorknob or in the air, or bacteria such as *Legionella* in water systems), although variations in human responses are common: microbial exposures may cause adverse health effects in one person while having minor or no effects in another.

---

[2] The report uses the term "building science" to refer to the field of knowledge that focuses on understanding physical and operational aspects of buildings and building systems and the impacts on performance; the term "building scientist" is used to refer to a broad range of integrated technical disciplines, including scientists, engineers, and architects who study this area.

[3] The study was sponsored by the Alfred P. Sloan Foundation, the National Aeronautics and Space Administration, the National Institutes of Health, and the U.S. Environmental Protection Agency, which asked the National Academies of Sciences, Engineering, and Medicine to convene a committee to address these multidimensional issues. In addition, the Gordon and Betty Moore Foundation supported travel awards for one of the committee's data-gathering workshops.

> **BOX S-1**
> **Key Terms Used in This Report**
>
> The term **built environment** encompasses many types of structures and related elements. This report focuses on residential, commercial, and mixed-use buildings, such as homes, offices, and schools, where most people spend extended periods of time. These types of environments have been the subject of recent research on understanding indoor microbial communities, yet there is enormous variability even within this subset of structures—not only among building designs, systems, and materials but also among the social and economic characteristics, densities, and behaviors of occupants and how buildings are related to each other and to external site and infrastructure. To illustrate selected points, the report sometimes touches on research conducted in other types of built environments, such as hospitals; however, it does not address a number of other specialized built environments, such as manufacturing facilities and transit systems.
>
> Paralleling the diversity of the built environment, the **microbial communities** discussed in this report encompass a heterogeneous group of organisms, including viruses, prokaryotes (bacteria and archaea), and microbial eukaryotes (fungi, microscopic algae, and protozoa) that exist in a particular location, including not only viable microorganisms along with their genetic material but also inactive or dormant microbes, proteins, metabolic products, and other cell components or fragments that may have an influence—positive, negative, or neutral—on human occupants. The report uses the term **microbiome** in this broad ecological context to include all of the microorganisms in an environment or a sample and their constituent parts. The report also discusses their metabolic products where relevant.

Microorganisms that are living and active in an indoor setting can also produce metabolites that may impact health. Metabolites of gut bacteria have been more well-characterized than those of indoor environmental bacteria or fungi, and further characterization of those metabolites and their influence on health will be needed. Exposure to a "dead" microbe or its component pieces, such as proteins and cell wall components of bacteria or fungi, also may cause irritant and allergic or nonallergic immune responses. Furthermore, humans participate with the indoor microbiome in a cycle of exposure, uptake, and shedding, interactions that can impact human health in myriad ways that remain imperfectly understood.

On the other hand, certain microorganisms, microbial compounds, and microbial communities are associated with beneficial health effects in ways that researchers are working to elucidate. Studies of exposure to "farm-type" microbiomes suggest, for instance, that children who grow up on farms in contact with livestock have a lower risk of developing asthma. Likewise, in urban U.S. settings, some studies suggest that exposure to dogs or to certain microbes early in life may protect against later allergic disease. Current research is investigating how these early-life microbial exposures correlate with subsequent health outcomes. Another line of investigation involves examining whether microorganisms in the built environment influence the populations of microorganisms on and in humans, such as on skin or in the gastrointestinal tract, and correlating such microbiome–microbiome interactions with health impacts. The difficulty of disentangling these pathways is compounded by the role of additional factors that can influence health and human behavior.

Questions remain as to whether favorable health outcomes are due to exposures to specific microbes, exposures to a greater diversity of microbes, the stages of life at which exposures occur, or other factors. Nor, beyond general advantages of moisture management and other practices, do researchers understand how characteristics of the built environment contribute to these favorable relationships. Nonetheless, advancing knowledge raises the possibility that future interventions to affect microbial communities in the built environment might be used both to reduce the risk of unfavorable outcomes and to promote beneficial or healthful outcomes. To develop effective interventions, researchers will need not only to define pathways that are relevant to human health but also to elucidate their mechanisms of action. Current findings provide a foundation for future directions in research that can yield this knowledge.

OTHER EFFECTS OF MICROORGANISMS

Interest in studying indoor microbial environments is also spurred by the desire to optimize energy performance and incorporate "green design" features into buildings while ensuring that the buildings maintain occupants' comfort and health. Managing microorganisms indoors may reduce biodegradation of building materials and finishes or reduce biofilm fouling to minimize energy losses. Achieving these objectives entails trade-offs. Increasing the temperature of water in building water heaters and pipes to a level that impedes growth of microorganisms can result in higher energy costs and an increased risk of scalding, or it may have limited effectiveness in inactivating certain microbes. Increased flow of outdoor air into buildings has been linked to occupant health and comfort and can promote exposure to a greater diversity of microorganisms, but it increases the energy consumption for building heating and cooling. Greater outdoor air ventilation also may increase occupants' exposure to allergens or other pollutants of outdoor origin. A better understanding of building design and use will yield a fuller understanding of the interplay among buildings, microbes, and humans.

RELATIONSHIP OF BUILDING CHARACTERISTICS TO INDOOR MICROBIAL COMMUNITIES

The composition and viability of indoor microbial communities are determined largely by characteristics of the buildings they inhabit, including the availability of water and nutrients for growth and survival; the buildings' occupants; and the external environment. These relationships affect microbial transport and removal and influence the formation and composition of indoor microbial reservoirs in air and water and on surfaces.

Air can enter buildings as a result of natural ventilation, such as through open windows; mechanical ventilation, such as through a heating, ventilation, and air conditioning (HVAC) system; or leakage into a building through uncontrolled infiltration through the building's envelope. When air enters buildings in a controlled manner via designed natural and mechanical ventilation systems, it can be filtered to remove particles of various sizes. A number of microorganisms fall into size ranges that can be captured by air filtration systems currently used in buildings, affecting the degree to which indoor microbial composition mimics or differs from that of the outdoor environment. However, the most commonly used filtration systems do not remove all microorganisms, nor do they remove gaseous contaminants, including the metabolic products from some microbes. Moreover, not only do microbes enter buildings through ventilation systems; these systems can also serve as microbial reservoirs, in part as a result of the presence of condensation. How HVAC systems are operated and maintained, the proportions of air drawn from outdoors or recirculated from occupied spaces, and whether systems include mechanisms to remove moisture all link the properties of HVAC systems with effects on indoor microbial communities.

Water systems serve as microbial reservoirs, and the development of microbial communities is affected by the composition of water piped into a building, as well as leaks, condensation, and the existence of other moisture sources. Although a building may appear to be dry, isolated locations of moisture can support microbial growth and activity. Microbes also may persist under arid conditions—for example, as spores.

Human behaviors contribute to how water impacts the indoor microbiome. Such practices as water temperature selection, occupant control of thermostats, whether toilets are flushed with the lid open or closed, and how indoor humidity levels are controlled can influence the development and maintenance of microbial communities. Moisture from the air that becomes adsorbed onto building surfaces or absorbed into materials can be an important anchor for indoor microbial communities, as can the availability of nutrients. It is also important to note that the role of humans in built environments varies. In addition to widely differing ages, health status, occupant density, and behaviors that facilitate or impede the transport and resuspension of indoor microorganisms, different occupants (e.g., homeowners versus renters; facilities managers versus office workers) vary in their ability to control or maintain features of the built environment.

TOOLS THAT FACILITATE ANALYSIS

To make targeted changes in built environments that positively impact microbial communities, building designers and managers and material scientists will need robust data on the relationships among the multiple factors relevant to microbiomes in the built environment. As noted, the existing base of research is starting to provide information that connects building characteristics with the composition and function of indoor microbial communities, while ongoing research is exploring associations with health and other outcomes. To move from research to application, it will be necessary to determine more fully the public health relevance of the relationships among built environments, indoor microbiomes, and humans, as well as how to demonstrate causal relationships in a clinically relevant framework. Although better quantitative information on microbial exposures is likely to be part of this framework, efforts to expand this knowledge face a number of technical and practical challenges.

Studies on the impacts associated with indoor microbiomes will need to use a variety of tools and data collection strategies to capture the dynamics involved. These tools include sensors to measure and monitor such building characteristics as temperature, moisture, and airflow. Identification tools for characterizing which microorganisms are in a sample collected from a given built environment (e.g., a sample taken from an air filtration system, a showerhead, or a carpet) need to be coupled to those that describe microbial functions (e.g., cellulose degradation, mycotoxin production, antibiotic production, production of immunosuppressants, or biofilm formation). Growing microbial samples in culture remains one technique for understanding whether the organisms collected in a building are viable. Since relatively few microorganisms can be cultured, "omics" technologies, including genomics, proteomics, and metabolomics,[4] have become increasingly important tools for moving research forward by providing a means to assess the composition, structure, and function of microbial communities. The detection of microbial products such as volatiles, toxins, or other microbial metabolites also may yield markers to provide exposure and

---

[4] Metagenomics is the study of the collections of genes present in a sample, which can be used to help identify the particular microorganisms a sample contains. Proteomics (measurement of the collection of proteins) and metabolomics (measurement of the collection of chemical metabolites) yield information on what the microorganisms were doing when the sample was collected and thus help provide a snapshot of the microorganisms' activity.

outcome data, although further development of techniques for measuring indoor microbial metabolites and development of biomarkers of human exposure will be important.

Despite the advantages these tools provide for characterizing microbial communities, persistent challenges hinder progress in understanding the interconnections among buildings, microorganisms, and health. When applied to built environment samples, sufficiently deep metagenomics sequencing is required to obtain resolution of community composition and abundance[5] at the species level, and many published reports provide less precise resolution. Advances in other "omics" areas, including the research infrastructure for sharing and analyzing data, are needed to provide fuller information on microbial activities within sampled communities. In addition, improved quantitative data on human exposures to microbes and evidence connecting indoor microbial activity to effects in humans will be needed to strengthen links between exposures and clinically relevant health outcomes. Scientists also need to be aware of the effects of sample collection and handling on downstream data and on the assumptions and limitations associated with the analytic methods that underpin "omics" tools. These tools rely on statistical methods to make sense of large amounts of data. For example, data from metagenomic analysis of a sample are in the form of many pieces of DNA from many different organisms, both living and dead. To assign the organisms in the sample to specific taxonomic groups requires comparing these pieces of DNA with databases to identify their similarities to reference genomes associated with specific microbial species. Researchers also may lack sufficient cultured representatives. Different theoretical assumptions can be made about how best to cluster these pieces of DNA into taxonomic units, which in turn can affect the microbial groups that are identified as being present. Understanding the assumptions associated with "omics" tools and improving ways to compare or standardize the results they produce would enable better comparisons across studies conducted by different groups.

There are additional challenges to facilitating cross-study comparisons. The research community has identified the need for a core set of building, environment, and occupant data to collect when studying indoor microbiomes; however, these elements (and the level of detail required) are not yet fully defined or agreed upon. Further work is needed on how to achieve a reasonable balance between collecting sufficiently detailed information and

---

[5]Relative abundance of a microorganism in a sample refers to the percentage of that type of microorganism that was identified compared with the total microorganisms identified in that sample. Absolute abundance, on the other hand, reflects the actual number of that microorganism that was in the substrate (surface, air, water, or bulk material) in the built environment from which the sample was collected. Several technical challenges arise in obtaining such data from samples.

the time and cost involved. The absence of a core of best practices and standardized dataset parameters creates challenges for comparing results across studies. The ability to draw larger conclusions will require multidisciplinary collaborations and consensus.

MOVING TOWARD PRACTICAL APPLICATION

An important aspect of this report is its emphasis on how future interventions in built environments may someday be able to change indoor microbiomes in ways that promote health, and on what practical steps can be taken to generate the data needed to support the development, assessment, and eventual implementation of these interventions. Relevant types of interventions include those focused on changes to characteristics of buildings, such as ventilation rates, air and water filtration efficiencies, and maintenance schedules. Other potential interventions need to take into account the effects of chemicals in indoor cleaning products, how human behaviors affect the use and effectiveness of such chemicals (e.g., cleaning frequency and methods), and the design and use of existing and novel building materials (e.g., materials designed to have antimicrobial properties).

One way in which scientists are trying to understand these complex relationships is by applying models that represent inputs and outflows of building environment and microbiological systems and that capture the relationships among these components; models that can help predict the effects when one or more factors are changed and generate information to inform the development of future interventions. One example is models of building airflow and contaminant dispersion, which can provide insight into how interventions are likely to impact microbial concentrations and transport in a building. Box S-2 lists knowledge gaps that need to be filled to support such efforts to move from research to practical application.

New data connecting the built environment, microbial communities, and human health could help inform building science and public health decision making. Nevertheless, it is important to remember that policy decisions will not be made in a vacuum. Given the data generated by future research, policy makers and others are likely to take into account additional considerations, including the economic and noneconomic costs of potential interventions, such as the burden of microbial illnesses, energy costs, and the possibility of unintended health effects. Models for designing and assessing built environment interventions will need to incorporate these important dimensions.

Additional challenges may also be associated with built environment interventions, health effects, or access for those in substandard housing who are of lower socioeconomic status. For example, people of low socioeconomic status, relative to those who are better off, may lack access to

> **BOX S-2**
> **Knowledge Gaps Identified in This Report**
>
> - Improve understanding of the transmission and impacts of infectious microorganisms within the built environment.
> - Clarify the relationships between microbial communities that thrive in damp buildings and negative allergic, respiratory, neurocognitive, and other health outcomes.
> - Elucidate the immunologic, physiologic, or other biologic mechanisms through which microbial exposures in built environments may influence human health.
> - Gain further understanding of the beneficial impacts of exposures to microbial communities on human health.
> - Develop an improved understanding of complex, mixed exposures in the built environment.
> - Design studies to test health-related hypotheses, drawing on the integrated expertise of health professionals, microbiologists, chemists, building scientists, and engineers.
> - Improve understanding of how building attributes are associated with microbial communities, and establish a common set of building and environmental data for collection in future research efforts.
> - Collect better information on air, water, and surface microbiome sources and reservoirs in the built environment.
> - Clarify the association of building attributes and conditions with the presence of indoor microorganisms that have beneficial effects.
> - Develop means to better monitor and maintain the built environment, including concealed spaces, to promote a healthy microbiome.
> - Deepen knowledge on the impact of climate and climate variations on the indoor environment.
> - Develop the research infrastructure in the microbiome–built environment–human field needed to promote reproducibility and enhance cross-study comparison.
> - Develop infrastructures and practices to support effective communication and engagement with those who own, operate, occupy, and manage built environments.
> - Improve understanding of "normal" microbial ecology in buildings of different types and under different conditions.
> - Further explore the concept of interventions that promote exposure to beneficial microorganisms, and whether and under what circumstances these might promote good health.
> - Obtain additional data necessary to support the use of a variety of quantitative frameworks for understanding and assessing built environment interventions.

information about the interactions between microbes and the built environment and may have more limited ability to make changes to their built environments, even when such guidance is available. A number of key dimensions, currently understudied, need to be informed by the social and behavioral sciences, such as effective communication about the results of scientific research on indoor environments; guidance in such areas as cleaning and maintenance practices; strategies for engaging with relevant sectors of the public, including owners, occupants, facilities managers, and others; and efforts to foster behavior changes where appropriate.

VISION FOR THE FUTURE: MICROBIOME-INFORMED BUILT ENVIRONMENTS

The built environment interacts with the indoor microbiome in multiple ways that impact humans. Microbial exchange between indoors and outdoors, microbial growth and persistence in indoor settings, and human exposures to indoor microbial communities are affected by building design, operation, and maintenance. Research that focuses only on one microbe, on a specific aspect of building design, or on a single human health outcome will not be sufficient to understand these multifactorial relationships. As noted earlier, integrating approaches from multiple disciplines and bodies of knowledge is essential to improve understanding of these relationships and apply the knowledge gleaned. In general, improved understanding of indoor environments holds promise that in the future, buildings can support a more productive, healthier population.

In this report's vision, future built environments will be informed by improved knowledge about the indoor microbiome. Researchers and building practitioners will have reached a deeper understanding of the effects of indoor microbial communities on human health. Detailed information about the growth, establishment, and evolution of indoor microbiomes and how these microbial communities relate to building characteristics, as well as greater insight into human exposures and responses, will be known.

New technologies will support building operation and maintenance. Examples include sensors to detect water penetration, filtration performance, occupant density, and air quality. Some of these technologies will require further development, while others are currently on the market but are not widely used in practice. Increased utilization of sensing and monitoring technologies can be coupled with a fuller understanding of microbial and environmental connections between the indoors and outdoors and what benefits these connections may provide. Where appropriate, this understanding can be incorporated into building design and operation. It may be hoped that a more purposeful approach to managing buildings and their microbiomes reflected in this vision for the future will result in building occupants who are more informed about and engaged in improving their indoor environments.

RESEARCH AGENDA

Gaining sufficient understanding of the relationships among microbial, physical, chemical, and human elements that contribute to the built environment and translating this knowledge into improved building design and operation is a long-term goal that will not be achieved quickly or easily. Its accomplishment will likely require partnerships among federal agencies, public entities, and private corporations. Steps that would fill the knowledge gaps identified in Box S-2 and advance progress toward this ultimate goal are reflected in the research agenda presented in Box S-3. The priorities the committee recommends for inclusion in this agenda build on current research, as well as on the questions and potential research directions presented throughout the report. They highlight steps that can be taken in generating the knowledge necessary to fully understand how microbiomes in built environments impact human health and what can be done to ensure that buildings and their occupants are, and remain, healthy into the future.

This research agenda reflects a need for significant additional research and tool and infrastructure development to sustain this field; will require time and support to accomplish; and is broad enough that many partners will need to be involved. Agencies such as the National Institute of Standards and Technology (NIST), the National Institutes of Health (NIH), the National Science Foundation (NSF), the U.S. Department of Energy (DOE), and the U.S. Environmental Protection Agency (EPA) can support, develop, and implement the foundational tools, data, and standards needed to support the field. Given the many types of built environments, occupants, and impacts, agencies including EPA, the General Services Administration (GSA), NIH, the U.S. Centers for Disease Control and Prevention (CDC), the U.S. Department of Defense (DOD), the U.S. Department of Housing and Urban Development (HUD), the U.S. Department of Veterans Affairs (VA), and others may pursue their own interests in exploring microbiome–built environment interactions—for example, health in public housing or in military facilities, vehicles, ships, and submarines. Specialized resources available to such agencies as the National Aeronautics and Space Administration (NASA), including the International Space Station, can provide laboratories in which to test hypotheses, given the close control over environmental features of these resources and the tradition of using such resources as experimental stations. Communities of practice, such as in the indoor air quality and HVAC fields, can follow these developments as they move closer to practical application. Further engagement of such disciplines as materials science and the social and behavioral sciences

> **BOX S-3**
> **A Research Agenda for Moving to Practical Application**
>
> A multidisciplinary research program on microbiomes of the built environment is needed to make progress toward understanding and predicting the interactions among microorganisms, built environments, and human occupants and their effects. Such a program will require integrating expertise from multiple scientific, health, and engineering disciplines, along with professional communities of practice in clinical medicine and in building design, operation, and maintenance.
>
> The microbiome–built environment field bridges the missions of multiple federal agencies without fitting directly into any one organization's portfolio. As a result, the implementation of a comprehensive and impactful research program will require an effort that cuts across individual agencies and foundations. Coordination of expertise from the United States, Europe, and other countries interested in the nexus of indoor environments, health, and microbial exposures would further strengthen the research effort by bringing diverse expertise to bear on the challenges involved, leveraging national research investments beyond the United States, enabling resources to be used most effectively while reducing duplication, and promoting international collaboration.
>
> The research agenda recommended in this report encompasses 12 priority areas, detailed below.
>
> ***Characterize interrelationships among microbial communities and built environment systems of air, water, surfaces, and occupants.*** Such efforts would build on the existing foundation of knowledge, address gaps in current understanding, and enable future progress.
>
> 1. Improve understanding of the relationships among building site selection, design, construction, commissioning, operation, and maintenance; building occupants; and the microbial communities found in built environments. Areas for further inquiry include fuller characterization of interactions among indoor microbial communities and materials and chemicals in built environment air, water, and surfaces, along with further studies to elucidate microbial sources, reservoirs, and transport processes.
> 2. Incorporate the social and behavioral sciences to analyze the roles of the people who occupy and operate buildings, including their critical roles in building and system maintenance.
>
> ***Assess the influences of the built environment and indoor microbial exposures on the composition and function of the human microbiome, on human functional responses, and on human health outcomes.*** This research could produce the knowledge base needed to inform the design and testing of future interventions to protect, promote, and improve health.
>
> 3. Use complementary study designs—human epidemiologic observational studies (with an emphasis on collection of longitudinal data), animal model studies (for hypothesis generation and validation of human observational findings), and intervention studies—to test health-specific hypotheses.

4. Clarify how timing (stage of life), dose, and differences in human sensitivity, including genetics, affect the relationships among microbial exposures and health. These relationships may be associated with protection or risk and are likely to have different strengths of effect, parameters that are important to understand further.
5. Recognize that human exposures in built environments are complex and encompass microbial agents, chemicals, and physical materials. Develop exposure assessment approaches to address how combinations of exposures influence functional responses in different human compartments (e.g., the lungs, the brain, the peripheral nervous system, and the gut) and downstream health outcomes at different stages of life.

Explore nonhealth impacts of interventions to manipulate microbial communities. The incorporation of data beyond health effects into the development of models would strengthen the assessment of potential interventions and inform future decision making.

6. Improve understanding of energy, environmental, and economic impacts of interventions that modify microbial exposures in built environments, and integrate the relevant data into existing built environment–microbial frameworks for assessing the effects of potential interventions.

Advance the tools and research infrastructure for addressing microbiome–built environment questions. The field relies on a diverse set of approaches aimed at understanding microbial communities, buildings, health, and other impacts. Improvements to this toolkit and infrastructure would support accelerated progress.

7. Refine molecular tools and methodologies for elucidating the identity, abundance, activity, and functions of the microbial communities present in built environments, with a focus on enabling more quantitative, sensitive, and reproducible experimental designs.
8. Refine building and microbiome sensing and monitoring tools, including those that enable researchers to develop building-specific hypotheses related to microbiomes and that assist in conducting intervention studies.
9. Develop guidance on sampling methods and exposure assessment approaches that are suitable for testing microbiome–built environment hypotheses.
10. Develop a data commons with data description standards and provisions for data storage, sharing, and knowledge retrieval. Creating and sustaining the microbiome–built environment research infrastructure would promote transparent and reproducible research in the field, increase access to experimental data and knowledge, support the development of new analytic and modeling tools, build on current benchmarking efforts, and facilitate improved cross-study comparison.

continued

> **BOX S-3 Continued**
>
> 11. Develop new empirical, computational, and mechanistic modeling tools to improve understanding, prediction, and management of microbial dynamics and activities in built environments.
>
> ***Translate research into practice.*** As the interconnections among microbiomes, built environments, and health become more clearly understood, this knowledge should be translated into guidance applicable to varied public and professional audiences.
>
> 12. Support the development of effective communication and engagement materials to convey microbiome–built environment information to diverse audiences, including guidance for professional building design, operation, and maintenance communities; guidance for clinical practitioners; and information for building occupants and homeowners. Social and behavioral scientists should be involved in creating and communicating these materials.

can result in significant contributions to these efforts. The integration of expertise across these many communities will be critical to achieving the vision of microbiome-informed design, operation, and maintenance of built environments.

1

Introduction

Human beings encounter microorganisms every day. In modern societies, most of these interactions occur inside the buildings where humans live and work. Data collected by the National Human Activity Pattern Survey (NHAPS), for example, revealed that Americans spent an average of nearly 69 percent of a 24-hour day in a residence, 5.4 percent in an office or factory, and nearly 13 percent in other indoor locations (Klepeis et al., 2001). Patterns were similar for Americans and Canadians, and the study found that spending significant time indoors had remained the case across several decades. Despite the age of the NHAPS data, it is unlikely that people are now spending less time inside. Thus, improving understanding of what humans encounter in their indoor environments, how these exposures affect them, and how their environments can be modified to affect these exposures has the potential to contribute to future health and well-being.

The microbial communities that surround humans are diverse and dynamic—consisting of ever-changing combinations of bacteria, viruses, and microbial eukaryotes (microorganisms whose cells have nuclei, such as fungi). They include microbes that may be actively persisting or proliferating; those that are inactive, dormant, or dying; and microbial molecules.[1]

---

[1] This Consensus Study Report considers built environment microbiomes as including a variety of microorganisms, including bacteria, viruses, and microbial eukaryotes (such as fungi, microscopic algae, and protozoa), and encompassing not only intact microorganisms but also microbial components and chemical products of microbial metabolism (such as mycotoxins and volatile organic compounds). The relevant microbial communities are found not only in the occupied spaces of built environments but also in indoor infrastructure, such as premise plumbing and ventilation ductwork, as well as in "unconditioned" indoor spaces, such as crawl spaces.

15

Microorganisms that live and die in built environments can have effects on human occupants. Prior to the advent of genomics and other molecular tools, studies of indoor microorganisms relied primarily on culture and microscopy to explore these microbial communities. Recent studies have taken advantage of the development of novel tools to quantify and sequence the DNA of communities of microbes and study their functions, revealing their abundance and complexity and leaving still more questions than answers about what these communities are doing and what functions they serve in people's indoor environments. Despite the wealth of research thus far, then, much remains unknown about the direct and indirect connections between humans and microorganisms in their built environments.

STUDY CHARGE

As more knowledge emerges about the communities of indoor microbes, the question arises as to whether this knowledge might be used to inform building design and operation to create and maintain more healthful indoor environments. This Consensus Study Report addresses this complex question. Because far more currently is unknown than known, the report does not provide a simple answer to this question. In accordance with the committee's statement of task (see Box 1-1), the report (1) summarizes the state of knowledge on microbiomes of the built environment and their relationships to human health, as well as their other potential impacts; (2) identifies knowledge gaps; and (3) outlines a research agenda for filling these gaps and moving the field forward.

STUDY FOCUS AND SCOPE

Microorganisms can impact the built environment and its occupants in a variety of ways. One of the key motivators for studying microbial communities of the built environment is to learn more about how microorganisms and their products affect human health. Such effects may arise from an infection causing a disease, as a result of allergic and other physiologic responses leading to symptoms, or by other means. For example, a number of studies have explored the transmission of pathogens, such as influenza virus or *Legionella* bacteria in indoor settings, as well as adverse associations among interior dampness, fungi, and respiratory health. Other potential effects are being explored as well, including whether exposure to environmental microorganisms affects the human microbiome. Alongside the consideration of indoor microbes as a potential source of risk, studies have suggested potential benefits from not living in too sterile an environment and from particular types of early-life exposures to microorganisms. The effects of exposures in the built environment likely have personal

INTRODUCTION 17

> **BOX 1-1**
> **Statement of Task**
>
> The National Academies of Sciences, Engineering, and Medicine will convene an ad hoc committee to examine the formation and function of microbial communities, or microbiomes, found in the interior of built environments. It will explore the implications of this knowledge for building design and operations to positively impact sustainability and human health. The committee will:
>
> - Assess what is currently known about the complex interactions among microbial communities, humans, and built environments, and their relationship to indoor environmental quality. Where knowledge is adequate, summarize implications for built environment design and operations and human health.
> - Articulate opportunities and challenges for the practical application of an improved understanding of indoor microbiomes, with an emphasis on how this knowledge might inform choices about built environment characteristics, both physical and operational, in order to promote sustainability and human health.
> - Identify a set of critical knowledge gaps and prioritized research goals to accelerate the application of knowledge about built environment microbiomes to improve built environment sustainability and human occupant health.
>
> The committee may discuss and recommend additional actions to advance understanding of microbiome-built environment interactions, including examples of the potential impacts of building and health-related policies and practices, and social or public engagement dimensions.

dimensions, influenced by such factors as age, genetics, and health status. Similar exposures to similar microorganisms may have beneficial, adverse, or neutral effects in different people, factors that will need to be considered in developing future guidance. In addition to effects on health, microorganisms affect building materials and building systems, including through degradation, corrosion, and fouling as a result of biofilm formation. These effects can have economic and sustainability costs, and the varied effects and trade-offs involved will influence the assessment of potential interventions to modify indoor microbiomes.

Both microbial communities and the built environments they inhabit are highly diverse. Myriad types of buildings and building uses exist in myriad climates, so clearly there is no "typical" building. Moreover, the built environment contains more than buildings. Used broadly, the term encompasses roads and transit systems, transportation vehicles such as ships and airplanes, space vehicles and space stations, manufacturing facilities, water supply systems and sources, wastewater treatment plants, and other components of the urban environments in which a majority of people in developed countries live.

This report does not address all of these built environments; it focuses primarily on single-family residential buildings (homes) and multioccupancy commercial and mixed-use buildings, including apartments, offices, and schools. These spaces represent environments in which humans spend large amounts of time. The report does not delve in detail into farm environments, factories, gathering spaces such as retail malls or amphitheaters, or transit systems, although it draws several comparisons with studies conducted in specialized living and working environments, such as hospitals and the International Space Station. The report also focuses on built environments found in temperate regions of the world and does not draw detailed comparisons among microbiomes and their impacts in nations with more widely varying climates and levels of economic development.

Even with these constraints, the building types considered in the report vary significantly in design, construction, operation, and occupant population. Residential buildings generally are designed so that occupants have control over environmental conditions such as temperature, humidity, and connection to the outdoor environment via operable doors and windows. However, the level of engagement of occupants in exercising such controls varies widely, with some being actively engaged in operating and maintaining the various building systems and others dealing with these systems only when the systems have become nonfunctional. The occupants of residential buildings generally are stable over time, and even transitory occupants tend to be related biologically to primary occupants. Occupants may span a range of ages, physical conditions, and overall health status, and occupant densities and occupancy characteristics can change over time. As the field moves closer to practice and to developing guidance on how indoor microorganisms affect health and how building factors affect these indoor microbiomes, it will also be important to consider challenges facing residents living in poor housing stock and of lower socioeconomic status, who may have less control over environmental conditions, may not be able to improve their residences, or may need information and resources to address indoor microbiome–built environment issues.

Commercial buildings, on the other hand, are necessarily transitory spaces. Most office buildings accept visitors every day, and the occupants are normally not related to one another. Commercial buildings generally are managed more intentionally relative to residences, typically with dedicated staff who operate and maintain the various building systems. Occupants usually have little to no control over their indoor environment. Even when individual occupants can control the temperature and lighting of their indoor environment or have operable windows, they still are subject to central management of many other aspects of that environment, and public spaces (e.g., lobbies and meeting rooms) are much less likely to allow for occupant control. Occupants in commercial buildings are also subject to the actions of other occupants and, as a result, may be exposed to temperatures, lighting, noise, cleaning products, or other conditions they would not necessarily choose for themselves.

EMERGING TOOLS THAT FACILITATE ANALYSIS

Although the percentage of microorganisms that can be cultured from an environmental sample varies, estimates as low as 1 percent have been reported (Amann et al., 1995; Kallmeyer et al., 2012; Quince et al., 2008). Thus, a reliance on culture-based surveys significantly limits understanding of microbial populations, including those present in buildings. New approaches for analyzing genetic material in environmental samples have revolutionized microbiology, and the tools available for detecting microbial DNA and RNA, along with associated bioinformatics approaches for connecting those sequences to microbial identification, have improved over the past decade. Simultaneously, databases containing reference microbial genomes have grown. Although these genomics technologies have brought new understanding of the composition and diversity of microbial communities in buildings, they also have limitations—for example, sequencing approaches do not distinguish between living and dead microorganisms—and having information on viability thus remains important. Additional high-throughput "omics" approaches with which to study microbial proteins and metabolic products provide valuable information not only on which species are present indoors but also on what these microorganisms are doing. These culture-independent techniques will need to be combined with culture-dependent methods, computer modeling, building characterization tools, and epidemiology to increase knowledge about the environments in which people live and work and how these building and microbial conditions are affecting them.

STUDYING THE INTERSECTION OF MICROBIAL COMMUNITIES, BUILT ENVIRONMENTS, AND HUMAN OCCUPANTS

Microorganisms within buildings are found in air, on surfaces, and in water systems. They are carried on and arise from living creatures that inhabit the environment, including human occupants, pets, plants, and pests, each of which has its own associated microbial communities. A simplified representation of the ecosystem built around these interactions among humans, the indoor environment, and indoor microorganisms is depicted

HUMAN OCCUPANTS

Physiology
Human-associated microbes
Behavior

BUILT ENVIRONMENT

Heating, cooling, and ventilating
Plumbing
Materials
Site

TRANSPORT MECHANISMS
Air
Water
Surfaces

MICROBIAL COMMUNITIES

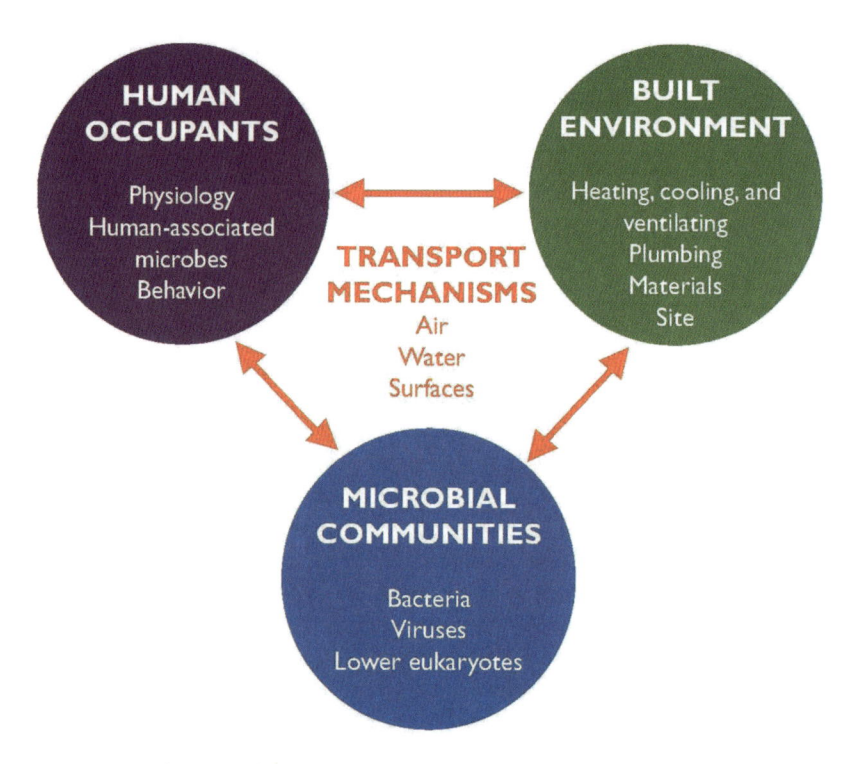

Bacteria
Viruses
Lower eukaryotes

FIGURE 1-1 The complex interactions among human occupants, built environments, and associated microbial communities.

in Figure 1-1. Examples of factors that affect community interactions are included in each circle, and the primary reservoirs of microbes and mechanisms of their transport are shown in the triangle connecting the circles. Not every factor is captured in this representation. For example, indoor building structures other than heating, ventilation, and air conditioning (HVAC) and plumbing systems and the outdoor environment beyond a building site may affect indoor microbiomes, as may building operation and maintenance practices that affect indoor building and building system conditions. The report's subsequent chapters explore these dimensions and their linkages in greater detail.

Ecosystems can be characterized in a number of ways, and an initial grounding in basic principles of ecological theory can aid in understanding how changes to one component of the system may affect other components. Some simple ecological relationships are briefly introduced below. The report turns to the idea of system models in more detail in Chapter 5, which explores the use of models in addressing specific questions, testing hypotheses, and assessing the impacts of different types of interventions on built environments, their associated microbiomes, and their human occupants.

The Built Environment as an Ecosystem: Fundamental Ecological Principles

A single building can be thought of as akin to an island that emerges from the ocean. It is a habitat patch that can be colonized by groups of microorganisms, all of which have the potential to proliferate further and interact. The resident community of microbes in a building will be amplified by colonization, modulated by in situ population dynamics, and depleted by extinction or depletion to below detectable levels.[2]

A mass balance (or material balance) equation is a technical way of describing the flows of a material into and out of a system. By identifying and accounting for what enters and leaves, this approach helps in understanding the properties and functions of a system and has widespread application across science and engineering. This concept is a useful way of describing the abundance (N) of a representative microbial population in an environment. The number of organisms grows through births (B) and immigration (I) into the habitat being studied, and decreases through deaths (D) and emigration (E). Expressing this mathematically, change in the species' abundance over time (dN/dt) is

$$dN/dt = B + I - D - E \qquad \text{(Equation 1.1)}$$

In a real environment, each of the terms of this simple equation encompasses a number of complex relationships—for example, accounting for multiple "immigration" sources of a particular microorganism in the built environment or accounting for multiple types of microorganisms, each with a different abundance. Examples of microbial inflows and outflows in a highly simplified built environment are depicted in Figure 1-2. As this simplified figure shows, microorganisms can enter a built environment from the outside through

- ventilation, such as through open doors and windows or through HVAC systems;
- through building water supply and premise plumbing systems through evaporation or aerosolization;
- infiltration, which includes moisture seepage or air leakage; and

---

[2]These dynamics generally refer to the processes that affect the composition, persistence, and proliferation of microbial communities.

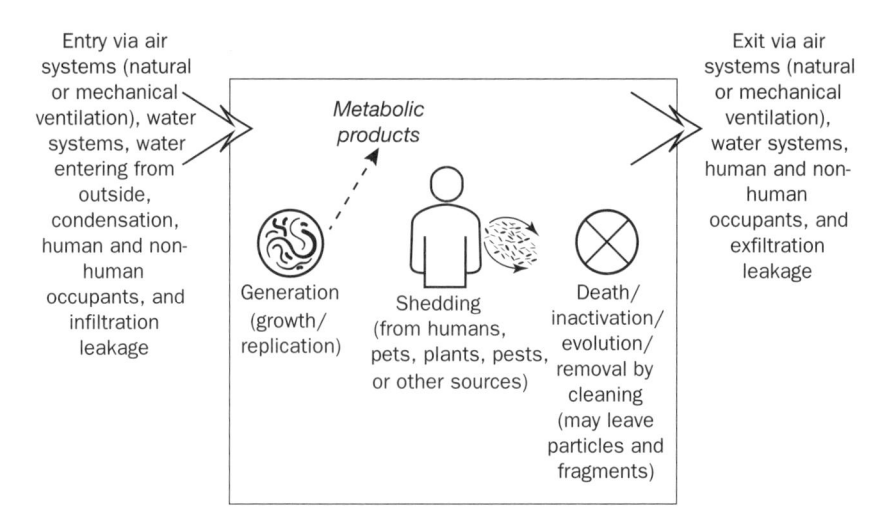

FIGURE 1-2 Transport and life cycle of indoor microbes. The abundance of a particular type of microorganism within a built environment can increase by transport from outside sources, by growth within the environment, and by shedding indoors from additional sources, such as humans. A population of specific microorganisms can decrease by being carried out of the environment, by being removed via cleaning, and by microbial death or evolution. Inactive cell components or microbially produced chemicals also may remain.

- transport on humans and other occupants, including pets and pests, and the microbes associated with soil and particulate matter these occupants track indoors.

Within the built environment, microorganisms can increase as a result of

- growth and replication if the environment has the right nutrients, water, and conditions for reproduction; and
- shedding from humans and other living occupants, each of which has its own associated microbiomes.

Microorganisms can leave the built environment through

- microbial death, although nonliving and dormant microbes and microbial fragments may remain;
- water via premise plumbing systems;
- air exfiltration and moisture seepage;
- outward transport by natural or mechanical ventilation;

- outward transport on humans and other occupants; and
- removal by cleaning, such as mopping, dusting, and vacuuming.

Permissive and Restrictive Environments

The habitat in an indoor environment also varies in terms of its "quality," in the sense of how well or poorly it suits an organism's life cycle. In the built environment, a *permissive environment* is one that has suitable conditions such that microorganisms can grow or persist. These conditions include such aspects as humidity/moisture, nutrients, temperature, pH, and lack of inhibitory compounds. Conversely, a *restrictive environment* is one in which suitable conditions are lacking and/or inhibitory substances are present such that microbial growth is limited by these substances and persistence is reduced, although persistence via dormancy may continue.

Making Use of Population Dynamics Equations

Equation 1-1 above, which can be used to assess the abundance of one microbial population, can also be utilized in different scenarios. These scenarios may include, for example, the microorganisms being inactive (in which case birth rate is zero), conditions in which immigration and emigration are equal, or a steady state in which there is no change in a species' abundance over time. Solving the equation for these and other conditions demonstrates that a relationship will exist between internal abundance (within the built environment) and concentration in the external "source" landscape. This relationship has two implications. First, it is important to note that microbial communities encompass not only viable microbes but also dormant microorganisms, as well as microbial fragments, which may play a role in impacting human health. The density of a microorganism in the built environment habitat will generally be affected by the density of microbes in the source landscape (e.g., in outdoor air), although indoor bacterial concentrations, for instance, can be higher than those outdoors as a result of indoor sources. Second, simple models such as this can also make predictions about transient dynamics and how long a system needs to settle into a new state following a perturbation. Box 1-2 provides additional examples of how mathematical representations help frame questions to enable better understanding of key system parameters and yield implications for studying population interactions in the built environment. Although these concepts may appear to be simplified and abstract, understanding the variables and processes behind how microbial communities develop, grow, and fail can guide models and predictions of how these communities will respond to features within the built environment (e.g., mechanical versus natural ventilation), as well as to perturbations (e.g., the occurrence of

BOX 1-2
Modeling Built Environment Ecology

A basic model for change in abundance of a microbial species in a built environment habitat can be developed from Equation 1-1, where N is microbial abundance (numbers of the microorganism in the BE habitat under study) and B, D, I, and E are microbial birth, death, immigration, and emigration rates, respectively.

$$dN/dT = B - D + I - E$$

The simple equation and mass balance concept can be expanded in a number of ways to represent more accurately the complexity of actual population dynamics. For example, if a representative microorganism is reproducing because it is in a permissive environment, and births exceed deaths, there will be a period of growth. In a constant environment, the population of microorganisms will grow until checked by some other dimension—a reflection of density dependence and resource availability. Such density dependence can arise, for example, as a result of resource depletion and waste product build-up that would inhibit other microorganisms in the population, or as a result of competition with other species.

The mass balance can also be adapted to reflect the concentrations of microorganisms in the air under equilibrium as expressed using contaminant species balance concepts, discussed in more detail in Chapter 5:

$$C_{in} = PC_{out} + G^* - L^*$$

In this case, the steady-state microbial concentration inside the building (C_{in}) is equal to the amount that penetrates indoors from outside (PC_{out}), plus indoor generation (G^*) and loss (L^*) terms.

All of the parameters that affect a representative microbial population in the built environment can be spatially heterogeneous and temporally variable. There may be multiple spatial "compartments" in a building, whose microbial populations are linked as organisms are transported between them and as microorganisms deposit and are resuspended. In many built environments with minimal moisture sources, there will be little growth and activity, and thus the dominant processes by which microbes are transported in buildings are likely to be physical. However, discrete disturbances to the environment may occur, and temporal variability in environmental conditions and in generation and loss can greatly affect the time-averaged abundance of a population, particularly for reproducing populations (Gonzalez and Holt, 2002; Holt et al., 2003). Moreover, members of any given microbial population will be heterogeneous as a result of genetic mutation, gene flow, and genetic recombination. Unless there is complete mixing with an external environment, adaptation to local conditions in the built environment via natural selection can occur. These factors can be repeated over the many species that make up microbial communities in built environments. Furthermore, buildings are not monolithic, and different buildings will vary in environmental conditions, building materials, and occupant activities.

The local abundance of a species compared with source populations can also vary greatly when species reproduce. Resources for reproduction can come from sources that may be easily overlooked. Walls in buildings can be covered with a thin layer of oily, squalene compounds that come from human skin shedding, and that can provide nutrients for bacterial growth. Particles can stick to the oily coating, bringing water to the surface and facilitating microbial growth. Condensation due to temperature gradients can also bring moisture to building surfaces. These interactions demonstrate the many ways in which microorganisms can interact with buildings and building surfaces.

Interactions among different microbial species can also come into play, and representative population models may require webs of interacting species. For example, certain types of viruses (phage) infect bacteria and may affect their populations. Environmental heterogeneity is common in the indoor environment. If such environmental heterogeneity occurs with distinct resources or gradients in abiotic conditions, then there can also be coexistence of competing species. Given interspecific interactions, a rich panoply of effects can spring into play. In competitive interactions, priority effects could be important—for instance, in biofilms, quorum-sensing processes can lead to dominance by initial colonists. Fungi colonizing a wall can provide a resource base for detrital food webs, including both bacteria and viruses.

Differences among species in local abundance will also reflect, in part, differences in loss rates, which will relate to environmental conditions such as moisture, light, temperature, cleaning practices, and presence of antimicrobial agents. These conditions can all vary within and among buildings and show temporal variability, affecting community composition. For instance, a rich microbiome may be maintained in a school by the constant influx of microbes from students entering and leaving during the school year, but this community may decay over the quiescent summer break.

Looking at the variables that form the basis for population models leads to a number of questions that need to be answered to better understand how microbial communities in built environments form and function:

- Where do the colonists come from? What is known about the sources that populate the microbiome of a particular building?
- What is known about the death, persistence, and removal rates of different types of microorganisms in the built environment?
- To what extent are local microbial communities and the individual species within those communities determined by dynamics within a building versus by coupling to the external environment?
- What is known about the kinds of resources available in the built environment?
- How well described are the environmental gradients that can define the "ecological niches" available for microbes in the built environment?

Research at this intersection of built environments, microorganisms, and occupants can help answer these questions.

flooding) or interventions (e.g., the inclusion of high-efficiency particulate air [HEPA] filters in a building's HVAC system).

The ability to apply such ecological analyses and modeling effectively remains preliminary in the indoor environment: many variables are still being characterized, while a number of additional factors (such as exposure to nonviable components and metabolites) may be relevant to understanding microbial impacts. As the field moves forward and research continues to resolve, or at least further clarify, these complexities, this modeling and prediction capability will contribute greatly to moving the field from basic research to application.

PRIOR EFFORTS ON WHICH THIS REPORT BUILDS

Research into the microbial populations associated with different built environments and the health effects of exposure to these microorganisms extends back for decades. In the public health community, many investigations have focused on how pathogens infect occupants within a building. Since the early 20th century, for example, research has explored the release of droplets into air through coughing and sneezing and the contributions of these aerosol and contact factors to the development of colds and flu (Dick et al., 1987; Lindsley et al., 2010; Marr, 2016; Stephens, 2016; Wells et al., 1939). Similarly, the proliferation and transmission of waterborne pathogens through municipal water systems and indoor plumbing have been studied for many years, drawing attention to the effects of such conditions as pipe degradation and water stagnation times in storage facilities and premise piping (NRC, 2006; Rhoads et al., 2015, 2016). For example, *Legionella* can colonize as biofilms in showers and be expelled in aerosols (Schoen and Ashbolt, 2011), and *mycobacterium avium* complex and *Pseudomonas* may colonize and persist in a similar manner (Falkinham et al., 2015). The recent water crisis in Flint, Michigan, provides one illustration of the importance of water quality, in this case impacting a subsequent outbreak of Legionnaires' disease (Schwake et al., 2016).

The positive and negative health consequences associated with the built environment also have been studied in the context of the relationship between indoor air quality and asthma. A report of the National Academies of Sciences, Engineering, and Medicine states that "major sources of indoor allergens in the United States are house dust mites, fungi and other microorganisms, domestic pets (cats and dogs), and cockroaches" (IOM, 1993, p. 2) that can be found in the buildings in which humans now spend more than 90 percent of their time (Klepeis et al., 2001). Increases in and exacerbation of asthma and allergies have been correlated with exposures to specific sources of indoor allergens (IOM, 1993, 2000). A 2004 Institute of Medicine report also "found evidence of an association between exposure to damp indoor environments and some respiratory health outcomes" (IOM, 2004, p. 10) (see also Mendell et al., 2011; WHO, 2009).

Other studies have addressed less-defined impacts of building conditions on human occupants. In the 1970s, for example, a number of environmental health investigations were prompted by symptoms associated with "sick building syndrome," in which people reported headaches, mucous-membrane irritation, and difficulty concentrating linked to arriving at and departing from a particular building without a clearly identified exposure or cause (Cox-Ganser et al., 2010, 2011; Hinkle and Murray, 1981; Hodgson and Kreiss, 1986; Kreiss, 1989; Levin, 1989; Melius et al., 1984; NRC, 1981; WHO, 1983). Around the same time, environmental awareness and concerns about energy efficiency led to changes in ventilation in public buildings, which in some cases included inoperable windows or HVAC systems that were undersized for the spaces they served or that operated contrary to their design intent. These changes sometimes resulted in decreased outdoor air ventilation rates and, at other times, in increased indoor dampness. The relative contributions of indoor microbes, their components, or their metabolic products to the development of symptoms of sick building syndrome are not well understood. Nevertheless, research into these symptoms has helped focus some members of the design, engineering, and construction fields on ways to promote occupant comfort and health.

The air quality of specialized environments, such as airplanes, also has received scrutiny, with one report recommending an increase in the number of cabin air changes per hour to "meet general comfort conditions, and dilute or otherwise reduce normally occurring odors, heat, and contaminants" (NRC, 2002, p. 4). In some circumstances, greater ventilation leads to more desirable outcomes. Similarly, studies on human exposure to particulate matter (PM) have noted that "levels of indoor PM have the potential to exceed outdoor PM levels," thus making indoor environments as critical to health as outdoor air, despite the fact that "the majority of studies have focused on outdoor PM levels and their impacts" (NASEM, 2016, p. 2). Concern about the health effects of indoor PM and lack of clarity regarding its components have led to renewed interest in understanding these indoor and outdoor sources, including microbial components. While increased ventilation with filtration of outdoor air can reduce the concentration of indoor PM (although not necessarily of gaseous pollutants), potential trade-offs, which could include economic and energy impacts, have yet to be fully studied.

In-depth studies to explore the connections among microbial communities, different environmental conditions in built environments, and such outcomes as health or illness need to integrate expertise from microbial ecology, building and building system design and operation, epidemiology and human health, materials science, and a number of other fields. During the past decade, the Alfred P. Sloan Foundation Microbiomes of the Built Environment (MoBE) program has catalyzed a number of complementary research collaborations at the intersection of microbial ecology and building science[3] to begin to tackle these challenges, foster greater cross-disciplinary collaboration, and bring attention to the idea of studying built environments as one would other types of ecosystems. The present report builds on the results of a number of studies supported by foundations, agencies, and societies, including the Sloan program, EPA, NIH, NSF, NASA, the American Society of Heating, Refrigerating and Air-Conditioning Engineers (ASHRAE), and others.

ORGANIZATION OF THE REPORT

The subsequent chapters of this report explore aspects of the dynamic interacting systems connecting humans, built environments, and environmental microbial communities.

A critical driver for better understanding these relationships will be the eventual ability to apply insights from microbiome research to the design and operation of buildings for human health and well-being. Chapter 2 emphasizes the human dimension of microbiome–built environment studies. The chapter reviews what is known about how indoor microbial exposures are associated with potential health outcomes and identifies knowledge gaps that need to be filled to advance the field.

Chapter 3 focuses on major components of buildings and reviews the state of knowledge on how buildings and building systems contain, enhance, or diminish microbial communities. It explores the transport of microorganisms into and within buildings associated with air, water, and surfaces and the influence of such factors as occupants and climate, and it concludes by identifying knowledge gaps in these areas.

Chapter 4 reviews tools and methods for studying the nexus of buildings, microbial communities, and human occupants, including tools with which to sample and characterize built environments and the microbial communities within them and approaches for studying health effects. It identifies needs to improve these tools and support further research in the field.

Chapter 5 examines opportunities and challenges for managing the microbiomes of built environments. It considers examples of interventions that could alter building microbiomes and approaches for assessing their potential benefits and trade-offs.

---

[3]The report uses the term "building science" to refer to the field of knowledge that focuses on understanding physical and operational aspects of buildings and building systems and the impacts on performance; the term "building scientist" is used to refer to a broad range of integrated technical disciplines, including scientists, engineers, and architects who study this area.

Finally, Chapter 6 lays out a vision for the future of buildings as informed by microbial understanding. It provides a research agenda to fill the knowledge gaps identified in the prior report chapters and to make progress toward achieving this vision.

REFERENCES

Amann, R. I., W. Ludwig, and K. H. Schleifer. 1995. Phylogenetic identification and in situ detection of individual microbial cells without cultivation. *Microbiological Reviews* 59(1):143-169.

Cox-Ganser, J. M., J. H. Park, and K. Kreiss. 2010. Office workers and teachers. In *Occupational and environmental lung diseases* (Ch. 23), edited by P. Cullinan, S. M. Tarlo, and B. Nemery. Chichester, UK: John Wiley & Sons, Ltd.

Cox-Ganser, J. M., J. H. Park, and R. Kanwal. 2011. Epidemiology and health effects in moisture-damaged damp buildings. In *Sick building syndrome and related illness: Prevention and remediation of mold contamination*, edited by W. Goldstein. Boca Raton, FL: CRC Press, Taylor & Francis Group. Pp. 11-22.

Dick, E. C., L. C. Jennings, K. A. Mink, C. D. Wartgow, and S. L. Inhorn. 1987. Aerosol transmission of rhinovirus colds. *The Journal of Infectious Diseases* 156(3):442-448.

Falkinham, J. O., A. Pruden, and M. Edwards. 2015. Opportunistic premise plumbing pathogens: Increasingly important pathogens in drinking water. *Pathogens* 4(2):373-386.

Gonzalez, A., and R. D. Holt. 2002. The inflationary effects of environmental fluctuations in source–sink systems. *Proceedings of the National Academy of Sciences of the United States of America* 99(23):14872-14877.

Hinkle, L. E., and S. H. Murray. 1981. The importance of the quality of indoor air. *Bulletin of the New York Academy of Medicine* 57(10):827-844.

Hodgson, M. J., and K. Kreiss. 1986. Building associated diseases: An update. In *Proceedings of IAQ '86: Managing Indoor Air for Health and Energy Conservation*, edited by J. E. Janssen. Atlanta, GA: American Society of Heating, Refrigerating and Air-Conditioning Engineers. Pp. 1-15.

Holt, R. D., M. Barfield, and A. Gonzalez. 2003. Impacts of environmental variability in open populations and communities: "Inflation" in sink environments. *Theoretical Population Biology* 64(3):315-330.

IOM (Institute of Medicine). 1993. *Indoor allergens: Assessing and controlling adverse health effects*. Washington, DC: National Academy Press.

IOM. 2000. *Clearing the air: Asthma and indoor air exposures*. Washington, DC: National Academy Press.

IOM. 2004. *Damp indoor spaces and health*. Washington, DC: The National Academies Press.

Kallmeyer, J., R. Pockalny, R. R. Adhikari, D. C. Smith, and S. D'Hondt. 2012. Global distribution of microbial abundance and biomass in subseafloor sediment. *Proceedings of the National Academy of Sciences of the United States of America* 109(40):16213-16216.

Klepeis, N. E., W. C. Nelson, W. R. Ott, J. P. Robinson, A. M. Tsang, P. Switzer, J. V. Behar, S. C. Hern, and W. H. Engelmann. 2001. The National Human Activity Pattern Survey (NHAPS): A resource for assessing exposure to environmental pollutants. *Journal of Exposure Analysis and Environmental Epidemiology* 11(3):231-252.

Kreiss, K. 1989. The epidemiology of building-related complaints and illness. *Occupational Medicine* 4(4):575-592.

Levin, H. 1989. Building epidemiology and investigations—approaches and results (U.S. experience). In *Chemical, microbiological, health and comfort aspect of indoor air quality—state of the art in SBS*, edited by H. Knöppel and P. Wolkoff. Boston, MA: Kluwer Academic Publishers.

Lindsley, W. G., F. M. Blachere, R. E. Thewlis, A. Vishnu, K. A. Davis, G. Cao, J. E. Palmer, K. E. Clark, M. A. Fisher, R. Khakoo, and D. H. Beezhold. 2010. Measurements of airborne influenza virus in aerosol particles from human coughs. *PLOS ONE* 5(11):e15100.

Marr, L. 2016. Viruses in the built environment. Presentation to the Committee on Microbiomes of the Built Environment: From Research to Application, October 18, 2017.

Melius, J., K. Wallingford, R. Keenlyside, and J. Carpenter. 1984. Indoor air quality—the NIOSH experience. *Annals of the American Conference of Governmental Industrial Hygienists* 10:3-7.

Mendell, M. J., A. G. Mirer, K. Cheung, M. Tong, and J. Douwes. 2011. Respiratory and allergic health effects of dampness, mold, and dampness-related agents: A review of the epidemiologic evidence. *Environmental Health Perspectives* 119(6):748-756.

NASEM (National Academies of Sciences, Engineering, and Medicine). 2016. *Health risks of indoor exposure to particulate matter: Workshop summary*. Washington, DC: The National Academies Press.

NRC (National Research Council). 1981. *Indoor pollutants*. Washington, DC: National Academy Press.

NRC. 2002. *The airliner cabin environment and the health of passengers and crew.* Washington, DC: National Academy Press.

NRC. 2006. *Drinking water distribution systems: Assessing and reducing risks.* Washington, DC: The National Academies Press.

Quince, C., T. P. Curtis, and W. T. Sloan. 2008. The rational exploration of microbial diversity. *The ISME Journal* 2(10):997-1006.

Rhoads, W. J., P. Ji, A. Pruden, and M. A. Edwards. 2015. Water heater temperature set point and water use patterns influence Legionella pneumophila and associated microorganisms at the tap. *Microbiome* 3(1):1-13.

Rhoads, W. J., A. Pruden, and M. A. Edwards. 2016. Convective mixing in distal pipes exacerbates Legionella pneumophila growth in hot water plumbing. *Pathogens* 5(1):E29.

Schoen, M. E., and N. J. Ashbolt. 2011. An in-premise model for Legionella exposure during showering events. *Water Research* 45(18):5826-5836.

Schwake, D. O., E. Garner, O. R. Strom, A. Pruden, and M. A. Edwards. 2016. Legionella DNA markers in tap water coincident with a spike in Legionnaires' disease in Flint, MI. *Environmental Science & Technology Letters* 3(9):311-315.

Stephens, B. 2016. What have we learned about the microbiomes of indoor environments? *mSystems* 1(4):e00083-e00116.

Wells, W. F., M. W. Wells, and S. Mudd. 1939. Infection of air: Bacteriologic and epidemiologic factors. *American Journal of Public Health and the Nations Health* 29(8):863-880.

WHO (World Health Organization). 1983. *Indoor air pollutants: Exposure and health effects.* EURO Reports and Studies No. 78. Geneva, Switzerland: WHO.

WHO. 2009. *WHO guidelines for indoor air quality: Dampness and mold.* http://www.euro.who.int/__data/assets/pdf_file/0017/43325/E92645.pdf (accessed April 27, 2017).

2

Microorganisms in Built Environments: Impacts on Human Health

Chapter Highlights

- There is a demonstrated link between exposures to infectious microorganisms present in built environments and human health. In a number of cases, the mechanisms of transmission are well understood, but more can be learned about how built environment design influences proliferation or transmission of such infectious microorganisms.
- There is evidence of a link between exposure to indoor microorganisms and the development of respiratory and allergic symptoms, particularly those arising from exposure to microorganisms that flourish in damp indoor settings.
- Preliminary evidence suggests that certain microbial exposures, including early-life exposures to diverse microorganisms associated with animals, may have beneficial health effects, such as protection from allergy and respiratory symptoms.
- A number of additional potential health impacts (beneficial or adverse) associated with exposures to indoor microorganisms are being explored. Impacts on nonrespiratory (e.g., neurologic) outcomes are less well understood. More investigation is needed to understand which exposures may be beneficial or adverse and by what mechanisms.

> - Additional studies will be needed to clarify causal links between microbial exposures in built environments and health impacts. These should include further longitudinal human studies with complementary animal and in vitro studies to assess how stage of life, exposure route, coexposures, dose, and genetic sensitivity influence the relation of individual indoor microbial exposures to health outcomes.

How do the microbiomes of the different indoor environments in which humans spend time for working, living, learning, and playing impact human health and well-being? What building conditions support microbial communities that benefit or harm human health and well-being? If most microorganisms do not infect humans, do those that thrive in indoor environments influence human health for good or bad, and if so, by what mechanisms? These questions are among those that motivate the study of microbiomes of the built environment. This chapter begins by laying the groundwork for understanding how microorganisms found in buildings may influence health. The chapter then addresses, in turn, infection transmission in indoor environments, noninfectious health outcomes associated with indoor microorganisms, and potential benefits of microbial exposures. The chapter concludes with summary observations and a discussion of knowledge gaps in the area of health impacts of built environment microbial exposures. Chapters 3–5 examine how building characteristics and occupants shape the indoor microbiome (see Chapter 3), tools that can be used in research on microbiomes of the built environment (see Chapter 4), and potential interventions that can alter these microbiomes (see Chapter 5).

INFLUENCE OF BUILDING MICROBIOMES ON HUMAN HEALTH: ECOLOGIC AND BIOLOGIC PLAUSIBILITY

Several considerations support the plausibility of the influence of building microbiomes on human health. First, in developed areas of the world, indoor environments are the primary ecosystems inhabited by people. Second, the environments people inhabit may influence the human microbiome, which may in turn impact human health. For example, microorganisms present in the environment may proliferate in niche-specific ecosystems of the human host—such as in airways, the gut, and on skin. Third, a wide array of microbial components and characteristics are known to impact human health. Finally, a number of sources of microorganisms within indoor environments impact human health.

Implications of Building Microbiomes for the Diversity of the Human Microbiome

In developed areas of the world, humans are born and spend the vast majority of their lives indoors, which may limit the diversity of microorganisms to which they are exposed. A building's envelope (foundation, walls, windows, and roofs) separates the indoor and outdoor environments, thus reducing exposure to microorganisms that thrive outdoors and potentially increasing exposure to organisms that thrive indoors.

The diversity of the microbiomes of the built environments in which humans live may impact the microbiomes of their bodies. Studies have shown that humans who spend significant time outdoors or live in dwellings with more open building envelope designs that result in high levels of unfiltered or minimally filtered air exchange with the outdoors have more diverse microbiomes relative to those who live in dwellings with less open designs (Clemente et al., 2015; Hanski et al., 2012).

The extent to which the indoor microbiome contributes to this diversity or lack thereof is not well understood, however. It has also been suggested that exposure to reduced microbial (especially bacterial) diversity may be less a function of the microbial content of buildings than a side effect of the modern human diet, which is less diverse than that of our ancestors: it varies little with the seasons; may be affected by the use of antibiotics; and may select for a limited number of human microbial taxa, particularly in the gut (Yatsunenko et al., 2012). This diversity may benefit human health as the microbiomes to which humans are exposed may be important for immune development and the processing of nutrients in the gut, which may not function as well when gut microbial communities change. People in economically disadvantaged and less developed societies who spend more time outdoors can have higher infectious disease risk and higher infant mortality (Clemente et al., 2015; Hanski et al., 2012). However, this may be due more to their health when exposed to infectious agents than to the diversity of the microorganisms to which they are exposed; as noted, some evidence suggests beneficial immune system effects from exposure to diverse microbes (see the section on "Beneficial Effects of Microbes"). Thus, the relative lack of microbial diversity may have positive or adverse effects on human physiologic and immune responses and health and may in turn influence risk of chronic noninfectious symptoms and diseases. By separating themselves from the outdoors, humans may have eroded the diversity of their own, as well as their environmental, microbiomes.

Environmental Influences on the Human Microbiome

Mounting evidence indicates that the human microbiome is influenced by the environment and that it is integral to human development. One of the most studied influences is the transmission of particular infectious microorganisms. For example, fomites are surfaces or objects on which microorganisms can deposit and that allow for transmission to a host (Julian, 2010). Fomites are well documented in the spread of infectious disease, and there is research associated with human exposure to indoor pathogens (Dick et al., 1987; Wong et al., 2010).

Another known influence of the environmental microbiome on the human microbiome is the process of birth. Each individual's microbiome is acquired both in utero and from the environment at birth. Babies delivered vaginally and by Cesarean delivery show differences in their microbiome composition (Bokulich et al., 2016; Hill et al., 2017; Rutayisire et al., 2016). Yet the human microbiome does not fully stabilize to adult patterns until 2–3 years of age (Dethlefsen et al., 2006). New technologies and bioinformatics techniques for genomic analysis of microbial DNA extracted from environmental samples are providing insights in this area not previously possible (see Box 2-1). One topic of interest is the nature of nonpathogenic interactions between indoor and human microbiomes.

Neonatal studies provide evidence that the microbes from the environment that are of human origin may influence the human microbiome. Other studies also have provided evidence that dogs and humans have bacteria in common (Song et al., 2013). Yet there is no concrete evidence that the human microbiome can be colonized by bacteria that originate from a building. For example, an in-depth study of the indoor and human microbiomes in which seven families were followed over 6 weeks indicated that the majority of the building microbiome measurable on home surfaces originated from the occupants. This study also found that the building microbiome did not appear to influence the occupants' skin microbial structure or composition (Lax et al., 2014). Further research will be needed to understand the reproducibility and generalizability of the findings of this study, and how temperature, humidity, building materials, and the integrity of the building structure impact the interchange between the indoor and human bacterial microbiomes.

Microbial Components Associated with Human Health Effects

Microorganisms can impact human health through a variety of mechanisms. The dominant microbial components linked to human health include pathogen-associated molecular patterns (PAMPs). PAMPs are molecules such as endotoxin and lipopolysaccharide (LPS) (a component of bacterial cell

> **BOX 2-1**
> **Transmission of Infectious and Noninfectious Organisms in Hospital Neonates**
>
> - Metagenomic technologies have supported new approaches to investigating whether or how hospital environments influence transmission of infectious or noninfectious microorganisms, colonization of full-term or premature infants, and subsequent infant health status. Studies employing these technologies are conducted in the delivery room, operating room, or neonatal intensive care unit (NICU).
> - Recent studies support transmission of skin, gut, or vaginal microorganisms (including many that are noninfectious) from mother to child during delivery, with potentially long-term immunomodulatory effects (Rutayisire et al., 2016).
> - Studies have also begun to characterize the NICU environment and its less diverse microbiome in order to evaluate its contribution to the immediate risk of infection and also to colonization of newborns (Aagaard et al., 2012; Banda, 2016; Bokulich et al., 2013, 2016; Cunnington et al., 2016; Forno et al., 2008; Groer et al., 2015; Hartz et al., 2015; Kitsios and Morris, 2016; Lax and Gilbert, 2015; Lokugamage, 2016; Ly et al., 2006; Martin et al., 2016; Neu, 2016; Pistiner et al., 2008; Prince et al., 2014, 2015; Shogan et al., 2013; Stokholm et al., 2016; Torrazza et al., 2013).

membranes), flagellin (from bacteria), and (1–3)-β-D glucans (also referred to as triple helical glucan, from fungi wall membranes). These molecules are associated with groups of microbes (bacteria or fungi) that may influence human innate immune system responses, interact with airway epithelial cell or irritant receptors (Lambrecht and Hammad, 2013, 2014), or have toxic effects. For example, many indoor fungi produce metabolites that can induce respiratory or systemic toxicity upon exposure (Kuhn and Ghannoum, 2003).

Mycotoxins, products of fungal metabolism (Robbins et al., 2000), can also provoke physiological responses. These fungal metabolites have been shown in mechanistic, toxicological studies to be relevant to human health, and more than 300 mycotoxins are potentially harmful with respect to food contamination (Alshannaq and Yu, 2017). This situation is much less clear for inhalation exposure to indoor-relevant mycotoxins. Only a handful of studies support a role of these compounds in inflammatory processes when combined with exposure to other microbial components (endotoxin, glucans) (Korkalainen et al., 2017). The potential airway epithelial toxicity of microbial components may be increased by the presence of tobacco smoke or other factors that disturb the epithelial barriers, tight junctions, antimicrobial production, or mucocilliary ability to clear bacteria (Lambrecht and Hammad, 2013, 2014).

A large respiratory and allergy literature that has evolved over two decades suggests complex positive, as well as negative, associations of various PAMPs with allergy and respiratory outcomes. In the observational epidemiologic literature, the directionality (whether the risk or protective factors) and the magnitude of associations with microbially produced molecules such as LPS/endotoxin appear to be dependent on dose, human body compartment (e.g., airway, gut), host, and stage of life (Perzanowski et al., 2006; Sordillo et al., 2010). Recent experimental literature also provides supporting evidence that PAMPs and other fungal (e.g., chitin) (Mohapatra et al., 2016; O'Dea et al., 2014) and bacterial components may "train" innate immune responses, with downstream effects on the body's ability to deal with either infection or allergic responses. While experimental models find direct effects of bacterial or fungal components (e.g., endotoxin or glucans), more recent observational birth cohort studies suggest that endotoxin and other PAMPs may be markers for complex communities of environmental bacteria and fungi, many of which have previously not been associated with disease (Manor et al., 2014).

Sources of Indoor Microbiomes That Are Relevant to Human Health

Recent observations suggest that occupants and outdoor microbes entering buildings through ventilation and tracked in through dust are the dominant origin of indoor environmental bacteria, particularly those that can be airborne (Adams et al., 2015; Prussin et al., 2015). Occupants contributing to the indoor microbiome include humans and nonhuman occupants, such as rodents and cockroaches, as well as pets, which are sources of bacteria and therefore can be direct sources of bacterial PAMPs (Thorne, 2015). In farm studies, likely sources for indoor PAMPs have included animal feed and farm animals. In urban environments, LPS/endotoxin sources include not only pets but also associations with moisture, such as that due to concrete basements, humidifiers (Park et al., 2001), and water damage in situations less extreme than flood conditions. Endotoxin on dirt and decaying plant material can be tracked into a house by its inhabitants. On the other hand, the relative contribution of endotoxin- or microbe-containing outdoor airborne particles to the suite of indoor microbial components (Hanson et al., 2016; Manzano-León, 2013) is not well understood. Although not causal or definitive, there is evidence that LPS/endotoxin exposures may be protective against the development of allergies in rural (Thorne, 2015) and U.S. urban environments (Park et al., 2001).

Sources of fungi indoors vary. Typically, in non-water-damaged buildings, fungi enter a building through leakage in the building envelope and through ventilation systems, are carried indoors by occupants, or may be brought indoors in association with building materials. The growth of fungi is generally moisture dependent. In cases of extreme water damage—as in flooding in New Orleans, Louisiana (Mitchell et al., 2012); in Cedar Rapids, Iowa (Hoppe et al., 2012); or in Boulder, Colorado (Emerson et al., 2015)—high levels of fungi not originating from building occupants have been measured on building surfaces and in air. Even in non-water-damaged buildings, fungi can grow in or on building materials when sufficient moisture is present (Adan and Samson, 2011; Macher et al., 2017), and their growth is affected by such factors as the chemical composition of building materials.

Less is currently understood about the origins of viruses in the built environment (beyond transmission of specific pathogens). Evidence suggests, however, either enhanced sources of bacteria relative to viruses indoors or preferential removal of viruses as air penetrates indoors (Prussin et al., 2015). See Chapter 3 for greater detail.

TRANSMISSION OF INFECTION IN INDOOR ENVIRONMENTS

Much prior work investigating the impact of environmental conditions on the survival of microorganisms and their transmission to and between humans has focused on infectious organisms. As noted earlier, fomites (Julian, 2010) are well documented in the spread of infectious disease, and there is research documenting aerosol transmission of pathogens in different nonresidential indoor environments (Dick et al., 1987; Wong et al., 2010). This section examines the modes of transmission and complex, mixed exposures or coinfections for selected pathogens.

Modes of Transmission

Some viruses, bacteria, fungi, protozoa, and algae in the indoor environment have long been known to be pathogens, with the potential to cause infectious disease or allergic illness (Burge, 1980). Potentially infectious organisms can vary in terms of their transmissibility (ease of spread), their mode of spread indoors, and their virulence (a quantitative measure of pathogenicity or potential to cause disease). Subspecies within a bacterial species also may have highly divergent health effects, highlighting the need for detailed microbial information (Ponnusamy et al., 2016). Transmissibility, virulence, and mode of spread all influence the mode and effectiveness of infection control, a topic generally beyond the scope of this report. Recent interventions to reduce transmission of tuberculosis[1] indoors, for example, have shown the benefit of advancing understand-

---

[1] Tuberculosis is caused by infection with the bacterium *Mycobacterium tuberculosis*.

ing of transmissibility and drug resistance patterns (Barrera et al., 2015; Dharmadhikari et al., 2014; Mphaphlele et al., 2015; Nardell, 2016).

Table 2-1 summarizes modes of transmission for selected pathogens that have been associated with infection due to exposure in the indoor environment. The organisms listed in this table vary widely in their transmissibility and virulence. Influenza A, for example, can be challenging to contain, as person-to-person transmission can begin on contact with asymptomatic individuals or 24 hours before symptoms are present. Severe acute respiratory syndrome (SARS), a virus more virulent than influenza A, may be somewhat easier to contain, in that symptoms are likely to be present when the virus is transmissible. Viral pathogens such as influenza also may have multiple modes of transmission—for example, by droplet inhalation or by fomite contact—that may vary in their relative contribution to transmission according to such building conditions as indoor temperature and humidity (Koep et al., 2013). Although some of the organisms listed in Table 2-1 have infectious potential, some may be present in buildings but are not highly virulent and rarely cause infection in people with healthy lungs and healthy immune systems. For example, many immunocompetent people breathe in a variety of Aspergillus species (including *Aspergillus fumigatis*) indoors on a regular basis without becoming infected.

Figure 2-1 illustrates various routes of transmission of infectious agents in the indoor environment. Chapter 3 explores indoor sources and reservoirs of microorganisms in greater detail, but one important pathway for human exposure is inhalation of microorganisms carried or resuspended in room air, as well as microorganisms found in building water systems that become aerosolized. Relevant exposure routes include both aerosolization of fine particles that can travel into the deep lung and inhalation of coarse particles that may be deposited in the upper respiratory system (Hatch, 1961) or be translocated to the gastrointestinal tract via the mucociliary escalator (Harada and Repine, 1985).

Another important route of transmission is via fomites, in which the microorganism is transferred from a surface. The initial deposition of microorganisms onto a fomite can occur by deposition of aerosols or dusts or by transference via contact from an individual. A susceptible host can become infected by touching a microbially laden fomite and subsequently touching the mouth, eye, nose, or other body surface. Many microorganisms can be transmitted in this manner, especially viruses, including rhinovirus, influenza virus, coronavirus, norovirus, rotavirus, hepatitis A virus, adenovirus, and astrovirus (Boone and Gerba, 2007). Multiple modes of transmission often occur for a particular microorganism (Nicas and Jones, 2009). For example, influenza may be transmitted by aerosolization or by contact with a virus-laden fomite, and the predominant mode of transmission may be related in part to the absolute humidity inside the building

TABLE 2-1 Mode of Transmission for Selected Pathogens Implicated in Infections Due to Inhalation or Fomite Interactions

Super Kingdom	Mode of Transmission	Examples
Bacteria	Inhalation	Bacillus anthracis Coxiella burnetii Chlamydia psittaci Legionella Mycobacterium tuberculosis Atypical mycobacteria
	Fomites	Clostridium difficile Staphylococcus aureus Enterococcus
Fungi	Inhalation	Cryptococcus neoformans Histoplasma capsulatum Aspergillus fumigatus
	Fomites	Trichophyton mentagrophytes Trichophyton rubrum
Protozoa	Inhalation	Acanthamoeba spp.
Viruses	Inhalation	Variola (smallpox) Rubella Norovirus Rotavirus Adenovirus Coxsackie virus Influenza Rhinovirus Coronaviruses (Middle East respiratory syndrome [MERS], severe acute respiratory syndrome [SARS])
	Fomites	Variola (smallpox) Rubella Norovirus Rotavirus Adenovirus Coxsackie virus Influenza Rhinovirus Coronaviruses (MERS, SARS)

SOURCES: Table created using data from Burrell (1991), Couch (1981), and Yu et al. (2004).

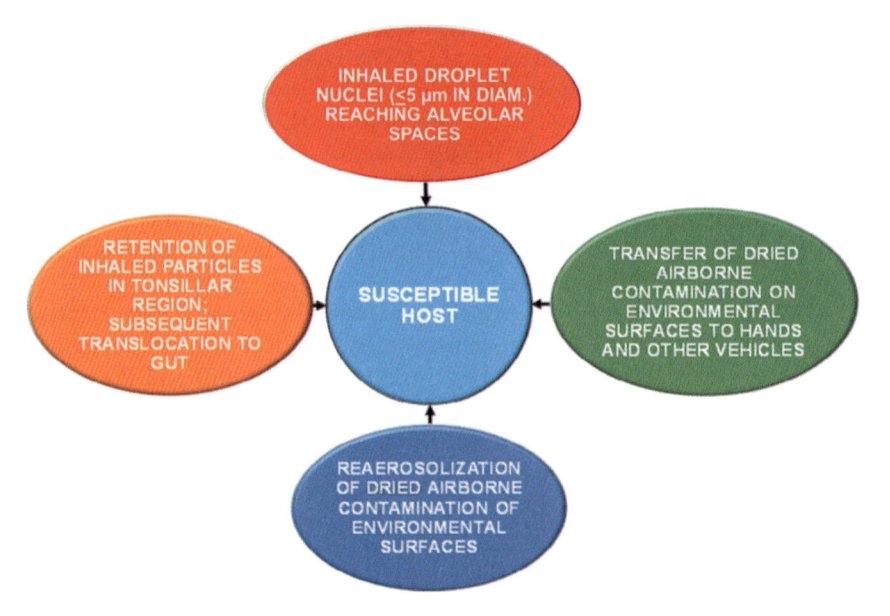

[Figure: Five ovals showing modes of transmission centered on "SUSCEPTIBLE HOST":
- INHALED DROPLET NUCLEI (≤5 μm IN DIAM.) REACHING ALVEOLAR SPACES
- RETENTION OF INHALED PARTICLES IN TONSILLAR REGION; SUBSEQUENT TRANSLOCATION TO GUT
- TRANSFER OF DRIED AIRBORNE CONTAMINATION ON ENVIRONMENTAL SURFACES TO HANDS AND OTHER VEHICLES
- REAEROSOLIZATION OF DRIED AIRBORNE CONTAMINATION OF ENVIRONMENTAL SURFACES]

FIGURE 2-1 Modes of transmission of microorganisms from the airborne environment.
SOURCE: Sattar, 2016.

(McDevitt et al., 2010; Tellier, 2009; Yang and Marr, 2011). Thus, transmission of microorganisms within the built environment is a complex process that is contingent not only on the class of microorganism itself but also the state of the building (e.g., its humidity or ventilation) and the number and behavior of building occupants.

A recently recognized class of pathogens, termed saprozoic (Ashbolt, 2015), is capable of amplifying on wetted surfaces such as those that may be found in built environments. In some cases, amplification is facilitated by growth of amoebae in biofilms that may harbor pathogens. A recent report suggests that fungi may also facilitate growth of saprozoites (Alum and Isaacs, 2016). Organisms of this class, which have been shown to be transmitted by indoor aerosolization or from fomites, include nontuberculosis *Mycobacterium* spp. *Legionella pneumophila* can similarly grow in association with biofilms, in this case in water systems. *L. pneumophila* can cause legionellosis and Pontiac fever, a nonfatal flu-like respiratory disease with a short incubation period and for which recovery usually occurs without medical intervention (OSHA, 2017; Principe et al., 2017). Locales in which amplification may occur along with subsequent aerosolization and infection include hot tubs and whirlpools (Falkinham, 2003); indoor fountains and architectural features (Haupt et al., 2012; O'Loughlin et al., 2007); shower heads and hoses (Feazel et al., 2009; Schoen and Ashbolt, 2011); heating, ventilation, and air conditioning (HVAC) systems and humidifiers (Stetzenbach, 2007); and toilets (Azuma et al., 2012).

Biofilms can also exist in piping systems and provide habitats for pathogen amplification (Wang et al., 2012). Corroded pipes may provide a more favorable environment for this microbial amplification, and this factor has been associated with a spike in legionellosis cases in Flint, Michigan (Schwake et al., 2016) (see Box 2-2).

Building design and operational characteristics can affect the relative importance of different modes of transmission (Li et al., 2007), as can cleaning and handwashing practices (Sandora et al., 2008) and the use of masks (Wei and Li, 2016). For example, in multiunit residential (and likely commercial) buildings, cross-transmission of aerosols via the ventilation system may serve as a conduit for disease transmission (Mao and Gao, 2015; Nardell et al., 1991).

Complex, Mixed Exposures

The potential exists in any environmental exposure for multiple agents to affect a single host concomitantly. Outbreaks of legionellosis have provided evidence that coinfection can adversely impact patient outcomes, although it is not clear whether the coinfections occurred following the initial *L. pneumophila* exposure (e.g., in a hospital) (Fernandez et al., 2002). Animal studies have shown that coinfections can modulate immune system response to one of the challenges (Redford et al., 2014). However, reports of multiple pathogen impacts in the context of indoor exposures in the built environment are currently unavailable.

Somewhat more information is available on the interaction between exposure to infectious agents and concomitant respiratory exposure to adverse chemical or physical agents. Most of this information is derived from animal models. Controlled rodent experiments showed that susceptibility to inhaled *Klebsiella* was increased by prior exposure to nitrogen dioxide (NO_2), as well as to aerosols of cadmium and nickel (as chloride) (Gardner, 1982). Intratracheal administration of combustion particles to mice was found to increase the lethality of inhaled *Streptococcus* (Hatch et al., 1985). Using a similar assay, Arany and colleagues (1986) showed that inhalation of a variety of volatile organic compounds either enhanced lethality or reduced the mice's ability to fight off the inhaled *Klebsiella* infection. It can be anticipated that many human exposures in the built environment will be complex, and future studies will need to explore in more detail the potentially interacting or modulating effects of combined chemical, physical particulate matter, and microbial exposures on health.

> **BOX 2-2**
> **Flint, Michigan: An Exemplar of the Role of the Municipal Water Supply in Triggering a Legionnaires' Disease Outbreak**
>
> In April 2014, the city of Flint, Michigan, switched its water supply from purchased water from Detroit, Michigan, to the local Flint River, in an effort to save money. A subsequent "man-made disaster" ensued because the much higher corrosivity of the Flint River water was overlooked and federal law requiring the addition of orthophosphate corrosion inhibitor was not followed, resulting in widespread contamination of tap water with lead and iron leaching from the city's pipes. At the same time, Legionnaires' disease incidence spiked in Genessee County, where Flint is located, with 45 cases in summer 2014, 43 cases in 2015, and 12 deaths overall. Prior to the water switch, the incidence of Legionnaires' disease was typically 4–6 cases per year (MDHHS, 2016; MDHHS and GCHD, 2015, 2016). Legionnaires' disease is caused by *Legionella* bacteria, usually *L. pneumophila* serogroup 1, which are known to inhabit the biofilms of potable water systems. Exposure usually occurs via inhalation of aerosols that can form when running the water, typically during showering, causing a severe and deadly form of pneumonia. Quantification of macrophage infectivity potentiator (mip) genes characteristic of *L. pneumophila* demonstrated that the copy numbers were up to 10,000 times higher at hospital taps than what is typically reported as background in potable water during the time Flint River water was being used (Schwake et al., 2016). Iron is an essential nutrient for *Legionella*, and iron leaching from pipes reacts with chlorine to decrease the residual disinfectant. Thus, the outbreak in Flint demonstrates that the quality of the water supplied by the municipality, in this case high corrosivity leading to increased iron and decreased chlorine, is an important factor in triggering Legionnaires' disease outbreak. Hot water heater temperature settings and maintenance and frequency of water use are believed to be important determinants of *L. pneumophila* growth and survival (Rhoads et al., 2015). There are currently no federal requirements for monitoring or controlling *Legionella* at the tap. Research efforts aimed at understanding the microbial ecological factors that trigger *L. pneumophila* growth in building plumbing systems and the effects of water chemistry provided by utilities can help inform future control measures, regulations, and monitoring approaches.
>
> SOURCE: Box courtesy of Amy Pruden and Marc Edwards, Department of Civil and Environmental Engineering, Virginia Polytechnic Institute and State University (used with permission).

In summary, the indoor environment can be a venue for exposure to a variety of infectious agents, including bacteria, protozoa, fungi, and viruses. These exposures may occur via inhalation (including of aerosols from premise plumbing) or contact with fomites. For bacteria, in particular, wetted surfaces can serve as a habitat for biofilms, which can amplify certain species, including several pathogens. These exposures are likely to be from a combination of multiple microorganisms, which can affect how the microorganisms impact the human host in a variety of ways, many of which have not yet been studied.

DAMP INDOOR ENVIRONMENTS, INDOOR MICROBIAL EXPOSURES, AND RESPIRATORY OR ALLERGIC DISEASE OUTCOMES

Another area in which there has been much previous work is the impacts of damp indoor environments on human health. This research has typically focused on a number of noninfectious[2] health effects from exposures to indoor microorganisms, with the greatest number of studies focusing on dampness, observations of microbial growth or detection of mold odors, and respiratory and allergic symptoms.

Damp building conditions promote the growth of mold, bacteria, and other microbial agents. Damp buildings may also contain other living organisms, such as dust mites and cockroaches (along with their associated microbial communities), which can potentially contribute to exacerbation of respiratory issues. Occupants in damp buildings can be exposed to pollutants in the air not only from biological contaminants but also from the deterioration of building materials, which can be accelerated by the presence of dampness inside the building.

Table 2-2 summarizes evidence from three review publications that damp buildings influence health (IOM, 2004; Mendell et al., 2011; WHO, 2009). There is a larger body of literature on upper and lower respiratory outcomes, with more limited attention to nonrespiratory outcomes. Over the years since the 2004 Institute of Medicine (IOM) report *Damp Indoor Spaces and Health* was issued, the evidence has become stronger for an association between damp buildings and the exacerbation of asthma and, importantly, the development of asthma. Furthermore, the 2011 review by Mendell and colleagues evaluates a number of health outcomes not considered by the IOM committee in 2004 (Mendell et al., 2011). Not all health effects that have been suggested as having possible associations with damp indoor environments have been the subject of sufficient published literature to enable evaluation. The 2004 IOM report lists a number of health outcomes as having inadequate or insufficient evidence with which to determine whether an association exists, including airflow obstruction (in otherwise healthy persons), skin symptoms, mucous membrane irritation

---

[2]The term "noninfectious" is used here to represent potential health associations that are not known to be due to specific infection, although conventional clinical testing through pathogen culture methods may not fully consider the breadth of diversity and community structure of indoor microbiomes that has been increasingly characterized.

TABLE 2-2 Associations Between Health Outcomes and Exposure to Damp Indoor Environments

Health Outcome	Strength of Association		
	IOM, 2004	WHO, 2009	Mendell et al., 2011
Upper respiratory (nasal and throat) tract symptoms	Sufficient Evidence	Sufficient Evidence	Sufficient Evidence
Wheeze	Sufficient Evidence	Sufficient Evidence	Sufficient Evidence
Cough	Sufficient Evidence	Sufficient Evidence	Sufficient Evidence
Shortness of breath	Limited or Suggestive Evidence	Sufficient Evidence	Sufficient Evidence
Exacerbation of existing asthma	Sufficient Evidence	Sufficient Evidence	Sufficient* Evidence
Development of asthma	Limited or Suggestive Evidence	Sufficient Evidence	Sufficient Evidence
Current asthma	Not Evaluated	Sufficient Evidence	Sufficient Evidence
Ever-diagnosed asthma	Not Evaluated	Not Evaluated	Sufficient Evidence
Bronchitis	Not Evaluated	Limited or Suggestive Evidence	Sufficient Evidence
Respiratory infections	Not Evaluated	Sufficient Evidence	Sufficient Evidence
Allergic rhinitis	Not Evaluated	Limited or Suggestive Evidence	Sufficient Evidence
Eczema	Not Evaluated	Not Evaluated	Sufficient Evidence
Common cold	Not Evaluated	Not Evaluated	Limited or Suggestive Evidence
Allergy/atopy	Not Evaluated	Inadequate/ Insufficient Evidence	Limited or Suggestive Evidence
Hypersensitivity pneumonitis	Clinical Evidence	Clinical Evidence	Clinical Evidence

*Evidence judged to be strongly suggestive of causation.

syndrome, gastrointestinal tract problems, chronic obstructive pulmonary disease, fatigue, inhalation fevers (nonoccupational exposures), neuropsychiatric symptoms, lower respiratory illness in otherwise healthy adults, cancer, acute idiopathic pulmonary hemorrhage in infants, reproductive effects, and rheumatologic and other immune diseases. In their 2011 review, Mendell and colleagues identify altered lung function as having inadequate evidence available.

These reviews indicate stronger evidence for adverse health effects due to signs of dampness relative to those due to measures of indoor environmental microorganisms, a point reiterated in a recent review on the associations of health effects with observational assessment of dampness and mold (Mendell and Kumagai, 2017). The potential benefits of reducing moisture problems in buildings have been discussed in the literature (Mendell, 2007; Mendell and Kumagai, 2017; Mendell et al., 2008; Mudarri and Fisk, 2007; WHO, 2009).

The rest of this section selectively discusses studies published in the past decade concerning indoor dampness and respiratory or allergic health outcomes, and it addresses the question "Can specific indoor microbial exposures account in part for the adverse respiratory effects of building dampness?" In summary, results have been inconsistent from study to study, and it is currently not understood which specific contaminants or combinations thereof in damp indoor environments cause the various health effects under what circumstances.

A review by Quansah and colleagues (2012) on dampness and mold in homes and asthma development found that associations with the presence of visible mold and with mold odor were evidence for mold-related causal agents for asthma (Quansah et al., 2012). In a recent review Sharpe and colleagues (2015) conclude that there is some evidence that in indoor environments *Penicillium*, *Aspergillus*, *Cladosporium*, and *Alternaria* species are associated with asthma development and with worsening of asthma symptoms, but that more work is needed on the role of fungal diversity. Another recent review on indoor exposures associated with the exacerbation of asthma found that many indoor exposures exacerbated asthma, including indoor dampness or dampness-related agents such as endotoxin, culturable *Penicillium*, and total fungi (Kanchongkittiphon et al., 2015). On the other hand, in an earlier paper, Hägerhed-Engman and colleagues (2009) report no association between concentrations of measured mold species and asthma in the Swedish Dampness in Buildings and Health study (Hägerhed-Engman et al., 2009). Similarly, a recent study of Danish schools found that high classroom dampness was associated with lower lung function and wheezing, but microbial components were not consistently associated with health outcomes (Holst et al., 2016).

Data from large cross-sectional studies of approximately 46,000 8- to 12-year-old children in 20 countries during phase two of the International Study of Asthma and Allergies in Childhood (ISAAC) found significant and consistent associations of current exposure to dampness or visible mold in homes with respiratory and allergic symptoms (Weinmayr et al., 2013). Associations with current exposure included wheezing, coughing up phlegm without a cold, rhinitis, rhinoconjunctivitis, and eczema. Children were similarly affected by dampness regardless of whether they had allergic sensitization or parents with allergy.

Studies of individual episodes of water-damaged buildings using detailed exposure measures and precise case definitions have subsequently produced stronger evidence that a variety of definable microbial groups may be partly responsible for associations between damp environments and respiratory health. In a water-damaged 20-story office building in the Northeast United States, linear associations were found between respiratory illnesses (particularly physician-diagnosed asthma) and hydrophilic fungi and ergosterol (a molecule in fungal cell membranes) (Park et al., 2008). In another study of a water-damaged U.S. building, associations were found between adverse respiratory functioning and thermophilic actinomycetes (a phylum of bacteria) and nontuberculosis mycobacteria (Park et al., 2017). While these cases are instructive, the range of microbial ecosystems in the built environment that may have long-term adverse effects on respiratory health is not well defined.

Higher fungal counts, specific fungal or bacterial species, and higher endotoxin/LPS levels have been associated with building dampness (Park et al., 2008). The absence of a gold standard for measuring mold (Chew et al., 2016) means that complementary measurement methods often yield a better perspective on which mold or mold component exposures linked to dampness may be relevant to health. As discussed by Cox-Ganser and colleagues (2011), mycotoxins have been posited as contributors to these symptoms, but their effects in the context of damp and poorly maintained buildings are not well understood.

Outright flooding, building-related water damage, indoor point sources of dampness and water, suboptimally maintained cooling towers, or poor maintenance of buildings leading to indoor moisture problems may lead to proliferation of microbial communities that can cause noninfectious adverse allergic and nonallergic respiratory responses. This is the case particularly in susceptible populations such as children and people with preexisting asthma.

Association with Asthma Development and Worsening of Asthma Control

Living, working, or attending school in damp indoor environments has been associated with onset or worsening and exacerbation of asthsma in children and adults. A meta-analysis of 33 studies estimates that exposure to dampness and mold in the home raises the risk for asthma development, history of asthma, and current asthma by about 30–50 percent (Fisk et al., 2007). It has been estimated that 21 percent of current asthma cases in the United States can be attributed to dampness and mold, which translates to 4.6 million of the 21.8 million U.S. asthma cases at the time of the estimate (Mudarri and Fisk, 2007). This study also estimates the annual cost of asthma attributable to exposure to dampness and mold at $3.5 billion.

Recent evidence suggests that both allergic and nonallergic asthma are more frequent in damp indoor environments. This evidence is important because it suggests that not all of the effects of dampness result from allergic responses to allergens from microbial and nonmicrobial sources (e.g., allergens on dust mites and cockroaches) that proliferate in damp conditions. As mentioned previously, microorganisms include components other than allergens. Fungal and bacterial components such as PAMPs, for example, can cause irritant or inflammatory symptoms through nonallergic biologic pathways.

While inheritance is presumed to play a role in how children or adults respond to exposures to indoor microorganisms such as fungi and bacteria, the genetics that might modulate either adverse or protective microbial responses resulting in the development or worsening of allergic or nonallergic respiratory disease is not well understood. Fungal components (e.g., chitin [Da Silva et al., 2008]) or bacterial components (e.g., LPS [Simpson and Martinez, 2010]) are known to stimulate innate pathways that are under genetic regulation, but the implications of this knowledge for characterization of susceptibility to disease are not well defined. Gene–environment interactions are likely to be complex and may even be sex-specific, since investigators of genes related to immununoglobulin E (IgE) or asthma have identified sex-specific polymorphisms of genes that regulate asthma or allergy responses, such as interleukin-17 receptor B (IL17RB) and thymic stromal lymphopoietin (TSLP) (Hunninghake et al., 2008, 2011). To complicate matters, a genetic polymorphism may be a risk factor for asthma, but it may increase the protective effects (on allergic asthma) of an environmental exposure such as the fungal and bacterial microbes on a farm (Loss et al., 2016).

A 9-year prospective follow-up study on onset of asthma was conducted on 7,104 young adults from 13 countries who had participated in the European Community Respiratory Health Survey (ECRHS) I and II and had not reported baseline respiratory symptoms or asthma (Norback et al., 2013). The findings strongly support the connection between indoor dampness and/or visible mold and new onset of asthma or bronchial hyperresponsiveness.

Various articles have reviewed the accumulating literature strengthening the links among dampness, dampness-related agents, and asthma exacerbation occurring in nonallergic as well as allergic children (Kanchongkittiphon et al., 2015). For example, a 2010 study of inner-city children indicated that indoor fungi originating from both indoor and outdoor sources could worsen asthma (Pongracic et al., 2010).

Prior work by investigators from the U.S. National Institute for Occupational Safety and Health showed that in occupants of a historically water-damaged office building, asthma was associated with concentrations of hydrophilic fungi in floor dust (Park et al., 2008) and that there was a synergistic effect of fungal and endotoxin exposure in respiratory health effects (Park et al., 2006). Other analyses of this office building population found evidence for exacerbation of building-related asthma. The onset of posthire asthma was associated with a lower prevalence of positive skin-prick reactions to common allergens, including indoor and outdoor mold mixes (Cox-Ganser et al., 2005). In summary, this study showed that building occupants who had developed asthma after working in the building were less allergic to common allergens, so the asthma must have developed directly in concert with something present in the building.

Fungi are a source of many different components that may have health effects. Allergens and other antigens on indoor fungi are one set of microbial components known to worsen respiratory symptoms in people who have established asthma and are allergic to the specific fungal allergen they inhale. Paradoxically, in early life, higher exposure to some nonmicrobial allergens (e.g., peanut allergen) may actually promote tolerance and protection from allergic responses (Du Toit et al., 2015). However, it is not known whether tolerance can occur with early-childhood exposure to fungal allergens. Overall, early-life fungal exposures worsen the health of young children, but many of the negative effects in early life may not be due to the allergens. Annex Table 2-1 at the end of this chapter summarizes epidemiologic studies that explore interventions aimed at reducing exposures and improving asthma outcomes.

Association with Rhinitis

Jaakkola and colleagues (2013) conducted a meta-analysis of the literature on the association between exposure to fungi in damp buildings and rhinitis (nasal inflammation). They concluded that there is evidence that dampness and mold exposures at home can cause or exacerbate rhinitis and its subcategories of allergic rhinitis and rhinoconjunctivitis (which produces nose and eye symptoms such as sneezing and itching). The strongest associations occurred with the presence of mold odor, which suggests that microbial agents were involved in some fashion in the development or exacerbation of this health effect.

The Role of Microbial Metabolic Compounds

The largest body of research on potential health associations of exposures to microbial metabolites focuses on fungi. Work by Miller and colleagues (2010) and Rand and colleagues (2011, 2013) on effects of fungal secondary compounds (mycotoxins) indicates that these compounds, in concentrations that may be found in indoor environments, can cause inflammation on a cellular level that points toward nonallergic asthma. With their models, they have shown that exposure to very low concentrations of these compounds precipitates such lung changes as secretion and modification of mucoids on lung surfaces, as well as changes in the respiratory cell composition obtained from sampling. This has been found with many pure compounds from a range of xerophilic, mesophilic, and hydrophilic fungi, as well as triple helical glucan. Moreover, it has been demonstrated that these compounds act on well-understood receptors (e.g., dectin-1 and NfkB). In addition, these compounds significantly modulate downstream gene transduction, transcription, and cytokine expression patterns associated with a variety of immune system pathways related to inflammatory and/or asthma provocation (including TH1, TH2, and TH3) in compound-, dose-, and time-dependent ways.

Association with Eczema

In a recent analysis of data on children in the ISAAC study, residential exposure to dampness and mold was found to be significantly associated with eczema, a type of skin inflammation, in the previous year. Dampness and mold in the first year of life also was associated with higher odds of the child's ever having had eczema, by parental report. Allergic sensitization of the child did not modify the association, suggesting that dampness or mold might increase the risk of eczema through mechanisms other than IgE-mediated allergy, although parental allergic disease did increase the odds of eczema with exposure (Tsakok et al., 2015). Occupational studies support the possibility of other mechanisms through which dampness-associated microbes of the built environment could increase the incidence of eczema, including contact dermatitis. However, little is known about mechanisms in the context of the indoor environments in which the children lived.

Association with Hypersensitivity Pneumonitis

Hypersensitivity pneumonitis, also known as allergic alveolitis, is a condition in which inhaled dust, fungi, or chemicals lead to inflammation in the lung. This condition has been recognized in relation to nonindustrial indoor environments for decades. Early publications on outbreaks usually implicated microbial dissemination from humidification and ventilation systems, although the specific organism(s) were rarely identified. Ventilation- and humidifier-related hypersensitivity pneumonitis may have decreased with recognition of the need for design changes and maintenance of cleanliness in these systems. More recent publications have implicated damp indoor environments, along with other exposures (e.g., chemicals), in causing hypersensitivity pneumonitis. Buildings with long-standing water damage from roof leaks, other building envelope water incursion, plumbing leaks, and below-grade moisture problems have all had reported clusters of hypersensitivity pneumonitis (Borchers et al., 2017).

NONAIRWAY AND NONALLERGY EFFECTS

The impact of bacterial or fungal communities in buildings on health outcomes other than those related to respiratory conditions or allergy is less well established. Epidemiologic and toxicologic data suggest that effects of environmental microbial exposures on human health are likely to be dependent on dose, stage of life, physiologic compartment (e.g., gut, nose, lung, skin), and/or host (including sex).

Endocrine Disruption and Child Development

Viable indoor microbes likely metabolize chemicals present in the built environment. A number of chemicals found in common household products or furnishings, such as bisphenol A, perfluorooctane sulfonate (PFOS), and perfluorooctanoate (PFOA), have been shown when ingested to influence endocrine function (e.g., glucose metabolism or diabetes risk) and child growth, including the risk of becoming overweight or obese (Heindel et al., 2017). Laboratory experiments show that environmental microbes can metabolize these and other chemicals with endocrine disruptive properties, sometimes creating more bioactive or bioavailable chemicals and sometimes reducing their toxicity (Blavier et al., 2016; Bradley et al., 2016; Gramec and Mašič, 2016; Janicki et al., 2016; Koestel et al., 2017; Vejdovszky et al., 2017). However, the chemical metabolites of indoor microbes have not been well characterized. Moreover, little is known about the ingestion, inhalation, transdermal, or other exposures of small children to chemical by-products of microbial metabolism and whether they influence endocrine function or growth. Fungi themselves can also produce mycotoxins that affect the production of estrogen, but the relevance of this finding to health and whether these fungi are found in buildings are similarly unknown (Vejdovszky et al., 2017).

In addition to producing active metabolites or metabolizing chemicals in the indoor environment, the indoor environmental microbiome itself may influence the human microbiome in ways that lead to health effects. The skin and its commensal microbial communities are instrumental in protection against the environment. These communities, which vary in composition by location on the body, are determined largely by environmental and physiologic conditions. The microbial communities on the skin modulate the health status of the body through immune responses that maintain health or, in specific situations, may promote disease (Barnard and Li, 2017). While knowledge has grown about the diversity and characteristics of skin microbial communities, relatively little is understood about the strains responsible for the function of these communities in maintaining protection and health or in promoting disease (Barnard and Li, 2017; Belkaid and Segre, 2014). Personal care products and other household chemicals that can act as endocrine disruptors have been shown to affect the skin and its microbiome (Bouslimani et al., 2015, 2016). However, there are no data to suggest that the indoor environmental microbiome influences the composition or structure of the skin microbiome (Lax et al., 2014).

Toxicologic and epidemiologic studies provide supporting evidence that specific groups of chemicals in plastics, furnishings, and personal care products may have endocrine-disrupting properties that increase the risk of hormone-related conditions such as diabetes or obesity, particularly in children exposed in early life. While it is known that microbes metabolize chemicals, and limited data show that building materials influence microbial degradation of those materials, it is not known whether metabolism of building materials, furnishings, or household chemicals by microbes results in active metabolites relevant to human health, specifically to endocrine disruption. These chemicals can also impact the skin microbiome and its ability to protect the body against disease, but modulation of the skin microbiome and its direct impact on human health is also not yet well characterized. While degradation or metabolism of building materials by microbes may be partly related to moisture, additional factors play into the health and safety aspects of these materials and the likelihood of their degradation by microbes. Further research in this area may inform the choice of "healthy" building materials and furnishings.

Future studies could focus on testing whether environmental microorganisms or by-products of household chemicals with endocrine-disrupting properties interact at major sites when encountering the human body. Research could elucidate whether these encounters lead to systematic changes in how the body functions and develops, with influences on endocrine or immune function or child growth. Such potential sites of interaction include the skin; airway; and gastrointestinal tract, including the oropharyngeal sites. Concurrently with the present study, a study by another committee of the National Academies of Sciences, Engineering, and Medicine (Advancing Understanding of the Implications of Environment–Chemical Interactions with Human Microbiomes) is under way, focused on developing a research agenda to guide the chemical risk assessment community in understanding how chemical exposures may modulate the human microbiome and how the human microbiome may modulate the effects of chemical exposures (via metabolism of chemicals) on human health outcomes. With that study's focus on the human microbiome and risk assessment for health outcomes related to indoor and other environmental chemical exposures, the proposed research agenda is likely to be of interest to the built environment community as it will likely suggest gaps in research needed to better understand the concerns outlined in this section.

Brain Health and Neurologic Outcomes

Studies in the United States and internationally have suggested that adult or child brain health may be influenced by aspects of homes or public buildings such as offices, schools, and hospitals. These studies have evaluated outcomes ranging from central nervous system (CNS) symptoms such as headaches, to mood (e.g., depression) and sleep disorders, to changes in neurocognitive or behavioral function. The findings of these studies have prompted investigators to posit that indoor microbial communities may mediate a portion of the observed associations of brain health outcomes with building characteristics or with other potential microbial sources within or proximal to buildings.

There are several potential pathways by which such effects could occur. Airborne indoor microbial components or metabolites may enter the brain directly via the olfactory bulb (Block et al., 2012). Alternatively, they may have indirect neurologic effects through airway autonomic stimulation or by causing airway or systemic inflammation. Microorganisms and their metabolites are also found in building water systems. Under certain circumstances, microbial deterioration of building or indoor plumbing materials may result in release of toxic chemicals into the indoor water system, and the absorption of these chemicals may lead to negative brain effects. To date, with specific exceptions, evidence to support these hypotheses is scant, in part because of methodologic challenges in exposure and outcome measurement and because of the potential complexity of the biologic response to exposures.

Another possible mechanism for the influence of indoor environmental microorganisms on brain development is through their potential interactions with the human gut microbiome. However, apart from the increasingly well-documented association of indoor dogs (Fujimura et al., 2014) or farm animals with the enrichment of certain beneficial microbes in the human gut and the diversification of home microbial communities, data to support a relationship between the indoor microbiome and the human microbiome are scant. As noted previously, the human microbiome can be influenced by the environment. For example, the ElderMet study found that subjects in long-term care facilities had different and less diverse gut microbiomes relative to those living in the community, although such factors as diet and health status appear to play important roles in explaining this variability (Claesson et al., 2012).

The question then arises of whether the environmental microbiome can influence the gut or respiratory microbiome. From a compartmental and stage-of-life point of view, infants and toddlers are more likely than adults to ingest environmental microbes, which may in turn influence the composition and function of the gut microbiome. A preliminary investigation using data on 20 infants from the Canadian Healthy Infant Longitudinal Development (CHILD) study, for example, found associations between house dust and fecal samples for several classes of bacteria (Konya et al., 2014). Other potential mechanisms by which environmental microbes might influence brain health might not require the ingestion or proliferation of microbes, and they could include responses to airway or skin encounters with microbial components or metabolites. Accumulating evidence suggests an association between microbiota present in the gut and brain function (Burokas et al., 2015), as the existence of bidirectional neural and immune interactions between the intestine and the brain has been proven (Keunen et al., 2015). While the exact mechanisms by which the microbiome can influence the development of the CNS are not completely understood, proposed communication between these systems is termed the microbiome–gut–brain axis. There are numerous complex interactions between the microbes that reside in the gastrointestinal tract and immune, endocrine, and neurologic systems. The vagus nerve directly connects the gastrointestinal nervous system to the brain. The immune system monitors the presence of microbes and reacts to changes in their structure and composition, transmitting this information to the CNS. Furthermore, human commensal organisms release metabolites that are precursors of important neurotransmitters such as gamma amino-butyric acid (GABA) and serotonin precursors (e.g., 5-hydroxytryptophan [5-HT]) or might induce the production of 5-HT by enteroendocrine cells, which in turn influences the nervous system and the brain (O'Mahony et al., 2015).

Altered microbial community structure, or dysbiosis, in the setting of stress and disease has been associated with alterations in behavior, cognition, emotion, and levels of inflammatory cells (Cryan and Dinan, 2012). Germ-free mice demonstrate altered risk-taking behavior, memory, and anxiety (Al-Asmakh et al., 2012; Cryan and Dinan, 2012; Neufeld et al., 2011). Diaz Heijtz and colleagues (2011) used an elegant experiment to demonstrate the role of the microbiome in brain and CNS development: germ-free mice had an altered neurologic response when subjected to stress tests, and this response was reversed only when bacteria were transplanted into the mouse caecum during infancy, rather than during adulthood.

Perturbing initial colonization and microbiome development has been shown to affect brain development and to pose a risk of developing neurologic disorders later in life (Borre et al., 2014; Diaz Heijtz et al., 2011). Therefore, environmental exposure to the built environment microbiome early in life could have much more significant effects on neurologic development than those that occur later in life.

Studies supporting the influence of gut microbial composition and function on a complex and bidirectional gut–brain axis were recently reviewed by Jasarevic and colleagues (2016). Microbial metabolites, such as short chain fatty acids and chemotactic peptides, may influence the brain directly or may bind to intestinal epithelial cell receptors to enable the secretion of peptide neurotransmitters. Microbes can also interact with gut immune cells, and resulting cytokines may influence brain function. Centrally activated neural circuits that may be activated by various stressors may also influence gut microbial composition or function. In addition to the gut, the nasal or lower airway epithelial layer, including its mucus interface, is a barrier, interactive site, or portal through which microbes or their components may influence immune function (and perhaps brain function) through multiple mechanisms. These mechanisms may include disruption of the mucous barrier, stimulation of toll-like receptors (TLRs) or other immune receptors (Davies, 2014), or direct translocation from the upper respiratory tract to the brain.

Moisture-Damaged Buildings, Poor-Quality Housing, and Brain Health

Brain health outcomes have also been studied for associations with moisture-damaged buildings, poor housing conditions in disadvantaged communities, outdoor traffic proximal to schools and homes, pets in homes, and the construction of what is colloquially termed "green" housing (see the section on "Beneficial Effects of Microbes"). It is important to note that the use of the term "green" encompasses a variety of design and building approaches, potential interventions, and actual success at achieving more healthful or energy-efficient buildings.

In addition to the large literature linking damp buildings, buildings with water damage, and housing in poor repair with respiratory symptoms, some studies also link these conditions to reduction in brain health, with symptoms of headache, nausea, mood disorders, difficulty concentrating, or sleep difficulties (Ansarin et al., 2013; Casas et al., 2013; Chambers et al., 2016; Cox-Ganser et al., 2010, 2011; Faber et al., 2015; Francisco et al., 2016; Jacobs et al., 2015; Oudin et al., 2016; Park et al., 2008, 2017; Schiffman et al., 2005, Shiue, 2015; Singh and Kenney, 2013; Tiesler et al., 2015). In circumstances such as the post–Hurricane Katrina experience, victims living in trailers suffer from stress and mood disorders, but it has been difficult to disentangle the trauma of the experience from responses to the physical environment, including molds and their products, as well as other airborne exposures. Mold odors and visible mold have been linked to sleep difficulties (Ansarin et al., 2013; Chambers et al., 2016; Faber et al., 2015;

Jacobs et al., 2015; Oudin et al., 2016; Shiue, 2015; Singh and Kenney, 2013; Tiesler et al., 2015), and it is unknown whether proinflammatory upper airway influences combine with the direct brain effects from mold to contribute to such difficulties, including sleep-disordered breathing.

The physical exposures linked to poverty, and thereby connected to lower socioeconomic status, are compounded by many other disparities, including reduced access to an adequate diet, an enriching environment, health care, and education, as well as the presence of environmental toxins. Children in poverty have a significantly increased risk for developmental delay, poor school performance, and behavioral problems (Blay et al., 2015; Chambers et al., 2016; Evans and Schamberg, 2009; Hanson et al., 2013; Saigal and Doyle, 2008). Poverty has even been linked to changes in brain structure (Hanson et al., 2013; Jednorog et al., 2012). Many of the aspects associated with poverty (e.g., poor diet and health, depression, and anxiety) are associated with alteration of the human microbiome (Myles, 2014; Rook et al., 2014). Even less obvious factors, such as the number of caregivers and the indoor and outdoor home environment (e.g., presence of animals, access to outdoor green spaces), can significantly impact the human microbiome (Lax et al., 2015). Providing clean, dry, well-maintained housing has recently been shown to improve respiratory health (Colton et al., 2014), but it is not known whether this could occur in part through changing microbial exposures.

Epidemiologic and toxicologic studies, including studies using rodent models, provide growing evidence that outdoor particulate and gaseous pollutants influence brain health. The Outdoor Air Pollution and Brain Health Workshop (Block et al., 2012) was followed by additional studies providing evidence that outdoor pollutants have effects on the brain, including children's neurodevelopment and neurocognitive and behavioral function in school (Basagana et al., 2016; Clifford et al., 2016; Dadvand et al., 2015; Harris et al., 2015; Kicinski et al., 2015; Sunyer et al., 2015), as well as cognitive function and decline in elders. A small but growing body of literature demonstrates the presence of microbes or microbial components on outdoor particles that may penetrate indoors and on indoor particles that may have both outdoor and indoor sources, which can be of local, regional, or even transoceanic origin (Frankel et al., 2012). Their contribution to effects on brain health is poorly understood. Although an in-depth discussion of outdoor environmental pollution is beyond the scope of this study, air, water, and other materials from the outdoor environment come indoors into built environments. The integration of disparate areas of knowledge that will underpin a clearer understanding of how indoor microbial exposures can lead to health outcomes and how this understanding could lead to practical application will include efforts to clarify the ties between the outdoors and the indoors.

While building conditions have been associated with adverse neurocognitive, behavioral, and other brain health outcomes, the specific contribution of the indoor microbiome to these adverse effects is unknown. The availability of evolving microbiologic, genomic, bioinformatics, and statistical technologies may facilitate a fuller assessment of whether and how brain health is influenced by indoor environmental microbes. An increased mechanistic understanding of how indoor microbes can act on neurologic outcomes would inform future environmental interventions designed to protect brain health. In addition, multiple adverse environmental exposures that are greater in disadvantaged populations and neighborhoods may add to or modify effects of indoor microbial exposures. Thus, it is important to conduct studies of indoor microbial effects on neurocognitive outcomes in built environments reflecting a range of socioeconomic circumstances and resources and to incorporate prospective longitudinal studies in future assessments.

BENEFICIAL EFFECTS OF MICROBES

Over the past two decades, interest in the potentially protective influences of indoor environmental microbes has been stimulated in part by farm community studies (see Box 2-3 and Annex Table 2-2 at the end of this chapter) that reproducibly demonstrate a reduced risk of allergic asthma with certain microbial exposures at home in early life. These exposures are estimated through measurement of bacteria components, microbial culture, or first-generation molecular biologic tools. Additionally, new metagenomics tools have opened a window into better understanding of the vast number of microbes that inhabit the human body and the microbial communities that are in the built environment.

Indoor microbes and their components and metabolites may have beneficial health effects in some circumstances and detrimental health effects in others. Characteristics of the built environment, the microbial community, and human behaviors within that environment may modulate the dose of the microorganism or the compartment exposed, and they may in turn influence whether the microorganism has a beneficial, adverse, or null effect on health. The same community of indoor microorganisms and their cell wall components may benefit human health in some circumstances and be detrimental in others depending on such circumstances as building characteristics, life stage of the person being exposed, exposure route, co-exposures, dose, and genetic sensitivity. For example, a baby who ingests microorganisms while crawling on the floor may respond differently from an adult with asthma who inhales the same microorganisms. Potentially beneficial microbes include primarily microorganisms that train or modulate the human immune system (Kelly et al., 2005), produce small molecules

> **BOX 2-3**
> **Asthma, Early-Life Exposures, and Farm-Type Environments**
>
> Evidence from multiple studies shows an association between children who grow up in traditional farm environments and a reduced risk of developing asthma (von Mutius and Vercelli, 2010). The "farm effect" is thought to be explained by the child's early-life contact with farm animals, in particular cattle, and their microbes (Ege et al., 2011; Illi et al., 2012; Wlasiuk and Vercelli, 2012). A recent study eloquently demonstrated the asthma-protective effects of traditional farming through observation of children in the Amish (AM) and Hutterite (HT) populations from Indiana and North Dakota in the United States (Stein et al., 2016). These two populations have similar genetic ancestries and lifestyles, but their farming practices are quite distinct: traditional among the AM, and industrialized among the HT. At the same time, AM and HT children show striking disparities in the prevalence of asthma, which is 4 times lower in the AM than in the HT, and allergic sensitization, which is 4.6 times lower among the AM. The environment in which the AM and the HT live is also quite different, because median levels of home endotoxin (a proxy of microbial exposure) were 6.7-fold higher among the AM, and the microbial composition of dust samples from AM and HT homes was also distinct. It should be noted that other exposures may be different among the AM and HT communities—for example, exposure to plant particulates and fecal microbes.
>
> These differences in the built environment exposure were paralleled by profound differences in the proportions, phenotypes, and functions of innate immune cells in the children; this points to strong environmental effects on the children's immune system. Finally, and notably, when instilled into the airways of ovalbumin-treated mice, extracts of AM but not HT house dust were sufficient to dramatically reduce airway hyperresponsiveness, broncho-alveolar eosinophilia, and immunoglobulin E (IgE), and these effects depended on innate immunity (Stein et al., 2016). In more recent unpublished data from the group, dust acquired from HT barns has also been shown to have asthma-protective properties in mouse models. However, HT children are not allowed in those barns in early life and therefore lack that protection during a critical developmental window (Gozdz et al., 2016). These findings underscore the fundamental role of environmental exposures in asthma protection and therefore highlight the importance of the farming built environment for immune-associated chronic diseases.

that mediate human health (Neish, 2009), or enable other functions that improve well-being in a human host (Reber et al., 2016; Rook and Lowry, 2008).

This knowledge has led to the identification of biomarkers that are associated with health benefit and an extensive body of research aimed at understanding the mechanisms of how microbial exposures could benefit human health (Heederik and von Mutius, 2012; Torow and Hornef, 2017; von Mutius, 2016; von Mutius and Vercelli, 2010; Wlasiuk and Vercelli,

2012). Annex Table 2-2 at the end of this chapter summarizes selected studies that have examined potential beneficial microbial exposures in the indoor environment.

This field of research needs many more targeted, longitudinal observational studies and intervention studies in order to pinpoint where beneficial tips into adverse, as well as the reverse, and to build the knowledge base needed to modulate built environments so as to positively impact human health.

Association of Microbial Exposures with Protection from Asthma and Respiratory Symptoms

While the adverse effects of microorganisms, their components, and their products have well-documented influence on the development, progression, or exacerbation of asthma and allergies (Eggleston et al., 1998; Lai et al., 2015; Quansah et al., 2012), there is also a substantive literature addressing *protection* from the development of asthma and allergy conditions through microbial exposures (Behbod et al., 2015; Celedón et al., 2007; Sordillo et al., 2010). Recent evidence suggests that the bacterial communities in dust in homes near farms may be reducing the incidence of asthma in certain populations (Stein et al., 2016) (see Box 2-3).

Dog-associated dust and the bacterial communities therein also have been shown to reduce atopy symptoms in mice (Fujimura et al., 2014), a finding suggesting that young children who live in homes with dogs may be less likely to develop asthma (Fall et al., 2015). The mechanism is hypothesized to be attributable to features of the microbial communities associated with animals. These microbes may act by shaping immune responses on skin, on airway mucosal surfaces, and in the gut (von Mutius, 2016). Differences in gut microbiota, including increased concentration of *Veillonella* spp., *Lachnospira* spp., and *Faecalibacterium* spp. from the phylum Firmicutes (Arrieta et al., 2015), in the first 3 months of life appear to play a role in asthma protection. Along with these taxa and Bacteroidetes (Lynch et al., 2014), increasing evidence suggests associations between exposure to high bacterial and fungal diversity in early life and protection from asthma and wheeze (Dannemiller et al., 2014; Ege et al., 2011; Tischer et al., 2016).

Exposure to insect and mammal stool within the first year of life has been shown to reduce the risk of development of preasthmatic wheeze (Dami and Bracken, 2016), while reduced exposure to certain species of Firmicutes and Bacteroidetes associated with house dust has been associated with an increase in atopy (Lynch et al., 2014). As these bacterial phyla are often associated with the mammalian gut, their absence suggests a reduction in stool in the environment. The observation that reducing mouse and cockroach stool can increase the probability of wheezing and asthma may be seen as paradoxical, but it accords with current hypotheses that immune system challenges can reduce atopy.

In addition to the effects of exposures to bacteria discussed above, a small but growing literature indicates selected beneficial effects of early-life exposures to fungi in relation to the development of allergy and respiratory disease (Behbod et al., 2015; Tischer et al., 2016). However, especially in the case of fungi (but also to some extent in the case of bacteria), researchers still know very little about what products specific microbes are making or dispersing that may benefit human health if encountered in early life. More longitudinal studies of early-life microbial effects on subsequent child health are needed that define specific taxa and microbial community structure and function. Observations in these studies will need to be further validated in animal models to elucidate the mode of transfer from the environment to human compartments and biologic mechanisms of immune, physiologic, and/or other health effects. These studies may not result in recommendations that suggest reproducing the lifestyle or building structure that is associated with protection (e.g., most people will not live in a house with cows inhabiting a barn below), but they may help define the components of microbial exposures that are of potential therapeutic benefit for some people. As has been shown with endotoxin, it is likely that various microbial components may be good for some people and bad for others, depending on dose, compartment (whether inhaled or ingested), stage of life at which they are exposed (early life or adulthood), and/or susceptibility factors (heredity and additional environmental factors covarying with poverty). This concept is not easily applied to building design, where the goal is to provide healthier buildings for all. Nonetheless, longitudinal studies that include microbial measurements can define conditions and specific exposures that are adverse or protective for specific groups of people.

Potential Beneficial Effects Associated with Green Buildings and Green Spaces

Green Building Design

"Green building" design aims to promote environmental and energy sustainability, and the concept has become adopted more widely within the architectural and design community: the U.S. Green Building Council's Leadership in Energy and Environmental Design (LEED) standard is one example. Green building designs reflect a mix of efforts focused on energy, water, and indoor environmental quality, but they offer no guarantees of meeting specific requirements for energy savings or healthful design. Green building design and what it entails is discussed in more detail in Chapter 3.

Green building design could potentially impact occupant health by improving air quality. Yet there have been no randomized trials of residents moving from conventional to green housing and how this change is associated with health outcomes. Findings of a few recent observational studies, some of them quite small, suggest that reducing home dampness through weatherization and by using green building approaches may improve asthma control.[3] These findings are based on the assumption that appropriate standards and guidelines on outdoor air ventilation rates and selection of building materials should be followed. To the committee's knowledge, there have been no published studies investigating ties among design, indoor microbiology, and health outcomes.

Researchers also have attempted to link green building design to neurocognitive outcomes. In one recent study, moving from poorly maintained or extremely aging housing to newly renovated or constructed green housing was found to result in fewer self-reported lost schooldays or workdays; less disturbed sleep, sadness, nervousness, and restlessness; and improved child behavior (Jacobs et al., 2015). Yet these improvements may have resulted from living in newer and better maintained housing, irrespective of green design features. Furthermore, it is unknown whether environmental microbial exposures contributed to the reported improvements.

While green buildings have repeatedly been cited as an approach to improving health (Allen et al., 2015, 2016; NRC, 2006), the specific attributes of green buildings that may contribute to improved health need to be broken down in order to understand how or why certain approaches to improved construction and operation under specific ecologic conditions may make such contributions. It is important to understand as well that many features of green buildings may make no contributions to improved indoor environmental quality or to the health of occupants.

Green Space

Outdoor green space surrounding buildings has been associated with improved patient outcomes (see Center for Health Care Design, 1995; IOM, 2007; Ulrich et al., 2004) and overall health status (Gong et al., 2016; Nieuwenhuijsen, 2016). Scientists have posited that exposure to green space may contribute to health benefits through exposure to plant-associated environmental microbiota. Studies in Finland, for example, have shown that living close to green space and agriculture rather than close to a town increases the biodiversity of the skin microbiota and correlates with reduced allergic sensitization (Hanski et al., 2012). For a review of potential

---

[3]See http://www.enterprisecommunity.org/solutions-and-innovation/green-communities (accessed April 25, 2017).

links among green space, the microbiome, and immune system function, see Rook (2013). Rook and Knight (2015) recently called for city planning and architectural designs that optimize the biodiversity of microbial exposure in urban settings, with an emphasis on green spaces.

SUMMARY OBSERVATIONS AND KNOWLEDGE GAPS

Summary Observations

The ability of microorganisms within the built environment to affect human health is supported by data for many types of infectious bacteria, viruses, fungi, and protozoa. Selected examples of microorganisms that can be encountered by humans in the built environment and can result in infections from inhalation or from contacts with fomites are provided in Table 2-1 presented earlier in this chapter.

A variety of health effects that are not infectious in nature have also been reported. Extensive research demonstrates that exposure to damp, water-damaged buildings and "sick buildings" results in negative respiratory health effects for building occupants. These respiratory effects often are not directly related to allergy and may also be caused by irritant or proinflammatory components of microbes. Connections between the built environment and a number of nonrespiratory health outcomes have been suggested—including effects on child development, brain health, and mental health—although less is known about whether these effects are due to exposures to indoor environmental microbes and through which physiologic mechanisms they occur.

Beneficial effects on health from exposure to microorganisms in built environments have also been reported, particularly for exposures that occur in early life. Evidence for mechanisms by which microbial exposures can have positive effects is starting to accumulate and can be built upon to better understand what constitutes microbial communities that have such effects and the potential mechanisms of action involved. In particular, a small but growing literature shows that selected early-life microbial exposures are associated with positive benefits in relation to the development of allergy and respiratory disease.

Research connecting microbial exposures in the built environment with health impacts draws on a number of study approaches, including epidemiologic observational studies, such as longitudinal cohort studies, as well as dose-response studies. For example, results of observational studies that suggest connections between built environment microbiomes and human health can benefit from further validation in animal models to elucidate the microbial communities or components responsible for protective or adverse health responses, the modes of transfer from the environment to humans, and the types and mechanisms of physiologic responses. There also may be opportunities to leverage data from existing health studies. These study designs are discussed in more detail in Chapter 4.

Knowledge Gaps

On the basis of the above summary observations and the information developed in this chapter connecting indoor microbial exposures to human health effects, the committee identified the following goals for research to address knowledge gaps and advance the field:

1. **Improve understanding of the transmission and impacts of infectious microorganisms within the built environment.** Continued elucidation of the transmission of infectious microorganisms in a variety of built environments would be useful, including studies on modes of transmission for emerging respiratory pathogens; for pathogens with evolving patterns of hosts (animal as well as human); and for pathogens with problematic characteristics, such as drug resistance (e.g., *Mycobacterium tuberculosis*, *Clostridium difficile* [Peng et al., 2017]). Improved understanding of the relationship between microbial transmission and the timing of symptom onset also would be useful in informing future strategies for minimizing exposures.
2. **Clarify the relationships between microbial communities that thrive in damp buildings and negative allergic, respiratory, neurocognitive, and other health outcomes.** A number of studies link human exposure to damp and water-damaged buildings with allergic and other respiratory health impacts. But further research is needed to identify how building conditions and maintenance result in dampness that leads to the proliferation of communities of microbes that can adversely affect respiratory health; to distinguish among the microbial and nonmicrobial effects of dampness; to understand the relationships among microbes, building materials, and chemicals within damp buildings; and to assess how human health is impacted when dampness is reduced.
3. **Elucidate the immunologic, physiologic, or other biologic mechanisms through which microbial exposures in built environments may influence human health.** A number of possible health impacts from microbial exposures have been suggested (beyond infectious disease and the association between dampness and respiratory health), including developmental and neurocognitive effects. Much remains unknown about how the composition of the microbial communities, stage of life, route of exposure, and other factors affect human biologic responses and potential health outcomes. For example, a

growing body of literature suggests that the human microbiome, particularly the microbial communities in the gut, can influence health. But questions remain about the extent to which indoor microbiomes influence the composition and function of the human microbiome (on the skin and in the gut, oral, or airway compartments) and what that may mean for health outcomes.

4. **Gain further understanding of the beneficial impacts of exposures to microbial communities on human health.** Several studies have documented associations between early-life microbial exposures and exposure to diverse microorganisms associated with animals and later protective health effects. Further longitudinal studies of the effects of early-life microbial exposures on subsequent child and adult health will be needed to understand these connections more fully. Also useful would be additional data with which to further explore the beneficial impacts of exposures to specific microbial communities and clarify such factors as the extent to which these impacts vary with the characteristics of a building's occupants, stage of life, and the routes through which the occupants are exposed.

5. **Develop an improved understanding of complex, mixed exposures in the built environment.** Responses to the complex and compound exposures that occur routinely in built environments, such as exposures to multiple microorganisms and to combinations of microorganisms and chemicals, have not been thoroughly elucidated to date.

6. **Design studies to test health-related hypotheses, drawing on the integrated expertise of health professionals, microbiologists, chemists, building scientists, and engineers.** Many of the studies investigating how human microbial exposures relate to health outcomes have been conducted in ways that make them difficult to reproduce in other buildings and make it difficult to understand how specific building attributes affect both the microbial exposures and the health outcomes. A variety of further studies will need to be developed and implemented to ensure that the experiments are reproducible and produce results that can be translated into actionable outcomes.

REFERENCES

Aagaard, K., K. Riehle, J. Ma, N. Segata, T. A. Mistretta, C. Coarfa, S. Raza, S. Rosenbaum, I. Van den Veyver, A. Milosavljevic, D. Gevers, C. Huttenhower, J. Petrosino, and J. Versalovic. 2012. A metagenomic approach to characterization of the vaginal microbiome signature in pregnancy. *PLOS ONE* 7(6):e36466.

Adams, R. I., A. C. Bateman, H. M. Bik, and J. F. Meadow. 2015. Microbiota of the indoor environment: A meta-analysis. *Microbiome* 3:49. doi:10.1186/s40168-015-0108-3.

Adan, O. C. G., and R. A. Samson. 2011. *Fundamentals of mold growth in indoor environments and strategies for healthy living.* Wageningen, The Netherlands: Wageningen Academic Publishers.

Al-Asmakh, M., F. Anuar, F. Zadjali, J. Rafter, and S. Pettersson. 2012. Gut microbial communities modulating brain development and function. *Gut Microbes* 3(4):366-373.

Allen, J. G., P. MacNaughton, J. G. Laurent, S. S. Flanigan, E. S. Eitland, and J. D. Spengler. 2015. Green buildings and health. *Current Environmental Health Reports* 2(3):250-258.

Allen, J. G., P. MacNaughton, U. Satish, S. Santanam, J. Vallarino, and J. D. Spengler. 2016. Associations of cognitive function scores with carbon dioxide, ventilation, and volatile organic compound exposures in office workers: A controlled exposure study of green and conventional office environments. *Environmental Health Perspectives* 124(6):805-812.

Alshannaq, A., and J. H. Yu. 2017. Occurrence, toxicity, and analysis of major mycotoxins in food. *International Journal of Environmental Research and Public Health* 14(6). doi:10.3390/ijerph14060632.

Alum, A., and G. Z. Isaacs. 2016. Aerobiology of the built environment: Synergy between Legionella and fungi. *American Journal of Infection Control* 44 (9 Suppl.):S138-S143.

Ansarin, K., L. Sahebi, and S. Sabur. 2013. Obstructive sleep apnea syndrome: Complaints and housing characteristics in a population in the United States. *Sao Paulo Medical Journal* 131(4):220-227.

Arany, C., J. A. Shea, W. J. O'Graham, and F. J. Miller. 1986. The effects of inhalation of organic chemical air contaminants on murine lung host defenses. *Fundamental and Applied Toxicology* 6(4):713-720.

Arrieta, M.-C., L. T. Stiemsma, P. A. Dimitriu, L. Thorson, S. Russell, S. Yurist-Doutsch, B. Kuzeljevic, M. J. Gold, H. M. Britton, D. L. Lefebvre, P. Subbarao, P. Mandhane, A. Becker, K. M. McNagny, M. R. Sears, T. Kollmann, W. W. Mohn, S. E. Turvey, and B. Brett Finlay. 2015. Early infancy microbial and metabolic alterations affect risk of childhood asthma. *Science Translational Medicine* 7(307):307ra152.

Ashbolt, N. J. 2015. Environmental (saprozoic) pathogens of engineered water systems: Understanding their ecology for risk assessment and management. *Pathogens* 4(2):390-405.

Azuma, K., I. Uchiyama, and J. Okumura. 2012. Assessing the risk of Legionnaires' disease: The inhalation exposure model and the estimated risk in residential bathrooms. *Regulatory Toxicology and Pharmacology* 65(1):1-6.

Banda, C. 2016. Offering women safer options than "vaginal seeding" for infants born by caesarean section. *British Medical Journal* 352:i1734.

Barnard, E., and H. Li. 2017. Shaping of cutaneous function by encounters with commensals. *The Journal of Physiology* 595(2):437-450.

Barrera, E., V. Livchits, and E. Nardell. 2015. F-A-S-T: A refocused, intensified, administrative tuberculosis transmission control strategy. *International Journal of Tuberculosis and Lung Disease* 19(4):381-384.

Basagana, X., M. Esnaola, I. Rivas, F. Amato, M. Alvarez-Pedrerol, J. Forns, M. Lopez-Vicente, J. Pujol, M. Nieuwenhuijsen, X. Querol, and J. Sunyer. 2016. Neurodevelopmental deceleration by urban fine particles from different emission sources: A longitudinal observational study. *Environmental Health Perspectives* 124(10):1630-1636.

Behbod, B., J. E. Sordillo, E. B. Hoffman, S. Datta, T. E. Webb, D. L. Kwan, J. A. Kamel, M. L. Muilenberg, J. A. Scott, G. L. Chew, T. A. E. Platts-Mills, J. Schwartz, B. Coull, H. Burge, and D. R. Gold. 2015. Asthma and allergy development: Contrasting influences of yeasts and other fungal exposures. *Clinical & Experimental Allergy* 45(1):154-163.

Belkaid, Y., and J. A. Segre. 2014. Dialogue between skin microbiota and immunity. *Science* 346(6212):954-959.

Birzele, L. T., M. Depner, M. J. Ege, M. Engel, S. Kublik, C. Bernau, G. J. Loss, J. Genuneit, E. Horak, M. Schloter, C. Braun-Fahrländer, H. Danielewicz, D. Heederik, E. von Mutius, and A. Legatzki. 2016. Environmental and mucosal microbiota and their role in childhood asthma. *Allergy* 72(1):109-119.

Blavier, J., G. Songulashvili, C. Simon, M. Penninckx, S. Flahaut, M. L. Scippo, and F. Debaste. 2016. Assessment of methods of detection of water estrogenicity for their use as monitoring tools in a process of estrogenicity removal. *Environmental Technology* 37(24):3104-3119.

Blay, S. L., A. J. Schulz, and G. Mentz. 2015. The relationship of built environment to health-related behaviors and health outcomes in elderly community residents in a middle income country. *Journal of Public Health Research* 4(2):548.

Block, M. L., A. Elder, R. L. Auten, S. D. Bilbo, H. Chen, J. C. Chen, D. A. Cory-Slechta, D. Costa, D. Diaz-Sanchez, D. C. Dorman, D. R. Gold, K. Gray, H. A. Jeng, J. D. Kaufman, M. T. Kleinman, A. Kirshner, C. Lawler, D. S. Miller, S. S. Nadadur, B. Ritz, E. O. Semmens, L. H. Tonelli, B. Veronesi, R. O. Wright, and R. J. Wright. 2012. The outdoor air pollution and brain health workshop. *Neurotoxicology* 33(5):972-984.

Bokulich, N. A., D. A. Mills, and M. A. Underwood. 2013. Surface microbes in the neonatal intensive care unit: Changes with routine cleaning and over time. *Journal of Clinical Microbiology* 51(8):2617-2624.

Bokulich, N. A., J. Chung, T. Battaglia, N. Henderson, M. Jay, H. Li, D. Lieber A, F. Wu, G. I. Perez-Perez, Y. Chen, W. Schweizer, X. Zheng, M. Contreras, M. G. Dominguez-Bello, and M. J. Blaser. 2016. Antibiotics, birth mode, and diet shape microbiome maturation during early life. *Science Translational Medicine* 8(343):343-382.

Boone, S. A., and C. P. Gerba. 2007. Significance of fomites in the spread of respiratory and enteric viral disease. *Applied and Environmental Microbiology* 7(6):1687-1696.

Borchers, A.T., C. Chang, and E.M. Gershwin. 2017. Mold and human health: A reality check. *Clinical Reviews in Allergy and Immunology* 52(3):305-322.

Borre, Y. E., R. D. Moloney, G. Clarke, T. G. Dinan, and J. F. Cryan. 2014. The impact of microbiota on brain and behavior: Mechanisms & therapeutic potential. *Advances in Experimental Medicine and Biology* 817:373-403.

Bouslimani, A., C. Porto, C. M. Rath, M. Wang, Y. Guo, A. Gonzalez, D. Berg-Lyon, G. Ackermann, G. J. Moeller Christensen, T. Nakatsuji, L. Zhang, A. W. Borkowski, M. J. Meehan, K. Dorrestein, R. L. Gallo, N. Bandeira, R. Knight, T. Alexandrov, and P. C. Dorrestein. 2015. Molecular cartography of the human skin surface in 3D. *Proceedings of the National Academy of Sciences of the United States of America* 112(17):E2120-E2129.

Bouslimani, A., A. V. Melnik, Z. Xu, A. Amir, R. R. da Silva, M. Wang, N. Bandeira, T. Alexandrov, R. Knight, and P. C. Dorrestein. 2016. Lifestyle chemistries from phones for individual profiling. *Proceedings of the National Academy of Sciences of the United States of America* 113(48):E7645-E7654.

Bradley, P. M., W. A. Battaglin, L. R. Iwanowicz, J. M. Clark, and C. A. Journey. 2016. Aerobic biodegradation potential of endocrine-disrupting chemicals in surface-water sediment at Rocky Mountain National Park, USA. *Environmental Toxicology and Chemistry* 35(5):1087-1096.

Breysse, J., D. E. Jacobs, W. Weber, S. Dixon, C. Kawecki, S. Aceti, and J. Lopez. 2011. Health outcomes and green renovation of affordable housing. *Public Health Reports* 126(Suppl 1):64-75.

Breysse, J., S. Dixon, J. Gregory, M. Philby, D. E. Jacobs, and J. Krieger. 2014. Effect of weatherization combined with community health worker in-home education on asthma control. *American Journal of Public Health* 104:e57-e64. doi:10.2105/AJPH.2013.301402.

Burge, H. 1980. "Bioaerosols" prevalence and health effects in the indoor environment. *Journal of Allergy and Clinical Immunology* 86(5):687-701.

Burokas, A., R. D. Moloney, T. G. Dinan, and J. F. Cryan. 2015. Microbiota regulation of the Mammalian gut-brain axis. *Advances in Applied Microbiology* 91:1-62.

Burrell, R. 1991. Microbiological agents as health risks in indoor air. *Environmental Health Perspectives* 95:29-34.

Casas, L., M. Torrent, J. P. Zock, G. Doekes, J. Forns, M. Guxens, M. Täubel, J. Heinrich, and J. Sunyer. 2013. Early life exposures to home dampness, pet ownership and farm animal contact and neuropsychological development in 4 year old children: A prospective birth cohort study. *International Journal of Hygiene and Environmental Health* 216(6):690-697.

Cavaleiro Rufo, J., J. Madureira, I. Paciencia, L. Aguiar, C. Pereira, D. Silva, P. Padrão, P. Moreira, L. Delgado, I. Annesi-Maesano, E. Oliveira Fernandes, J. P. Teixeira, and A. Moreira. 2017. Indoor fungal diversity in primary schools may differently influence allergic sensitization and asthma in children. *Pediatric Allergy and Immunology* 28(4):332-339.

Celedón, J. C., D. K. Milton, C. D. Ramsey, A. A. Litonjua, L. Ryan, T. A. Platts-Mills, and D. R. Gold. 2007. Exposure to dust mite allergen and endotoxin in early life and asthma and atopy in childhood. *Journal of Allergy and Clinical Immunology* 120(1):144-149.

Center for Health Care Design. 1995. *Gardens in healthcare facilities: Uses, therapeutic benefits, and design recommendations.* https://www.healthdesign.org/sites/default/files/Gardens%20in%20HC%20Facility%20Visits.pdf (accessed July 13, 2017).

Chambers, E. C., M. S. Pichardo, and E. Rosenbaum. 2016. Sleep and the housing and neighborhood environment of urban Latino adults living in low-income housing: The AHOME Study. *Behavioral Sleep Medicine* 14(2):169-184.

Chew, G. L., J. Wilson, F. A. Rabito, F. Grimsley, S. Iqbal, T. Reponen, M. L. Muilenberg, P. S. Thorne, D. G. Dearborn, and R. L. Morley. 2006. Mold and endotoxin levels in the aftermath of Hurricane Katrina: A pilot project of homes in New Orleans undergoing renovation. *Environmental Health Perspectives* 114:1883-1889.

Chew, G. L., W. E. Horner, K. Kennedy, C. Grimes, C. S. Barnes, W. Phipatanakul, D. Larenas-Linnemann, J. D. Miller, J. Portnoy, E. Levetin, P. B. Williams, S. Baxi, and J. Scott. 2016. Procedures to assist health care providers to determine when home assessments for potential mold exposure are warranted. *The Journal of Allergy and Clinical Immunology: In Practice* 4(3):417-422.

Claesson, M. J., I. B. Jeffery, S. Conde, S. E. Power, E. M. O'Connor, S. Cusack, H. M. Harris, M. Coakley, B. Lakshminarayanan, O. O'Sullivan, G. F. Fitzgerald, J. Deane, M. O'Connor, N. Harnedy, K. O'Connor, D. O'Mahony, D. van Sinderen, M. Wallace, L. Brennan, C. Stanton, J. R. Marchesi, A. P. Fitzgerald, F. Shanahan, C. Hill, R. P. Ross, and P. W. O'Toole. 2012. Gut microbiota composition correlates with diet and health in the elderly. *Nature* 488(7410):178-184.

Clemente, J. C., E. C. Pehrsson, M. J. Blaser, K. Sandhu, Z. Gao, and B. Wang. 2015. The microbiome of uncontacted Amerindians. *Science Advances* 1(3):e1500183. doi:10.1126/sciadv.1500183.

Clifford, A., L. Lang, R. Chen, K. J. Anstey, and A. Seaton. 2016. Exposure to air pollution and cognitive functioning across the life course—A systematic literature review. *Environmental Research* 147:383-398.

Colton, M. D., P. MacNaughton, J. Vallarino, J. Kane, M. Bennett-Fripp, J. D. Spengler, and G. Adamkiewicz. 2014. Indoor air quality in green vs. conventional multifamily low-income housing. *Environmental Science & Technology* 48(14):7833-7841.

Colton, M. D., J. G. Laurent, P. MacNaughton, J. Kane, M. Bennett-Fripp, J. Spengler, and G. Adamkiewicz. 2015. Health benefits of green public housing: Associations with asthma morbidity and building-related symptoms. *American Journal of Public Health* 105:2482-2489.

Couch, R. B. 1981. Viruses and indoor air pollution. *Bulletin of the New York Academy of Medicine* 57(10):907-921.
Cox-Ganser, J. M., S. K. White, R. Jones, K. Hilsbos, E. Storey, P. L. Enright, C. Y. Rao, and K. Kreiss. 2005. Respiratory morbidity in office workers in a water-damaged building. *Environmental Health Perspectives* 113(4):485-490.
Cox-Ganser, J. M., J.-H. Park, and K. Kreiss. 2010. Office workers and teachers. In *Occupational and environmental lung diseases*, edited by P. Cullinan, S. Tarlo, and B. Nemery. Chichester, UK: John Wiley & Sons, Ltd.
Cox-Ganser, J. M., J.-H. Park, and R. Kanwal. 2011. Epidemiology and health effects in moisture-damaged damp buildings. In *Sick building syndrome and related illness: Prevention and remediation of mold contamination*, edited by W. Goldstein. Boca Raton, FL: CRC Press, Taylor & Francis Group. Pp. 11-22.
Cryan, J. F., and T. G. Dinan. 2012. Mind-altering microorganisms: The impact of the gut microbiota on brain and behaviour. *Nature Reviews Neuroscience* 13(10):701-712.
Cunnington, A. J., K. Sim, A. Deierl, J. S. Kroll, E. Brannigan, and J. Darby. 2016. "Vaginal seeding" of infants born by caesarean section. *British Medical Journal* 352:i227.
Da Silva, C. A., D. Hartl, W. Liu, C. G. Lee, and J. A. Elias. 2008. TLR-2 and IL-17A in chitin-induced macrophage activation and acute inflammation. *Journal of Immunology* 181(6):4279-4286.
Dadvand, P., M. J. Nieuwenhuijsen, M. Esnaola, J. Forns, X. Basagana, M. Alvarez-Pedrerol, I. Rivas, M. Lopez-Vicente, M. De Castro Pascual, J. Su, M. Jerrett, X. Querol, and J. Sunyer. 2015. Green spaces and cognitive development in primary schoolchildren. *Proceedings of the National Academy of Sciences of the United States of America* 112(26):7937-7942.
Dami, A. J., and S. J. Bracken. 2016. Breathing better through bugs: Asthma and the microbiome. *The Yale Journal of Biology and Medicine* 89(3):309-324.
Dannemiller, K. C., M. J. Mendell, J. M. Macher, K. Kumagai, A. Bradman, and N. Holland. 2014. Sequencing reveals that low fungal diversity in house dust is associated with childhood asthma development. *Indoor Air* 24(3):236-247.
Davies, D. E. 2014. Epithelial barrier function and immunity in asthma. *Annals of the American Thoracic Society* 11(Suppl. 5):S244-S251.
Dethlefsen, L., P. B. Eckburg, E. M. Bik, and D. A. Relman. 2006. Assembly of the human intestinal microbiota. *Trends in Ecology & Evolution* 21(9):517-523.
Dharmadhikari, A. S., M. Mphahlele, K. Venter, A. Stoltz, R. Mathebula, T. Masotla, M. van der Walt, M. Pagano, P. Jensen, and E. Nardell. 2014. Rapid impact of effective treatment on transmission of multidrug-resistant tuberculosis. *International Journal of Tuberculosis and Lung Disease* 18(9):1019-1025.
Diaz Heijtz, R., S. Wang, F. Anuar, Y. Qian, B. Bjorkholm, A. Samuelsson, M. L. Hibberd, H. Forssberg, and S. Pettersson. 2011. Normal gut microbiota modulates brain development and behavior. *Proceedings of the National Academy of Sciences of the United States of America* 108(7):3047-3052.
Dick, E. C., L. C. Jennings, K. A. Mink, C. D. Wartgow, and S. L. Inborn. 1987. Aerosol transmission of rhinovirus colds. *The Journal of Infectious Diseases* 156(3):442-448.
Du Toit, G., G. Roberts, P. H. Sayre, H. T. Bahnson, S. Radulovic, A. F. Santos, H. A. Brough, D. Phippard, M. Basting, M. Feeney, V. Turcanu, M. L. Sever, M. Gomez Lorenzo, M. Plaut, and G. Lack. 2015. Randomized trial of peanut consumption in infants at risk for peanut allergy. *New England Journal of Medicine* 372:808-813.
Ege, M. J., M. Mayer, A. C. Normand, J. Genuneit, W. O. Cookson, C. Braun-Fahrlander, D. Heederik, R. Piarroux, and E. von Mutius. 2011. Exposure to environmental microorganisms and childhood asthma. *New England Journal of Medicine* 364(8):701-709.

Eggleston, P., D. Rosenstreich, H. Lynn, P. Gergen, D. Baker, M. Kattan, K. M. Mortimer, H. Mitchell, D. Ownby, R. Slavin, and F. Malveaux. 1998. Relationship of indoor allergen exposure to skin test sensitivity in inner-city children with asthma. *Clinical Immunology* 102(4 Pt. 1):563-570.

Emerson, J. B, P. B. Keady, T. E. Brewer, N. Clements, E. E. Morgan, J. Awerbuch, S. L. Miller, and N. Fierer. 2015. Impacts of flood damage on airborne bacteria and fungi in homes after the 2013 Colorado Front Range flood. *Environmental Science & Technology* 49(5):2675-2684.

Evans, G. W., and M. A. Schamberg. 2009. Childhood poverty, chronic stress, and adult working memory. *Proceedings of the National Academy of Sciences of the United States of America* 106(16):6545-6549.

Faber, S., G. M. Zinn, A. Boggess, T. Fahrenholz, J. C. Kern, II, and H. M. Kingston. 2015. A cleanroom sleeping environment's impact on markers of oxidative stress, immune dysregulation, and behavior in children with autism spectrum disorders. *BMC Complementary and Alternative Medicine* 15:71.

Falkinham, J. O. 2003. Mycobacterial aerosols and respiratory disease. *Emerging Infectious Diseases* 9(7):763-767.

Fall, T., C. Lundholm, K. Fall, F. Fang, A. Hedhammar, O. Kämpe, E. Ingelsson, and C. Almqvist. 2015. Early exposure to dogs and farm animals and the risk of childhood asthma. *JAMA Pediatrics* 169(11):e153219.

Feazel, L. M., L. K. Baumgartner, K. L. Peterson, D. N. Frank, J. K. Harris, and N. R. Pace. 2009. Opportunistic pathogens enriched in showerhead biofilms. *Proceedings of the National Academy of Sciences of the United States of America* 106(38):16393-16399.

Fernandez, J. A., P. Lopez, D. S. Orozco, and J. Merino. 2002. Clinical study of an outbreak of Legionnaire's disease in Alcoy, Southeastern Spain. *European Journal of Clinical Microbiology and Infectious Diseases* 21(10):729-735.

Fisk, W. J., Q. Lei-Gomez, and M. J. Mendell. 2007. Meta-analyses of the associations of respiratory health effects with dampness and mold in homes. *Indoor Air* 17:284-296.

Forno, E., A. B. Onderdonk, J. McCracken, A. A. Litonjua, D. Laskey, M. L. Delaney, A. M. Dubois, D. R. Gold, L. M. Ryan, S. T. Weiss, and J. C. Celedón. 2008. Diversity of the gut microbiota and eczema in early life. *Clinical and Molecular Allergy* 6:11.

Francisco, P. W., D. E. Jacobs, L. Targos, S. L. Dixon, J. Breysse, W. Rose, and S. Cali. 2016. Ventilation, indoor air quality, and health in homes undergoing weatherization. *Indoor Air* 27(2):463-477.

Frankel, M., G. Beko, M. Timm, S. Gustavsen, E. W. Hansen, and A. M. Madsen. 2012. Seasonal variations of indoor microbial exposures and their relation to temperature, relative humidity, and air exchange rate. *Applied and Environmental Microbiology* 78(23):8289-8297.

Fujimura, K. E., T. Demoor, M. Rauch, A. A. Faruqi, S. Jang, and C. C. Johnson. 2014. House dust exposure mediates gut microbiome Lactobacillus enrichment and airway immune defense against allergens and virus infection. *Proceedings of the National Academy of Sciences of the United States of America* 111(2):805-810.

Gardner, D. E. 1982. Use of experimental airborne infections for monitoring altered host defenses. *Environmental Health Perspectives* 43:99-107.

Gold, D. R., G. Adamkiewicz, S. Hasan Arshad, J. C. Celedón, M. D. Chapman, G. L. Chew, D. N. Cook, A. Custovic, U. Gehring, J. E. Gern, C. C. Johnson, S. Kennedy, P. Koutrakis, B. Leaderer, H. Mitchell, A. A. Litonjua, G. A. Mueller, G. T. O'Connor, D. Ownby, W. Phipatanakul, V. Persky, M. S. Perzanowski, C.D. Ramsey, P. M. Salo, J. M. Schwaninger, J. E. Sordillo, A. Spira, S. F. Suglia, A. Togias, D. C. Zeldin, and E. C. Matsui. 2017. NIAID, NIEHS, NHLBI, and MCAN Workshop Report: The indoor environment and childhood asthma—implications for home environmental intervention in asthma prevention and management. *Journal of Allergy and Clinical Immunology* [Epub ahead of print].

Gong, Y., S. Palmer, J. Gallacher, T. Marsden, and D. Fone. 2016. A systematic review of the relationship between objective measurements of the urban environment and psychological distress. *Environment International* 96:48-57.

Gozdz, J., C. Ober, and D. Vercelli. 2016. Innate immunity and asthma risk. *New England Journal of Medicine* 375:1897-1899.

Gramec, S. D., and L. P. Mašič. 2016. Bisphenol A and its analogs: Do their metabolites have endocrine activity? *Environmental Toxicology and Pharmacology* 47:182-199.

Groer, M. W., K. E. Gregory, A. Louis-Jacques, S. Thibeau, and W. A. Walker. 2015. The very low birth weight infant microbiome and childhood health. *Birth Defects Research Part C: Embryo Today* 105(4):252-264.

Hägerhed-Engman, L., C.-G. Bornehag, and J. Sundell. 2009. Building characteristics associated with moisture related problems in 8,918 Swedish dwellings. *International Journal of Environmental Health Research* 19(4):251-265.

Hanski, I., L. von Hertzen, N. Fyhrquist, K. Koskinen, K. Torppa, T. Laatikainen, P. Karisola, P. Auvinen, L. Paulin, M. J. Mäkelä, E. Vartiainen, T. U. Kosunen, H. Alenius, and T. Haahtela. 2012. Environmental biodiversity, human microbiota, and allergy are interrelated. *Proceedings of the National Academy of Sciences of the United States of America* 109(21):8334-8339.

Hanson, B., Y. Zhou, E. Bautista, B. Urch, M. Speck, F. Silverman, M. Muilenberg, W. Phipatanakul, E. Sodergren, D. R. Gold, and J. E. Sordillo. 2016. Characterization of the bacterial and fungal microbiome in indoor dust and outdoor air samples: A pilot study. *Environmental Science: Processes & Impacts* 18(6):713-724.

Hanson, J. L., N. Hair, D. G. Shen, F. Shi, J. H. Gilmore, B. L. Wolfe, and S. D. Pollak. 2013. Family poverty affects the rate of human infant brain growth. *PLOS ONE* 8(12):e80954.

Harada, R. N., and J. E. Repine. 1985. Pulmonary host defense mechanisms. *CHEST Journal* 87(2):247-252.

Harris, M. H., D. R. Gold, S. L. Rifas-Shiman, S. J. Melly, A. Zanobetti, B. A. Coull, J. D. Schwartz, A. Gryparis, I. Kloog, P. Koutrakis, D. C. Bellinger, R. F. White, S. K. Sagiv, and E. Oken. 2015. Prenatal and childhood traffic-related pollution exposure and childhood cognition in the project viva cohort (Massachusetts, USA). *Environmental Health Perspectives* 123(10):1072-1078.

Hartz, L. E., W. Bradshaw, and D. H. Brandon. 2015. Potential NICU environmental influences on the neonate's microbiome: A systematic review. *Advances in Neonatal Care* 15(5):324-335.

Hatch, G. E., E. Boykin, J. A. Graham, J. Lewtas, F. Pott, K. Loud, and J. L. Mumford. 1985. Inhalable particles and pulmonary host defense: In vivo and in vitro effects of ambient air and combustion particles. *Environmental Research* 36(1):67-80.

Hatch, T. F. 1961. Distribution and deposition of inhaled particles in respiratory tract. *Bacteriological Reviews* 25(3):237-240.

Haupt, T. E., R. T. Heffernan, J. J. Kazmierczak, H. Nehls-Lowe, B. Rheineck, C. Powell, K. K. Leonhardt, A. S. Chitnis, and J. P. Davis. 2012. An outbreak of Legionnaires disease associated with a decorative water wall fountain in a hospital. *Infection Control and Hospital Epidemiology* 33(2):185-191.

Heederik, D., and E. von Mutius. 2012. Does diversity of environmental microbial exposure matter for the occurrence of allergy and asthma? *Journal of Allergy and Clinical Immunology* 130(1):44-50.

Heindel, J. J., L. A. Skalla, B. R. Joubert, C. H. Dilworth, and K. A. Gray. 2017. Review of developmental origins of health and disease publications in environmental epidemiology. *Reproductive Toxicology* 68:34-48.

Hill, C. J., D. B. Lynch, K. Murphy, M. Ulaszewska, I. B. Jeffery, C. A. O'Shea, C. Watkins, E. Dempsey, F. Mattivi, K. Tuohy, R. P. Ross, C. A. Ryan, P. W. O'Toole, and C. Stanton. 2017. Evolution of gut microbiota composition from birth to 24 weeks in the INFANTMET Cohort. *Microbiome* 5(1):4.

Holst, G. J., A. Høst, G. Doekes, H. W. Meyer, A. M. Madsen, K. B. Plesner, and T. Sigsgaard. 2016. Allergy and respiratory health effects of dampness and dampness related agents in schools and homes: A cross-sectional study in Danish pupils. *Indoor Air* 26:880-891.

Hoppe, K. A., N. Metwali, S. S. Perry, T. Hart, P. A. Kostle, and P. S. Thorne. 2012. Assessment of airborne exposures and health in flooded homes undergoing renovation. *Indoor Air* 22(6):446-456.

Hunninghake, G. M., J. Lasky-Su, M. E. Soto-Quirós, L. Avila, C. Liang, S. L. Lake, T. J. Hudson, M. Spesny, E. Fournier, J. S. Sylvia, N. B. Freimer, B. J. Klanderman, B. A. Raby, and J. C. Celedón. 2008. Sex-stratified linkage analysis identifies a female-specific locus for IgE to cockroach in Costa Ricans. *American Journal of Respiratory and Critical Care Medicine* 177(8):830-836.

Hunninghake, G. M., J. H. Chu, S. S. Sharma, M. H. Cho, B. E. Himes, A. J. Rogers, A. Murphy, V. J. Carey, and B. A. Raby. 2011. The CD4+ T-cell transcriptome and serum IgE in asthma: IL17RB and the role of sex. *BMC Pulmonary Medicine* 11:17.

Illi, S., M. Depner, J. Genuneit, E. Horak, G. Loss, C. Strunz-Lehner, G. Buchele, A. Boznanski, H. Danielewicz, P. Cullinan, D. Heederik, C. Braun-Fahrlander, and E. von Mutius. 2012. Protection from childhood asthma and allergy in Alpine farm environments: The GABRIEL Advanced Studies. *Journal of Allergy and Clinical Immunology* 129(6):1470-1477.

IOM (Institute of Medicine). 2004. *Damp indoor spaces and health*. Washington, DC: The National Academies Press.

IOM. 2007. *Green healthcare institutions: Health, environment, and economics: Workshop summary*. Washington, DC: The National Academies Press.

Jaakkola, M. S., R. Quansah, T. T. Hugg, S. A. Heikkinen, and J. J. Jaakkola. 2013. Association of indoor dampness and molds with rhinitis risk: A systematic review and meta-analysis. *Journal of Allergy and Clinical Immunology* 132(5):1099-1110.

Jacobs, D. E., E. Ahonen, S. L. Dixon, S. Dorevitch, J. Breysse, J. Smith, A. Evens, D. Dobrez, M. Isaacson, C. Murphy, L. Conroy, and P. Levavi. 2015. Moving into green healthy housing. *Journal of Public Health Management and Practice* 21(4):345-354.

Janicki, T., M. Krupiński, and J. Długoński. 2016. Degradation and toxicity reduction of the endocrine disruptors nonylphenol, 4-tert-octylphenol and 4-cumylphenol by the non-ligninolytic fungus Umbelopsis isabellina. *Bioresource Technology* 200:223-229.

Jasarevic, E., K. E. Morrison, and T. L. Bale. 2016. Sex differences in the gut microbiome-brain axis across the lifespan. *Philosophical Transactions of the Royal Society B: Biological Sciences* 371(1688):20150122. doi:10.1098/rstb.2015.0122.

Jednorog, K., I. Altarelli, K. Monzalvo, J. Fluss, J. Dubois, C. Billard, G. Dehaene-Lambertz, and F. Ramus. 2012. The influence of socioeconomic status on children's brain structure. *PLOS ONE* 7(8):e42486.

Julian, T. R. 2010. *Fomites in infectious disease transmission: A modeling, laboratory, and field study on microbial transfer between skin and surfaces*. Ph.D. dissertation. Stanford, CA: Stanford University.

Kanchongkittiphon, W., M. J. Mendell, J. M. Gaffin, G. Wang, and W. Phipatanakul. 2015. Indoor environmental exposures and exacerbation of asthma: An update to the review by the Institute of Medicine. *Environmental Health Perspectives* 123(1):6-20.

Kelly, D., S. Conway, and R. Aminov. 2005. Commensal gut bacteria: Mechanisms of immune modulation. *Trends in Immunology* 26(6):326-333.
Kercsmar, C. M., D. G. Dearborn, M. Schluchter, L. Xue, H. L. Kirchner, J. Sobolewski, S. J. Greenberg, S. J. Vesper, and T. Allan. 2006. Reduction in asthma morbidity in children as a result of home remediation aimed at moisture sources. *Environmental Health Perspectives* 114:1574-1580.
Keunen, K., R. M. van Elburg, F. van Bel, and M. J. Benders. 2015. Impact of nutrition on brain development and its neuroprotective implications following preterm birth. *Pediatric Research* 77(1-2):148-155.
Kicinski, M., G. Vermeir, N. Van Larebeke, E. Den Hond, G. Schoeters, L. Bruckers, I. Sioen, E. Bijnens, H. A. Roels, W. Baeyens, M. K. Viaene, and T. S. Nawrot. 2015. Neurobehavioral performance in adolescents is inversely associated with traffic exposure. *Environment International* 75:136-143.
Kitsios, G. D., and A. Morris. 2016. Mode of delivery to the brave new (microbial) world: A defining moment for the respiratory microbiome? *EBioMedicine* 9:25-26.
Koep, T. H., F. T. Enders, C. Pierret, S. C. Ekker, D. Krageschmidt, K. L. Neff, M. Lipsitch, J. Shaman, and W. C. Huskins. 2013. Predictors of indoor absolute humidity and estimated effects on influenza virus survival in grade schools. *BMC Infectious Diseases* 13:71.
Koestel, Z. L., R. C. Backus, K. Tsuruta, W. G. Spollen, S. A. Johnson, A. B. Javurek, M. R. Ellersieck, C. E. Wiedmeyer, K. Kannan, J. Xue, N. J. Bivens, S. A. Givan, and C. S. Rosenfeld. 2017. Bisphenol A (BPA) in the serum of pet dogs following short-term consumption of canned dog food and potential health consequences of exposure to BPA. *Science of the Total Environment* 579:1804-1814.
Konya, T., B. Koster, H. Maughan, M. Escobar, M. B. Azad, D. S. Guttman, M. R. Sears, A. B. Becker, J. R. Brook, T. K. Takaro, A. L. Kozyrskyj, J. A. Scott, and CHILD Study Investigators. 2014. Associations between bacterial communities of house dust and infant gut. *Environmental Research* 131:25-30.
Korkalainen, M., M. Täubel, J. Naarala, P. Kirjavainen, A. Koistinen, A. Hyvärinen, H. Komulainen, and M. Viluksela. 2017. Synergistic proinflammatory interactions of microbial toxins and structural components characteristic to moisture-damaged buildings. *Indoor Air* 27(1):13-23.
Krieger, J. W., T. K. Takaro, L. Song, and M. Weaver. 2005. The Seattle-King County Healthy Homes Project: A randomized, controlled trial of a community health worker intervention to decrease exposure to indoor asthma triggers. *American Journal of Public Health* 95:652-659.
Kuhn, D. M., and M. A. Ghannoum. 2003. Indoor mold, toxigenic fungi, and stachybotrys chartarum: Infectious disease perspective. *Clinical Microbiology Reviews* 16(1):144-172.
Lai, P. S., W. J. Sheehan, J. M. Gaffin, C. R. Petty, B. A. Coull, D R. Gold, and W. Phipatanakul. 2015. School endotoxin exposure and asthma morbidity in inner-city children. *Chest* 148(5):1251-1258.
Lambrecht, B. N., and H. Hammad. 2013. Asthma: The importance of dysregulated barrier 782 immunity. *European Journal of Immunology* 43(12):3125-3137.
Lambrecht, B. N., and H. Hammad. 2014. Allergens and the airway epithelium response: 780 gateway to allergic sensitization. *Journal of Allergy and Clinical Immunology* 134(3):499-507.
Lax, S., and J. A. Gilbert. 2015. Hospital-associated microbiota and implications for nosocomial infections. *Trends in Molecular Medicine* 21(7):427-432.
Lax, S., D. P. Smith, J. Hampton-Marcell, S. M. Owens, K. M. Handley, N. M. Scott, S. M. Gibbons, P. Larsen, B. D. Shogan, S. Weiss, J. L. Metcalf, L. K. Ursell, Y. Vazquez-Baeza, W. Van Treuren, N. A. Hasan, M. K. Gibson, R. Colwell, G. Dantas, R. Knight, and J. A. Gilbert. 2014. Longitudinal analysis of microbial interaction between humans and the indoor environment. *Science* 345(6200):1048-1052.

Lax, S., C. R. Nagler, and J. A. Gilbert. 2015. Our interface with the built environment: Immunity and the indoor microbiota. *Trends in Immunology* 36(3):121-123.

Li, Y., G. M. Leung, J. W. Tang, X. Yang, C. Y. Chao, J. Z. Lin, J. W. Lu, P. V. Nielsen, J. Niu, H. Qian, A. C. Sleigh, H. J. Su, J. Sundell, T. W. Wong, and P. L. Yuen. 2007. Role of ventilation in airborne transmission of infectious agents in the built environment—a multidisciplinary systematic review. *Indoor Air* 17(1):2-18.

Lokugamage, A. 2016. Study provides evidence that "vaginal seeding" of infants born by caesarean partially restores microbiota. *British Medical Journal* 352:i1737.

Loss, G. J., M. Depner, A. J. Hose, J. Genuneit, A. M. Karvonen, A. Hyvärinen, C. Roduit, M. Kabesch, R. Lauener, P. I. Pfefferle, J. Pekkanen, J. C. Dalphin, J. Riedler, C. Braun-Fahrländer, E. von Mutius, and M. J. Ege. 2016. The early development of wheeze: Environmental determinants and genetic susceptibility at 17q21. *American Journal of Respiratory and Critical Care Medicine* 193(8):889-897.

Ly, N. P., B. Ruiz-Perez, A. B. Onderdonk, A. O. Tzianabos, A. A. Litonjua, C. Liang, D. Laskey, M. L. Delaney, A. M. DuBois, H. Levy, D. R. Gold, L. M. Ryan, S. T. Weiss, and J. C. Celedón. 2006. Mode of delivery and cord blood cytokines: A birth cohort study. *Clinical and Molecular Allergy* 4:13.

Lynch, S. V., R. A. Wood, H. Boushey, L. B. Bacharier, G. R. Bloomberg, M. Kattan, G. T. O'Connor, M. T. Sandel, A. Calatroni, E. Matsui, C. C. Johnson, H. Lynn, C. M. Visness, K. F. Jaffee, P. J. Gergen, D. R. Gold, R. J. Wright, K. Fujimura, M. Rauch, W. W. Busse, and J. E. Gern. 2014. Effects of early-life exposure to allergens and bacteria on recurrent wheeze and atopy in urban children. *Journal of Allergy and Clinical Immunology* 134(3):593-601.

Macher, J. M., M. J. Mendell, W. Chen, and K. Kumagai. 2017. Development of a method to relate the moisture content of a building material to its water activity. *Indoor Air* 27(3):599-608.

Manor, O., R. Levy, and E. Borenstein. 2014. Mapping the inner workings of the microbiome: Genomic- and metagenomic-based study of metabolism and metabolic interactions in the human microbiome. *Cell Metabolism* 20(5):742-752.

Manzano-León, N. 2013. Variation in the composition and in vitro proinflammatory effect of urban particulate matter from different sites. *Journal of Biochemical and Molecular Toxicology* 27(1):87-97.

Mao, J., and N. Gao. 2015. The airborne transmission of infection between flats in high-rise residential buildings: A review. *Building and Environment* 94(Part 2):516-531.

Martin, R., H. Makino, A. Cetinyurek Yavuz, K. Ben-Amor, M. Roelofs, E. Ishikawa, H. Kubota, S. Swinkels, T. Sakai, K. Oishi, A. Kushiro, and J. Knol. 2016. Early-life events, including mode of delivery and type of feeding, siblings and gender, shape the developing gut microbiota. *PLOS ONE* 11(6):e0158498.

McDevitt, J., S. Rudnick, M. First, and J. Spengler. 2010. Role of absolute humidity in the inactivation of influenza viruses on stainless steel surfaces at elevated temperatures. *Applied and Environmental Microbiology* 76(12):3943-3947.

MDHHS (Michigan Department of Human and Health Services). 2016. *MDHHS issues update to 2015 Legionnaires' disease report for Genesee County.* http://www.michigan.gov/mdhhs/0,5885,7-339--379243--,00%20.html (accessed January 30, 2017).

MDHHS and GCHD (Genesee County Health Department). 2015. *Legionellosis outbreak-Genesee County, May 2015–November 2015 summary analysis.* https://www.michigan.gov/documents/mdhhs/Updated_5-15_to_11-15_Legionellosis_Analysis_Summary_511707_7.pdf (accessed January 30, 2017).

MDHHS and GCHD. 2016. *Legionellosis outbreak-Genesee County, June 2014–March 2015 full analysis.* https://www.michigan.gov/documents/mdhhs/6-14_to_3-15_Legionellosis_Report_Full_Analysis_Results_511708_7.pdf (accessed January 30, 2017).

Mendell, M. J. 2007. Indoor residential chemical emissions as risk factors for respiratory and allergic effects in children: A review. *Indoor Air* 17(4):259-277.
Mendell, M. J., and K. Kumagai. 2017. Observation-based metrics for residential dampness and mold with dose-response relationships to health: A review. *Indoor Air* 27(3):506-517.
Mendell, M. J., Q. Lei-Gomez, A. G. Mirer, O. Seppänen, and G. Brunner. 2008. Risk factors in heating, ventilating, and air-conditioning systems for occupant symptoms in U.S. office buildings: The U.S. EPA BASE study. *Indoor Air* 18(4):301-316.
Mendell, M. J., A. G. Mirer, K. Cheung, M. Tong, and J. Douwes. 2011. Respiratory and allergic health effects of dampness, mold, and dampness related agents: A review of the epidemiologic evidence. *Environmental Health Perspectives* 119:748-756.
Miller, J. D., M. Sun, A. Gilyan, J. Roy, and T. G. Rand. 2010. Inflammation-associated gene transcription and expression in mouse lungs induced by low molecular weight compounds from fungi from the built environment. *Chemico-Biological Interactions* 183(1):113-124.
Mitchell, H., R. D. Cohn, J. Wildfire, E. Thornton, S. Kennedy, J. M. El-Dahr, P. C. Chulada, M. M. Mvula, L. F. Grimsley, M. Y. Lichtveld, L. E. White, Y. M. Sterling, K. U. Stephens, and W. J. Martin. 2012. Implementation of evidence-based asthma interventions in post-Katrina New Orleans: The Head-off Environmental Asthma in Louisiana (HEAL) study. *Environmental Health Perspectives* 120(11):1607-1612.
Mohapatra, A., S. J. Van Dyken, C. Schneider, J. C. Nussbaum, H. E. Liang, and R. M. Locksley. 2016. Group 2 innate lymphoid cells utilize the IRF4-IL-9 module to coordinate epithelial cell maintenance of lung homeostasis. *Mucosal Immunology* 9(1):275-286.
Mphaphlele, M., A. S. Dharmadhikari, P. A. Jensen, S. N. Rudnick, T. H. van Reenen, M. A. Pagano, W. Leuschner, T. A. Sears, S. P. Milonova, M. van der Walt, A. C. Stoltz, K. Weyer, and E. A. Nardell. 2015. Institutional tuberculosis transmission. Controlled trial of upper room ultraviolet air disinfection: A basis for new dosing guidelines. *American Journal of Respiratory and Critical Care Medicine* 192(4):477-484.
Mudarri, D., and W. J. Fisk. 2007. Public health and economic impact of dampness and mold. *Indoor Air* 17(3):226-235.
Myles, I. A. 2014. Fast food fever: Reviewing the impacts of the Western diet on immunity. *Nutrition Journal* 13:61.
Nardell, E. A. 2016. Indoor environmental control of tuberculosis and other airborne infections. *Indoor Air* 26(1):79-87.
Nardell, E. A., J. Keegan, S. A. Cheney, and S. C. Etkind. 1991. Airborne infection: Theoretical limits of protection achievable by building ventilation. *American Review of Respiratory Diseases* 144(2):302-306.
Neish, A. S. 2009. Microbes in gastrointestinal health and disease. *Gastroenterology* 136(1):65-80.
Neu, J. 2016. The microbiome during pregnancy and early postnatal life. *Seminars in Fetal and Neonatal Medicine* 21(6):373-379.
Neufeld, K. M., N. Kang, J. Bienenstock, and J. A. Foster. 2011. Reduced anxiety-like behavior and central neurochemical change in germ-free mice. *Neurogastroenterology & Motility* 23(3):255-264.
Nicas, M., and R. M. Jones. 2009. Relative contributions of four exposure pathways to influenza infection risk. *Risk Analysis* 29(9):1292-1303.
Nieuwenhuijsen, M. J. 2016. Urban and transport planning, environmental exposures and health: New concepts, methods and tools to improve health in cities. *Environmental Health* 15(Suppl. 1):38.

Norback, D., J.-P Zock, E. Plana, J. Heinrich, C. Svanes, J. Sunyer, N. Kunzli, S. Villani, M. Olivieri, and A. Soon. 2013. Mould and dampness in dwelling places, and onset of asthma: The population-based cohort ECRHS. *Occupational and Environmental Medicine* 70(5):325-331.

NRC (National Research Council). 2006. *Green schools: Attributes for health and learning.* Washington, DC: The National Academies Press.

O'Dea, E. M., N. Amarsaikhan, H. Li, J. Downey, E. Steele, S. J. Van Dyken, R. M. Locksley, and S. P. Templeton. 2014. Eosinophils are recruited in response to chitin exposure and enhance Th2-mediated immune pathology in Aspergillus fumigatus infection. *Infection and Immunity* 82(8):3199-3205.

O'Loughlin, R. E., L. Kightlinger, M. C. Werpy, E. Brown, V. Stevens, C. Hepper, T. Keane, R. F. Benson, B. S. Fields, and M. R. Moore. 2007. Restaurant outbreak of Legionnaires' disease associated with a decorative fountain: An environmental and case-control study. *BMC Infectious Diseases* 7:93.

O'Mahony, S. M., G. Clarke, Y. E. Borre, T. G. Dinan, and J. F. Cryan. 2015. Serotonin, tryptophan metabolism and the brain-gut-microbiome axis. *Behavioural Brain Research* 277:32-48.

OSHA (Occupational Safety and Health Administration). 2017. *OSHA technical manual; Section III, Chapter 7; Legionnaires' disease.* https://www.osha.gov/dts/osta/otm/otm_iii/otm_iii_7.html (accessed March 5, 2017).

Oudin, A., J. C. Richter, T. Taj, L. Al-Nahar, and K. Jakobsson. 2016. Poor housing conditions in association with child health in a disadvantaged immigrant population: A cross-sectional study in Rosengård, Malmö, Sweden. *BMJ Open* 6(1):e007979.

Park, J.-H., D. L. Spiegelman, D. R. Gold, H. A. Burge, and D. K. Milton. 2001. Predictors of airborne endotoxin in the home. *Environmental Health Perspectives* 109(8):859-864.

Park, J.-H., J. Cox-Ganser, C. Rao, and K. Kreiss. 2006. Fungal and endotoxin measurements in dust associated with respiratory symptoms in a water-damaged office building. *Indoor Air* 16(3):192-203.

Park, J.-H., J. M. Cox-Ganser, K. Kreiss, S. K. White, and C. Y. Rao. 2008. Hydrophilic fungi and ergosterol associated with respiratory illness in a water-damaged building. *Environmental Health Perspectives* 116(1):45-50.

Park, J.-H., J. M. Cox-Ganser, S. K. White, A. S. Laney, S. M. Caulfield, W. A. Turner, A. D. Sumner, and K. Kreiss. 2017. Bacteria in a water-damaged building: Associations of actinomycetes and nontuberculous mycobacteria with respiratory health in occupants. *Indoor Air* 27(1):24-33.

Peng, Z., D. Jin, H. B. Kim, C. W. Stratton, B. Wu, W. Tang, and Z. Sun. 2017. An update on antimicrobial resistance in *Clostridium difficile*: Resistance mechanisms and antimicrobial susceptibility testing. *Journal of Clinical Microbiology* 55:7. doi:10.1128/JCM.02250-16.

Perzanowski, M. S., R. L. Miller, P. S. Thorne, R. G. Barr, A. Divjan, B. J. Sheares, R. S. Garfinkel, F. P. Perera, I. F. Goldstein, and G. L. Chew. 2006. Endotoxin in inner-city homes: Associations with wheeze and eczema in early childhood. *Journal of Allergy and Clinical Immunology* 117(5):1082-1089.

Pistiner, M., D. R. Gold, H. Abdulkerim, E. Hoffman, and J. C. Celedón. 2008. Birth by cesarean section, allergic rhinitis, and allergic sensitization among children with a parental history of atopy. *Journal of Allergy and Clinical Immunology* 122(2):274-279.

Pongracic, J. A., G. T. O'Connor, M. L. Muilenberg, B. Vaughn, D. R. Gold, M. Kattan, W. J. Morgan, R. S. Gruchalla, E. Smartt, and H. E. Mitchell. 2010. *Journal of Allergy and Clinical Immunology* 125(3):593-599.

Ponnusamy D., E. V. Kozlova, J. Sha, T. E. Erova, S. R. Azar, E. C. Fitts, M. L. Kirtley, B. L. Tiner, J. A. Andersson, C. J. Grim, R. P. Isom, N. A. Hasan, R. R. Colwell, and A. K. Chopra. 2016. Cross-talk among flesh-eating Aeromonas hydrophila strains in mixed infection leading to necrotizing fasciitis. *Proceedings of the National Academy of Sciences of the United States of America* 113(3):722-727.

Prince, A. L., K. M. Antony, D. M. Chu, and K. M. Aagaard. 2014. The microbiome, parturition, and timing of birth: More questions than answers. *Journal of Reproductive Immunology* 104-105:12-19.

Prince, A. L., D. M. Chu, M. D. Seferovic, K. M. Antony, J. Ma, and K. M. Aagaard. 2015. The perinatal microbiome and pregnancy: Moving beyond the vaginal microbiome. *Cold Spring Harbor Perspectives in Medicine* 5(6):a023051. doi:10.1101/cshperspect.a023051.

Principe, L., P. Tomao, and P. Visca. 2017. Legionellosis in the occupational setting. *Environmental Research* 152:485-495.

Prussin, A. J., E. B. Garcia, and L. C. Marr. 2015. Total virus and bacteria concentrations in indoor and outdoor air. *Environmental Science & Technology Letters* 2(4):84-88.

Quansah, R., M. S. Jaakkola, T. T. Hugg, S. A. M. Heikkinen, and J. J. K. Jaakkola. 2012. Residential dampness and molds and the risk of developing asthma: Review and meta-analysis. *PLOS ONE* 7(11):e47526.

Rand, T. G., J. Dipenta, C. Robbins, and J. D. Miller. 2011. Effects of low molecular weight fungal compounds on inflammatory gene transcription and expression in mouse alveolar macrophages. *Chemico-Biological Interactions* 190(2-3):139-147.

Rand, T. G., C. Robbins, D. Rajaraman, M. Sun, and J. D. Miller. 2013. Induction of Dectin-1 and asthma-associated signal transduction pathways in RAW 264.7 cells by a triple-helical (1, 3)-β-D glucan, curdlan. *Archives of Toxicology* 87(10):1841-1850.

Reber, S. O., P. H. Siebler, N. C. Donner, J. T. Morton, D. G. Smith, J. M. Kopelman, K. R. Lowe, K. J. Wheeler, J. H. Fox, J. E. Hassell, B. N. Greenwood, C. Jansch, A. Lechner, D. Schmidt, N. Uschold-Schmidt, A. M. Füchsl, D. Langgartner, F. R. Walker, M. W. Hale, G. Lopez Perez, W. Van Treuren, A. González, A. L. Halweg-Edwards, M. Fleshner, C. L. Raison, G. A. Rook, S. D. Peddada, R. Knight, and C. A. Lowry. 2016. Immunization with a heat-killed preparation of the environmental bacterium Mycobacterium vaccae promotes stress resilience in mice. *Proceedings of the National Academy of Sciences of the United States of America* 11(22):E3130-E3139.

Redford, P. S., K. D. Mayer-Barber, E. F. W. McNab, E. Stavropoulos, A. Wack, A. Sher, and A. O'Garra. 2014. Influenza A virus impairs control of Mycobacterium tuberculosis coinfection through a type I interferon receptor-dependent pathway Influenza A virus impairs control of Mycobacterium tuberculosis coinfection through a type I interferon receptor-dependent pathway. *Journal of Infectious Diseases* 209(2):270-274.

Rhoads, W. J., P. Ji, A. Pruden, and M. A. Edwards. 2015. Water heater temperature set point and water use patterns influence Legionella pneumophila and associated microorganisms at the tap. *Microbiome* 3(1):1-13.

Robbins, C. A., L. J. Swenson, M. L. Nealley, R. E. Gots, and B. J. Kelman. 2000. Health effects of mycotoxins in indoor air: A critical review. *Applied Occupational and Environmental Hygiene* 15(10):773-784.

Rook, G. A. 2013. Regulation of the immune system by biodiversity from the natural environment: An ecosystem service essential to health. *Proceedings of the National Academy of Sciences of the United States of America* 110(46):18360-18367.

Rook, G. A. W., and R. Knight. 2015. Environmental microbial diversity and noncommunicable diseases. In *Connecting global priorities: Biodiversity and human heath. A state of knowledge review* (Ch. 8). http://www.grahamrook.net/resources/Rook-and-Knight_SOK-biodiversity_15.pdf (accessed July 13, 2017).

Rook, G. A. W., and C. A. Lowry. 2008. The hygiene hypothesis and psychiatric disorders. *Trends in Immunology* 29(4):150-158.

Rook, G. A. W., C. L. Raison, and C. A. Lowry. 2014. Microbial "old friends," immunoregulation and socioeconomic status. *Clinical & Experimental Immunology* 17(1):1-12.

Rutayisire, E., K. Huang, Y. Liu, and F. Tao. 2016. The mode of delivery affects the diversity and colonization pattern of the gut microbiota during the first year of infants' life: A systematic review. *BMC Gastroenterology* 16(1):86.

Saigal, S., and L. W. Doyle. 2008. An overview of mortality and sequelae of preterm birth from infancy to adulthood. *The Lancet* 371(9608):261-269.

Sandora, T. J., M. C. Shih, and D. A. Goldmann. 2008. Reducing absenteeism from gastrointestinal and respiratory illness in elementary school students: A randomized, controlled trial of an infection-control intervention. *Pediatrics* 121(6):e1555-e1562.

Sattar, S. A. 2016. Indoor air as a vehicle for human pathogens: Introduction, objectives, and expectation of outcome. *American Journal of Infection Control* 44(9 Suppl.):S95-S101.

Schiffman, S. S., C. E. Studwell, L. R. Landerman, K. Berman, and J. S. Sundy. 2005. Symptomatic effects of exposure to diluted air sampled from a swine confinement atmosphere on healthy human subjects. *Environmental Health Perspectives* 113(5):567-576.

Schoen, M. E., and N. J. Ashbolt. 2011. An in-premise model for Legionella exposure during showering events. *Water Research* 45(18):5826-5836.

Schwake, D. O., E. Garner, O. R. Strom, A. Pruden, and M. A. Edwards. 2016. Legionella DNA markers in tap water coincident with a spike in Legionnaires' disease in Flint, MI. *Environmental Science & Technology Letters* 3(9):311-315.

Sharpe, R. A., N. Bearman, C. R. Thornton, K. Husk, and N. J. Osborne. 2015. Indoor fungal diversity and asthma: A meta-analysis and systematic review of risk factors. *Journal of Allergy and Clinical Immunology* 135(1):110-122.

Shiue, I. 2015. Indoor mildew odour in old housing was associated with adult allergic symptoms, asthma, chronic bronchitis, vision, sleep and self-rated health: USA NHANES, 2005-2006. *Environmental Science and Pollution Research* 22(18):14234-14240.

Shogan, B. D., D. P. Smith, A. I. Packman, S. T. Kelley, E. M. Landon, S. Bhangar, G. J. Vora, R. M. Jones, K. Keegan, B. Stephens, T. Ramos, B. C. Kirkup, Jr., H. Levin, M. Rosenthal, B. Foxman, E. B. Chang, J. Siegel, S. Cobey, G. An, J. C. Alverdy, P. J. Olsiewski, M. O. Martin, R. Marrs, M. Hernandez, S. Christley, M. Morowitz, S. Weber, and J. Gilbert. 2013. The Hospital Microbiome Project: Meeting report for the 2nd Hospital Microbiome Project, Chicago, USA, January 15(th), 2013. *Standards in Genomic Sciences* 8(3):571-579.

Simpson, A., and F. D. Martinez. 2010. The role of lipopolysaccharide in the development of atopy in humans. *Clinical and Experimental Allergy* 40(2):209-223.

Singh, G. K., and M. K. Kenney. 2013. Rising prevalence and neighborhood, social, and behavioral determinants of sleep problems in U.S. children and adolescents, 2003-2012. *Sleep Disorders* 2013:394320. doi:10.1155/2013/394320.

Song, S. J., C. Lauber, E. K. Costello, C. A. Lozupone, G. Humphrey, D. Berg-Lyons, J. G. Caporaso, D. Knights, J. C. Clemente, S. Nakielny, J. I. Gordon, N. Fierer, and R. Knight. 2013. Cohabiting family members share microbiota with one another and with their dogs. *Elife* 2:e00458. doi:10.7554/eLife.00458.

Sordillo, J. E., E. B. Hoffman, J. C. Celedón, A. A. Litonjua, D. K. Milton, and D. R. Gold. 2010. Multiple microbial exposures in the home may protect against asthma or allergy in childhood. *Clinical & Experimental Allergy* 40:902-910.

Stein, M. M., C. L. Hrusch, J. Gozdz, C. Igartua, V. Pivniouk, S. E. Murray, J. G. Ledford, M. M. dos Santos, R. Anderson, N. Metwali, J. Neilson, R. Maier, J. Gilbert, M. Holbreich, P. Thorne, F. Martinez, E. von Mutius, D. Vercelli, C. Ober, and A. I. Sperling. 2016. Innate immunity and asthma risk in Amish and Hutterite farm children. *New England Journal of Medicine* 375(5):411-421.

Stetzenbach, L. D. 2007. Airborne bacteria in indoor environments. *Sampling and Analysis of Indoor Microorganisms* 123-132.

Stokholm, J., J. Thorsen, B. L. Chawes, S. Schjorring, K. A. Krogfelt, K. Bonnelykke, and H. Bisgaard. 2016. Cesarean section changes neonatal gut colonization. *Journal of Allergy and Clinical Immunology* 138(3):881-889.

Sunyer, J., M. Esnaola, M. Alvarez-Pedrerol, J. Forns, I. Rivas, M. Lopez-Vicente, E. Suades-Gonzalez, M. Forarter, R. Garcia-Esteban, X. Basagana, M. Viana, M. Cirach, T. Moreno, A. Alastuey, N. Sebastian-Galles, M. Nieuwenhuijsen, and X. Querol. 2015. Association between traffic-related air pollution in schools and cognitive development in primary school children: A prospective cohort study. *PLOS Medicine* 12(3):e1001792.

Tellier, R. 2009. Aerosol transmission of influenza A virus: A review of new studies. *Journal of the Royal Society Interface* 6(Suppl. 6):S783-S790.

Thorne, P. S. 2015. Endotoxin exposure: Predictors and prevalence of associated asthma outcomes in the United States. *American Journal of Respiratory and Critical Care Medicine* 192(11):1287-1297.

Tiesler, C. M., E. Thiering, C. Tischer, I. Lehmann, B. Schaaf, A. von Berg, and J. Heinrich. 2015. Exposure to visible mould or dampness at home and sleep problems in children: Results from the LISAplus study. *Environmental Research* 137:357-363.

Tischer, C., F. Weikl, A. J. Probst, M. Standl, J. Heinrich, and K. Pritsch. 2016. Urban dust microbiome: Impact on later atopy and wheezing. *Environmental Health Perspectives* 124(12):1919-1923.

Torow, N., and M. W. Hornef. 2017. The neonatal window of opportunity: Setting the stage for life-long host-microbial interaction and immune homeostasis. *The Journal of Immunology* 198(2):557-563.

Torrazza, R. M., M. Ukhanova, X. Wang, R. Sharma, M. L. Hudak, J. Neu, and V. Mai. 2013. Intestinal microbial ecology and environmental factors affecting necrotizing enterocolitis. *PLOS ONE* 8(12):e83304.

Tsakok, T., G. Weinmayr, A. Jaensch, D. P. Strachan, H. C. Williams, and C. Flohr. 2015. Eczema and indoor environment lessons from the International Study of Asthma and Allergies in Childhood ISAAC Phase 2. *The Lancet* 385:S99.

Ulrich, R. P., X. Quan, C. P. Zimring, A. Joseph, and R. Choudhary. 2004. *The role of the physical environment in the hospital of the 21st century: A once-in-a-lifetime opportunity.* https://www.healthdesign.org/chd/knowledge-repository/role-physical-environment-hospital-21st-century-once-lifetime-opportunity-0 (accessed July 13, 2017).

Vejdovszky, K., K. Hahn, D. Braun, B. Warth, and D. Marko. 2017. Synergistic estrogenic effects of Fusarium and Alternaria mycotoxins in vitro. *Archives of Toxicology* 91(3):1447-1460.

von Mutius, E. 2016. The microbial environment and its influence on asthma prevention in early life. *Journal of Allergy and Clinical Immunology* 137(3):680-689.

von Mutius, E., and D. Vercelli. 2010. Farm living: Effects on childhood asthma and allergy. *Nature Reviews Immunology* 10(12):861-868.

Wang, H., S. Masters, Y. Hong, J. Stallings, J. O. Falkinham, III, M. A. Edwards, and A. Pruden. 2012. Effect of disinfectant, water age, and pipe material on occurrence and persistence of Legionella, mycobacteria, Pseudomonas aeruginosa, and two amoebas. *Environmental Science & Technology* 4(21):11566-11574.

Wei, J., and Y. Li. 2016. Airborne spread of infectious agents in the indoor environment. *American Journal of Infection Control* 44(9 Suppl.):S102-S108.

Weinmayr, G., U. Gehring, J. Genuneit, G. Buchele, A. Kleiner, R. Siebers, K. Wickens, J. Crane, B. Brunekreef, D. P. Strachan, and ISAAC Phase Two Study Group. 2013. Dampness and moulds in relation to respiratory and allergic symptoms in children: Results from Phase Two of the International Study of Asthma and Allergies in Childhood (ISAAC Phase Two). *Clinical and Experimental Allergy* 43(7):762-774.

WHO (World Health Organization). 2009. *WHO guidelines for indoor air quality: Dampness and mold*. http://www.euro.who.int/__data/assets/pdf_file/0017/43325/E92645.pdf (accessed April 27, 2017).

Wlasiuk, G., and D. Vercelli. 2012. The farm effect, or: When, what and how a farming environment protects from asthma and allergic disease. *Current Opinion in Allergy and Clinical Immunology* 12(5):461-466.

Wong, B. C. K., N. Lee, Y. Li, P. K. S. Chan, H. Qiu, and Z. Luo. 2010. Possible role of aerosol transmission in a hospital outbreak of influenza. *Clinical Infectious Diseases* 51(10):1176-1183.

Yang, W., and L. C. Marr. 2011. Dynamics of airborne influenza A viruses indoors and dependence on humidity. *PLOS ONE* 6(6):e21481.

Yatsunenko, T., F. E. Rey, M. J. Manary, I. Trehan, M. G. Dominguez-Bello, and M. Contreras. 2012. Human gut microbiome viewed across age and geography. *Nature* 486(7402):222-227.

Yu, I. T. S., Y. Li, T. W. Wong, W. Tam, A. T. Chan, J. H. W. Lee, D. Y. C. Leung, and T. Ho. 2004. Evidence of airborne transmission of the severe acute respiratory syndrome virus. *New England Journal of Medicine* 350(17):1731-1739.

ANNEX TABLE 2-1 Selected Studies on Building/Home-Based Exposure Reduction and Asthma Outcomes in Children (2000–2017)

Reference	Population	Study Design	Exposure Focus	Intervention	Exposure Outcome	Asthma Outcome	Comments
Krieger et al., 2005	274 4- to 12-year-old children with asthma from low-income families (Seattle-King County, Washington)	RCT	Multiple asthma "triggers"	Intervention (n = 138): Multifaceted—5–8 home visits by community health worker over 1 year, including home assessment, education, support for behavior change, and resources to reduce exposures Control (n = 136): 1 visit, limited resources	N/A	Increased parent/caregiver actions to reduce exposures Decreased urgent visits, and increased caregiver QOL No differences in asthma symptoms between groups	Separate effects of each component of intervention unknown Intervention not tailored to child's sensitivities Exposures not measured Projected 4-year savings: $189–$721
Chew et al., 2006	Three uninhabited water-damaged homes after a major hurricane (New Orleans, Louisiana)	Pre–post treatment comparison	Mold (spore counts, cultures, PCR analysis, glucan), endotoxin, and PM	Intervention: Removal of drywall, carpet, insulation, and all water-damaged furnishings	Reductions in mold and endotoxin pre–post, but high levels during cleanup	N/A	Pre- and during treatment mold and endotoxin levels orders of magnitudes above those in homes without severe water damage Adequate respirator use recommended during cleanup

continued

ANNEX TABLE 2-1 Continued

Reference	Population	Study Design	Exposure Focus	Intervention	Exposure Outcome	Asthma Outcome	Comments
Kercsmar et al., 2006	62 2- to 17-year-old children with asthma in homes with mold (Cuyahoga County, Illinois)	RCT	Mold scores; allergen levels	Intervention (n = 29) and control (n = 33): asthma action plan, education, individualized problem solving Intervention group only: + household repairs and modifications	At 6 months, but not at 12 months, greater reduction in mold scores in intervention group compared with control	Decreased asthma symptom days and prevalence of exacerbations in intervention group compared with control	Low sample size, limited power frequency of families moving complexity of applying for household repairs and working with landlords

| Breysse et al., 2011 | 49 adults, 29 children from 31 units in a low-income 3-building, 60-unit apartment complex (Minnesota) | Cross-sectional health survey of pre-/immediately post renovation health, followed by survey 12–18 months post renovation | Green specifications targeting ventilation, moisture, mold, pests, radon | Intervention: Renovation according to Enterprise Green Communities green specifications, using "healthy housing" features; new mechanical ventilation installed (94) | Reduction in energy use (45 percent) Tightening of building envelope Functional exhaust fans Fresh air at 70 percent of ASHRAE standard Lower radon Annual average indoor CO_2 982 ppm | Immediately post renovation: Self-reports of cleaner, more comfortable, safer housing Improvement in overall adult health, in nonasthma respiratory health (adults + children), and in asthma health (adults) | Potential recall bias Nonrandomized, unblinded study design Nonindependence of health reports from residents in same apartment Potential communication problems with non-English-speaking residents Potential selection bias toward healthier residents Some retrofitting required as not all renovations worked Reports of health benefits appear fewer in follow-up |

continued

ANNEX TABLE 2-1 Continued

Reference	Population	Study Design	Exposure Focus	Intervention	Exposure Outcome	Asthma Outcome	Comments
Mitchell et al., 2012	182 4- to 12-year-old children with moderate to severe asthma living in post–Hurricane Katrina flooded areas (New Orleans, Louisiana)	Observational, pre-post intervention study	Indoor allergens, moisture and mold	Intervention: Individually tailored multifaceted environmental intervention plus asthma counselor (timing of introduction of counselor varied)	Reduction in bedroom mold spores and in Alternaria in settled dust	Reduction (45 percent) in asthma symptom days Children with asthma counselor had greater symptom decrease	Separate effects of individual interventions unknown Unclear whether mold decrease occurred because of intervention
Hoppe et al., 2012	Families living in 73 flood/water-damaged homes (Cedar Rapids, Iowa)	Cross-sectional assessment of homes and health at two levels of remediation (in progress [n = 24] or complete [n = 49])	Extensive (e.g., mold, bacteria, endotoxin, PM, allergens)	Intervention: Removal of drywall, carpet, insulation, and all water-damaged furnishings	Levels of mold, bacteria, endotoxin, PM, glucan higher in homes with remediation in progress compared with homes with remediation complete	Compared with before the flood, residents of in-progress homes reported more allergies All residents reported more wheeze and meds for breathing problems	Cross-sectional Stage of in-progress cleanup variable Many in-progress families not moved back full time Potential participation bias

| Breysse et al., 2014 | 102 low-income households in rental properties with 1 or more children with not-well-controlled asthma (King County, Washington) | Observational, pre–post intervention study with historical comparison group | Intervention (n = 34): Weatherization plus community health worker (CHW) education Historical comparison group (n = 68): CHW education without weatherization | Reduction in evidence of water damage greater with intervention group But no consistent evidence for greater improvement in intervention versus comparison group in other environmental exposures | Increased asthma control Increased caregiver QOL | Separate effects of weatherization and CHW not demonstrated Small study size IPM not used |

continued

ANNEX TABLE 2-1 Continued

Reference	Population	Study Design	Exposure Focus	Intervention	Exposure Outcome	Asthma Outcome	Comments
Colton et al., 2014	31 low-income households in rental housing	Observational comparison of exposures and health in green versus conventional housing, including among those who moved between housing types		Intervention (n = 18): Move from conventional to new buildings designed to green standards; smoke-free policies and IPM practices employed Control 1 (n = 6): Move from conventional to conventional housing Control 2 (n = 7): Live in conventional housing (61 visits, including pre and post for 24 who moved)	Green versus conventional housing: Lower $PM_{2.5}$, NO_2, and nicotine Fewer reports of mold, pests, inadequate ventilation, and stuffiness	Fewer sick building syndrome symptoms	Suggested benefits of move to green housing need further assessment Number of controls limit pre–post analysis

| Colton et al., 2015 | 235 households in 3 Boston public housing 188 residents (80 percent) with 2 visits | Observational comparison of conditions and health in green versus conventional housing. Visits included home inspection and questionnaire | Visits to green units (n = 201) and conventional public housing units (n = 222) | Fewer reports and observations of mold, pests, inadequate ventilation, and secondhand smoke in green compared with conventional housing | Fewer asthma symptoms, hospital visits, school absences for children in green compared with conventional public housing | Suggested benefits of move to green housing Effects observed only for children with asthma; effects on adults not certain |

NOTES: ASHRAE = American Society of Heating, Refrigerating and Air-Conditioning Engineers; IPM = integrated pest management; N/A = not available; NO_2 = nitrogen dioxide; PCR = polymerase chain reaction; PM = particulate matter; QOL = quality of life; RCT = randomized controlled trial. References (2000–2016) selected by participants from the National Institute of Allergy and Infectious Diseases (NIAID); National Institute of Environmental Health Sciences (NIEHS); National Heart, Lung, and Blood Institute (NHLBI); and Merck Childhood Asthma Network (MCAN) Workshop on the Indoor Environment and Childhood Asthma as representative and illustrative of asthma management intervention studies in children.

SOURCE: Adapted from Gold et al., 2017.

ANNEX TABLE 2-2 Beneficial Associations of Indoor Microbiota with Asthma or Allergy Outcomes in Selected Studies Using Metagenomics, Molecular Biologic, or Culture Methods to Measure Indoor Environmental Microbiota

Reference	Population	Study Design	Microbiota Exposure Source and Measurement Method	Protective Associations	Adverse Associations	Comments
Ege et al., 2011	Two European studies comparing children on farms with reference group: PARSIFAL (n = 489); GABRIELA (n = 444)	Cross-sectional	Mattress dust PARSIFAL: screened for bacterial DNA with SSCP* GABRIELA: bacterial and fungal culture	Less asthma associated with: Greater diversity Specific taxa: fungi = eurotium, penicillium; bacteria = Listeria moncytogenecs, bacillus species, coryne-bacterium species and others	N/A	Diversity as outcome has limitations Families of species, but not specific microbes, identified as potentially protective
Fujimura et al., 2014	Two households: one with (D) and one without dog (ND)	Murine model validation of epidemiologic human study	Experiment 1: Mouse dust from D and ND homes Experiment 2: Lactobaccillus johnsonii	Experiment 1: Reduced Th2 (allergy)-related airway responses, cecal enrichment with L. johnsonii Experiment 2: Reduced Th2 (allergic) cytokine expression	N/A	Suggests GI microbiome manipulation may protect individuals against allergic airway disease Dust from two homes Murine model

Lynch et al., 2014	104 inner-city children (Baltimore, Boston, New York, St. Louis)	Longitudinal birth cohort followed from birth to age 3	At 3 months of age: Living room dust assessed with 16S rRNA-based phylogenetic microarray	Reduced atopy or atopy/recurrent wheeze with more bacterial richness; protective taxa primarily in Bacteroidetes and Firmicutes phyla (Prevotellaceae, Lachnospiraceae, Ruminococcaceae families)	N/A	Suggests children with highest exposures to multiple bacteria and allergens may get protection from allergy and wheeze Sample of children limited Taxonomic definition limited
Behbod et al., 2015	408 children (Boston, Massachusetts)	Longitudinal birth cohort followed from birth to age 13	At 2–3 months of age: Bedroom and outdoor air and bedroom dust; bacterial and fungal culture	Reduced wheeze, fungal sensitization, and asthma by age 13 associated with bedroom floor dust	Increased rhinitis associated with outdoor air Cladosporium, dust Aspergillus at 2–3 months	Suggests responses vary by fungi taxon and mode of exposure Culture misses nonculturable taxa
Tischer et al., 2016	189 children (Munich, Germany)	Longitudinal birth cohort followed from birth to age 10	At 3 months of age: Living room floor dust bacterial and fungal diversity assessed with tRFLP**	Reduced aeroallergen sensitization and ever wheezing associated with higher fungal (not bacterial) diversity by age 10	N/A	Protective associations may attenuate with age Diversity as outcome has limitations

continued

87

ANNEX TABLE 2-2 Continued

Reference	Population	Study Design	Microbiota Exposure Source and Measurement Method	Protective Associations	Adverse Associations	Comments
Stein et al., 2016	60 children aged 7–14 from the Amish and Hutterite communities	Cross-sectional	Bedroom and living room dust assessed with 16S rRNA V4-5 amplicon sequencing; bacterial and archaeal diversity and compositional structure; endotoxin levels; animal exposure validation studies	Reduced asthma with increased endotoxin levels and increased microbial diversity Sensitized animals protected from eosinophilia with exposure to dust from Amish homes	N/A	Protective associations may be related only to immune activation in the upper airway Microbial organisms likely associated with bovine animals
Birzele et al., 2016	86 school-age children	Cross-sectional	Mattress dust and nasal samples assessed with 454 pyrosequencing of 16S rRNA; bacterial diversity and composition based on operational taxonomic units (OTUs)	Reduced asthma with diversity/richness—stronger for dust than for nasal mucosal samples	N/A	Protective associations may relate not only to upper-airway colonization
Cavaleiro Rufo et al., 2017	858 8- to 10-year-old children from 20 primary schools (Portugal)	Cross-sectional	Classroom air samples for bacterial and fungal culture and endotoxin measures	Reduced allergic sensitization but not asthma with increased diversity	Higher sensitization with higher air *Penicillium spp.* and endotoxin	Culture misses nonculturable taxa

NOTES: GI = gastrointestinal; N/A = not available; SSCP = single-strand conformation polymorphism analysis; tRFLP = terminal restriction fragment length polymorphism.
Diversity: small numbers of microbial exposures may be sufficient to stimulate all pattern-recognition receptors.
*Same pattern-recognition receptors on multiple microbial taxa.
**Overgrowth of harmful bacteria might be achieved by a limited number of bacterial species.

3

The Built Environment and Microbial Communities

Chapter Highlights

- The composition and viability of indoor microbial communities are determined by the characteristics and dynamic interactions of the building they inhabit, the building's occupants, and the surrounding external environment.
- Air, water, and surfaces are the primary reservoirs for microbes found indoors. Microbes enter the indoor environment primarily through occupant shedding, through being carried from the outdoors through the air and water, and through microbial growth that occurs indoors.
- Exposure to microbes is affected by how buildings exchange air with the environment that surrounds them. This exchange can take place through infiltration (unintentional air leakage) and other forms of ventilation, including natural (such as opening windows) and mechanical (heating, ventilation, and air conditioning [HVAC] systems). It is also strongly affected by human activity.
- Microbes are found throughout buildings' plumbing systems and in places with standing water or moisture. Indoor humidity influences the airborne survival and virulence of infectious agents and the surface moisture characteristics that affect mold growth.

- Surfaces can serve as sources for microbes exchanged from human to human through shared contact with doorknobs, keyboards, and the like; for microbes in or on dust on indoor surfaces, such as carpets and furnishings, and resuspended into the air by such activities as walking or cleaning; and for microbes from indoor plants, pests, and pets.
- An appreciation of how buildings are designed and used in different climates is essential to improving understanding of harmful and beneficial microbial environments.

The composition and viability of indoor microbial communities and their metabolic products are determined by the characteristics of the building they inhabit, the building's occupants and their behaviors, and the surrounding external environment. The intricate, dynamic interplay of these elements affects human health in both positive and negative ways that are, at least currently, poorly understood.

This chapter focuses on buildings and how their characteristics and occupants shape the indoor microbiome. The chapter characterizes indoor microbial sources and reservoirs associated with air, water, and building surfaces; examines how features of the building and the environment impact indoor microbial communities and occupant exposures; considers how microbial communities affect material degradation and energy use; and identifies research needs. Subsequent chapters address how changes in one component of the built environment–microbiome–occupant nexus, intentional or otherwise, affect the others (see Chapter 5) and describe the research toolkit available for studying these interactions (see Chapter 4). This chapter is focused on microbes that have demonstrated adverse health effects; however, much of the discussion is likely applicable as well to (the far less-studied) potentially beneficial microbes.

INTRODUCTION TO MICROBIAL RESERVOIRS IN COMMERCIAL AND RESIDENTIAL BUILDINGS

There are three primary reservoirs[1] for microbes found indoors: air, water, and surfaces. The primary sources for these microbes are outdoor

---

[1] A reservoir is any person, animal, plant, material, or particle on which a microbe lives and multiplies. The reservoir typically harbors the microbe without injury to itself and serves as a source from which it may be spread (adapted from MedicineNet.com, 2017).

microbes carried indoors carried by air, water, or occupants and microbial growth that occurs indoors.

The sources and reservoirs of microbes that can be found in the **air** include

- mechanical heating, ventilation, and air conditioning (HVAC) systems;
- airborne particles that have been aerosolized via HVAC operation or occupant activities, such as walking or cleaning;
- outdoor air that enters through infiltration and natural or mechanical ventilation; and
- reservoirs in unfinished spaces such as crawl spaces, basements and attics, and concealed spaces (defined in Box 3-1) that are linked to occupied spaces via a range of airflow pathways.

Microbes are also found in **water** sources and on moist surfaces and materials, including

- municipal or well water supplies, harvested rainwater, recycled water, and drinking fountain water;
- roof, foundation, and plumbing leaks;
- condensation on or in walls and on cold water pipes;

BOX 3-1
Concealed Spaces

Buildings contain many spaces that are typically hidden from view or inaccessible. These spaces have been described as follows:

Concealed spaces or voids are non-occupied spaces created by building construction. Areas that are occupied or used for storage would not be considered a concealed space. A concealed space is not visible, with limited or no access to it. (Quimby, 2016)

Concealed spaces include vertical chases for pipes, ducts, and mechanical/electrical systems; spaces above suspended ceilings; spaces within framed walls; and inaccessible components in heating, ventilation, and air conditioning (HVAC) systems. While they can play an important role in the indoor microbiome—serving as reservoirs for mold and other microbes that exchange air with occupied spaces—they are often unmonitored, difficult to clean, and neglected by those who do not think of a building as a system. Concealed spaces thus present a great, but largely unexamined challenge for those seeking to establish and maintain a healthy indoor environment.

- mechanical equipment drain pans, coils, insulation, and filters;
- cooling towers and ponds;
- whole-house or room humidifiers;
- hot water storage tanks, with subsequent aerosolization through plumbing fixtures;
- moisture generated by household appliances and food preparation;
- aerosolized water from personal hygiene practices (showering, bathing, and the like) and toilet flushing; and
- water features, including fountains, pools, hot tubs, whirlpool baths, and spas.

Finally, microbes and microbial products linked to human health may exist on **surfaces** of objects and materials that serve as transmission sources (called "fomites"), including microbes found in or on

- dust from floors, mattresses, furniture, and other surfaces that is resuspended in the air and inhaled;
- surfaces (doorknobs, faucets, remote control devices, keyboards, light switches) that are touched and thereby allow for dermal absorption or ingestion via hand-to-mouth;
- materials and objects that are used (toothbrushes), touched (pillows, textiles), or worn (clothing), leading to ingestion either directly or via hand to mouth; and
- soil floors and rain- or pest-damaged surfaces in crawl spaces or basements.

In addition, indoor plants, pests (such as rodents and cockroaches), and pets carry bacteria, fungi, and viruses that are then inhaled, ingested, or transmitted dermally through touch. Building occupants also may bring in or store food and beverage products that contain associated microorganisms or can support microbial growth. This report does not attempt to draw a sharp distinction between flourishing microbial communities and transient microbial presence in air, in water, and on surfaces. Built environments contain both established and transient microorganisms in different spaces and at different times, and both are affected by such practices as cleaning and remediation. However, there are still significant gaps in knowledge about the factors that shape their persistence, growth, evolution, transmission, and dynamics.

This chapter distinguishes the indoor air, water, and surface microbial reservoirs and transmission routes and details how they may lead to human exposure and are affected by building design and operation and by occupant actions. Where appropriate, it notes the distinctive features of building systems and management that lead to unique issues for residential buildings and for small and large commercial buildings.

THE DIVERSITY OF BUILDINGS AND ITS IMPACT ON THEIR MICROBIOMES

Buildings are as diverse as living things: they are differentiated by geographic location and the associated climatic conditions; their type, age, and occupancy; their HVAC and other systems; the investments made in their ongoing operations and maintenance; and the expertise of their operations staff. Like living things, moreover, they change over time. The commonalities and differences among the various types of buildings and how elements of their design, construction, and operation affect the indoor microbial environment[2] are briefly discussed below.

In **single-family residential buildings** there is little systematic or institutional control over the introduction and management of sources of microbes, and the indoor microbiome tends to reflect the actions or inactions of the building's occupants within the context of geographic location and seasonal variations. Mechanical, plumbing, and other systems reflect the choices of the designers, builders, and occupants; the use, maintenance, and condition of this equipment depend primarily on the owners and occupants and are highly variable. A wide range of biocides and antifungal and antibacterial chemicals of highly varying composition and inadequately studied efficacy may be used to control microbial presence in such buildings.

HVAC systems in U.S. single-family residential buildings rarely incorporate outdoor air intake but instead recirculate interior air primarily for temperature control, typically with low-efficiency particle filtration. These buildings generally have local exhaust fans in kitchens and bathrooms to remove moisture and odor, but the effectiveness and use of these systems vary greatly. As a result, these residences are ventilated primarily by weather-driven infiltration through unintentional building leakage, supplemented by the opening of windows based on outdoor weather conditions and occupant preferences. **High-rise, multifamily residential buildings** are more likely to incorporate some amount of mechanical outdoor air intake, often supplied to hallways, but the HVAC and outdoor air systems in such buildings vary greatly.

While the microbial environments of **commercial buildings**—including offices, schools, and other nonindustrial workplaces—share some similarities with residences, there are a number of important differences. Relative to residences, for example, commercial buildings tend to have

---

[2]The discussion in this chapter is focused on building types and climate conditions found in North America and, to a lesser extent, Europe, reflecting the environments where much of the English-language research in this field has been conducted. Its applicability to building types in other parts of the world and in different climates will vary.

- greater density of occupants in contact with shared surfaces through which microbes can be transferred via doorknobs, handrails, faucets, remotes, keyboards, counters, light switches, elevator buttons, and the like;
- central HVAC systems with particle filtration and with liquid and aerosolized water associated with air conditioning coils and humidification systems that may be sources of microbes distributed to occupied spaces via mechanical ventilation;
- rooftop HVAC components (including outdoor air intakes and cooling towers) that may be exposed to standing water that supports microbial growth;
- intentional outdoor air intakes in HVAC systems in larger commercial buildings[3]; and
- higher pressure differentials than those in low-rise residential buildings, created by outdoor wind, the propensity for warmer air to rise ("stack effect"), and mechanical exhausts, all of which may increase the entry of outdoor air and microbes, as well as air and microbial migration between interior spaces.

Both residential and commercial buildings are characterized by widespread use of carpeting and textiles—known reservoirs of microbes—as well as maintenance and cleaning practices that may limit or promote the accumulation of microbial material and microbial growth, depending on frequency and the methods and materials used.

Characteristics of commercial buildings vary considerably by the structure's size and use. Small and medium-sized buildings constitute the vast majority of U.S. commercial and institutional building stock, although they do not contain the majority of occupants or floor area. Such buildings—restaurants, office parks, gas stations, hair salons, bodegas, and dental offices, for example—seldom have on-site engineering staff and often rely on sometimes-distant owners or real estate management companies to manage HVAC, water intrusion, cleaning, and indoor air quality. Smaller commercial buildings tend to be designed similarly to single-family residences in terms of the building envelope and HVAC systems and are given a similar level of attention to operations and maintenance. A field study of small and medium-sized buildings in California prepared for the California Energy Commission (Bennett et al., 2011, p. 3) found that

> sixteen of the thirty-seven buildings [examined in the study] did not have mechanically supplied outdoor air, including all the buildings built before

---

[3]In contrast, many small commercial buildings rely on a combination of infiltration, local exhaust, and windows, similar to single-family residences.

1980 and 19 percent of the buildings built after 1980. In some cases, the air handling unit was generally a residential model rather than a commercial model, and thus did not have the capability to bring outdoor air inside. Air filters used in the buildings' ventilation systems generally had low efficiency, with 56 percent having a Minimum Efficiency Reporting Value rating of four or lower. Only a quarter of the buildings had a ventilation maintenance contractor that inspected regularly. Buildings with regular contractor visits had HVAC systems that were better maintained than buildings that did not have regular inspections.

Larger commercial buildings are more likely to have dedicated in-house operations and maintenance staff. The tendency of larger buildings to have more complex HVAC systems makes the systems' presence important, as their performance will degrade over time if they are not properly maintained, increasing the likelihood of poor control of temperature, humidity, and ventilation. However, it is not enough to have an operations and maintenance staff. A separate question is whether the budget allotted to operations and maintenance allows such tasks as filter changing, system inspections and repair, and control sensor calibration to be carried out in the manner recommended by system manufacturers.

Schools are a significant exposure environment for those 6–18 years of age. American children spend an average of 180 days per year in school, and a little more than one-quarter of each of those days is spent in the school environs (6.64 hours/day on average) (ED, 2008). A 2006 National Research Council report on "green schools" includes the following findings:

- A robust body of scientific evidence indicates that the health of children and adults can be affected by indoor air quality. A growing body of evidence suggests that teacher productivity and student learning may also be affected by indoor air quality.
- Well-designed, -constructed, and -maintained building envelopes are critical to the control and reduction of excess moisture and mold growth. (NRC, 2006, p. 6)

This report offers a number of recommendations regarding these findings, including that "future green school guidelines should emphasize the control of excess moisture, dampness, and mold to protect the health of children and adults in schools," and that "such guidelines should specifically address moisture control as it relates to the design, construction, operation, and maintenance of a school building's envelope (foundations, walls, windows, and roofs) and related items such as siting and landscaping" (NRC, 2006, p. 6). The report further notes that "the survival, dispersal, and removal of airborne pathogens are affected by relative humidity, ventilation rate, and the percentage of recirculated air in the air supply" and recommends "additional research . . . to determine the optimal infection-control interventions in terms of measurable outcomes such as absenteeism and academic achievement" (p. 119).

In **multiunit residential** (apartments, dormitories) **and mixed-use buildings** with domiciliary activities (hotels, firehouses), all of the sources and building features detailed above affect the overall microbiome. Within these categories of buildings, hospitals, nursing homes, and other inpatient health care facilities represent particular challenges because the sources of microbes—viruses in particular—and occupant vulnerabilities are greatest.

Only a limited literature addresses the microbiome of buildings—commercial or mixed-use—that contain retail operations. Hoisington and colleagues (2016) examined HVAC filter dust in a total of 13 electronics, furniture, grocery, home improvement, office supply, and general merchandise stores located in Pennsylvania and Texas. They found that, for this admittedly limited sample,

> the indoor environment in retail stores may offer a variety of niches for microbial populations that support a diverse community as compared to other built environment studies.…The microbiome was significantly influenced by several parameters including human microbiota (most notably to oral and skin bacterial communities) and the outdoor environment [but only a] tangential relationship between the bacterial community present and factors such as season, store location, and store type. (p. 685)

There are also specialized types of living and working environments, such as aircraft, submarines, and spacecraft, that pose special challenges for the management of microbes; Box 3-2 provides an example.

In some cases, buildings fail to deliver acceptable indoor environmental quality even at the time of their initial occupation. This issue has led, especially in the case of larger-scale commercial and other buildings, to a growing commitment to "building commissioning"[4]—a process intended to ensure that "systems are designed, installed, functionally tested, and capable of being operated and maintained according to the owner's operational needs" (DOE, 1999, p. 9). In the past, testing, adjusting, and balancing (TAB) of HVAC components were performed only once, after construction was completed. In commissioning, this process is expanded to include dynamic testing of multiple systems, including plumbing, lighting, and the building envelope, in all modes of building operation to capture seasonal changes. Most recently, this process has been extended to include "retrocommissioning"—the same systematic process applied to existing

---

[4]A more complete discussion of commissioning may be found in the Transportation Research Board report *Optimizing Airport Building Operations and Maintenance Through Retrocommissioning: A Whole-Systems Approach* (TRB, 2015).

BOX 3-2
Challenges for the Management of Microbes:
The International Space Station

Humans occupy a range of built environments that share characteristics with homes, schools, and offices but also differ in important ways, which may have an effect on the indoor microbiome as well as human health. An extreme example of such an environment is the International Space Station (ISS). The ISS is a closed habitat with carefully controlled conditions of humidity, temperature, and air circulation. Unlike most built environments on Earth, it provides no way to exchange air with the outdoor environment in space aside from the replenishment of pressurized oxygen during resupply missions. Other differences include, but are not limited to, the need to rely exclusively on recirculated air passed through a heat-exchange process in the high-efficiency particulate air (HEPA) filter system; the use of fans to move air through the microgravity environment; and the highly specialized functions each of the work areas of the ISS serves, potentially creating different microenvironments (Checinska et al., 2015). Aside from the distinct built environment characteristics on the ISS, it has been demonstrated that bacteria exhibit enhanced virulence, antibiotic resistance, and increased biofilm formation in space (Mayer et al., 2016). Therefore, detecting and monitoring microbes, particularly pathogens, in the air-handling systems has been a major concern of the National Aeronautics and Space Administration (NASA) in its efforts to maintain the health of astronauts.

The ISS provides a surrogate highly controlled environment to study. NASA has a long history of examining the growth of microbes in clean rooms and clean room air systems, although recent studies revealed that these environments host different microbiomes, likely because of extensive cleaning and maintenance procedures conducted for clean rooms that are not feasible on the ISS (Checinska et al., 2015). NASA has designed and created test chambers for training astronauts for their missions that replicate all aspects of the ISS (excluding microgravity and radiation). These chambers, as well as the ISS, can serve as specialized test chambers to examine potential built environment interventions and their effect on the indoor microbiome.

As the field of research examining microbiomes of built environments has grown, NASA has undertaken more detailed examination of the microbiome of the ISS, demonstrating that the microbial burden consists of more bacterial than fungal species, and that these microbes are most likely of human origin (Venkateswaran, 2016). At the time this report was written, no studies had been published directly comparing the microbiomes of astronauts with those of the spaces they occupy in the ISS, nor had associations been identified with particular health outcomes.

buildings that have never been commissioned; "recommissioning"—the reevaluation of a previously commissioned building, sometimes in conjunction with changes in use or renovations involving upgrades to the physical plant; and "continuous commissioning"[5]—an ongoing process that uses technology to identify and address performance problems, enhance occupant comfort, and optimize energy use via monitoring and dynamic adjustment of building systems (DOE, 2007; TEES, 2017). The negative consequences of understaffing and underfunding building operations and maintenance activities and the growing complexity of both building systems and environmental factors of concern highlight the importance of ongoing commissioning and draw attention to the need to incorporate into the commissioning process consideration of the effects of the indoor microbiome on occupants. Such knowledge may one day lead to the use of measures of the state of the indoor microbiome as part of the commissioning process.

The next three sections of this chapter identify indoor air, water, and surface microbial reservoirs and transmission routes and detail how they may lead to human exposure and how they are affected by occupant actions.

INDOOR AIR SOURCES AND RESERVOIRS OF MICROBES

Air is a critical transport vehicle of microbes and their metabolites in the built environment because it connects surfaces, water, and dust to what occupants inhale, inadvertently ingest, or absorb through their skin. The sources of the microbial communities that make up the microbiome of the built environment include indoor and outdoor sources of bacteria, fungi, and viruses, and the movement of air is a significant factor in their distribution. Indeed, the impact of these microbes on human health is often a function of how they become airborne and subsequently move through a building. This section reviews four key factors in the air transport of indoor and outdoor microbes: air leakage through unintentional openings in the building air envelope, internal migration of air between zones, mechanical ventilation, and natural ventilation.

In practice, most residential and commercial buildings are ventilated through a combination of envelope infiltration and mechanical and natural ventilation. HVAC systems in commercial buildings are designed and operated to maintain temperature and humidity within a comfortable range and to ensure the delivery of outdoor air for ventilation. In residential buildings, until recently, mechanical systems rarely incorporated outdoor air intake, relying on infiltration and operable windows to provide outdoor air. Engineered natural ventilation systems involving, for example, thermal chimneys and stack effect (discussed below) with carefully designed and located

---

[5] Also known as "ongoing commissioning" or "monitoring-based commissioning."

inlet vents are now becoming prevalent in northern Europe and Asia. The combination of these natural ventilation strategies with mechanical ventilation with or without mechanical cooling is termed "hybrid" or "mixed-mode" ventilation, and interest has been growing in using this approach to achieve energy savings while maintaining a healthy and comfortable indoor environment (Chenari et al., 2016; Heiselberg, 2006).

Increased attention has been paid to indoor air chemistry in recent years, revealing many important mechanisms affecting the fate and transport of airborne chemical contaminants within buildings (Morrison, 2015; Nazaroff and Goldstein, 2015; Weschler, 2011, 2016). This work, however, has to date not focused on impacts of indoor microbiomes. Accordingly, one of the areas for future investigation suggested by Adams and colleagues (2016, p. 227) is "how does the microbiome affect indoor chemistry, and how do chemical processes and the composition of building materials influence the indoor microbiome?" The authors' summary of available information on the topic notes that while the chemical metabolites produced by microbes may affect indoor chemistry, evidence suggests that their impact may be weak. Chemical agents in the environment may influence the microbiome, though, with research finding that differences in growth substrate lead to differences in microbial composition and metabolite production on wetted materials, and source strength may drive microbial community structure (Adams et al., 2016). A greater understanding of the extent to which indoor chemistry influences and is influenced by the building microbiome should result from the research recommended in the present report and from work being conducted under the Alfred P. Sloan Foundation's Chemistry of Indoor Environments[6] initiative.

Air Leakage Through Unintentional Openings in the Building Envelope

Airflow into and within a building can bring outdoor microbes indoors and transport them throughout the structure, and it can also transport microbes from interior sources from one part of a building to another. Although certain airflows are established by design (e.g., via mechanical ventilation systems), others are unintended and uncontrolled. Air leakage through unintentional openings in the building envelope[7]—known as "air infiltration"—is an important pathway for bioaerosols to enter buildings. Leaky building envelopes in both residential and commercial buildings lead

---

[6] See https://sloan.org/programs/science/chemistry-of-indoor-environments (accessed July 14, 2017).

[7] "Building envelope" is the collective term for the physical separators between the interior and exterior of a building, comprising such components as walls, floors, roofs, windows, skylights, and doors.

to considerable outdoor air entry, and even very "tight" buildings have non-negligible infiltration levels (Ng et al., 2015). Leakage alone can result in outdoor air ventilation rates equal to the lower range of mechanical ventilation rates (Grot et al., 1989). Research suggests that in naturally ventilated buildings nearly all particles in the diameter range 0.1–10 µm can flow through leaks in the building envelope with no significant losses (Liu and Nazaroff, 2001; Nazaroff, 2016).

Infiltration is driven by differences between indoor and outdoor air pressures, which vary with outdoor weather conditions, including wind velocity and direction relative to building exterior surfaces and their exposure to the wind, as well as differences between outdoor and indoor air temperatures. It is also affected by the operation of building equipment, including furnaces and boilers, local exhaust fans in bathrooms and kitchens, and mechanical ventilation systems that may have an imbalance between outdoor air intake and exhaust. Such infiltration and exhaust airflows often are highly complex, even in apparently simple buildings, and may be subject to significant short-term variations. These effects need to be considered to understand the impacts of airflow on microbial transport and growth.

Internal Migration of Air Between Zones

Building envelope construction, mechanical infrastructure, and interior layout define important physical features of a building that impact how airborne microbes may come into contact with occupants. These features define a building's major space conditioning (heating and cooling) zones. Pressure differences among zones combined with airflow pathways lead to interzone airflow and movement of airborne contaminants. In turn, these features drive airflow and surface and material moisture levels in relation to the available sites for potential microbial growth.

Building height plays an important role in creating indoor–outdoor pressure differences, especially when it is colder outside the building than inside. A phenomenon known as the stack or chimney effect results in significant airflow into lower floors of a building that then moves upward and flows out of the building at higher levels. During summer cooling periods, when it is colder inside than out, these airflow directions are reversed. The stack effect in a building is enhanced by its vertical shafts—elevators, stairwells, plumbing, and other service chases—providing important paths for airflows that can transport microbes. Variations in space or zone temperatures and their humidity levels—whether they are conditioned or unconditioned, occupied or unoccupied, or directly connected to the outdoors—play a role in microbial growth and movement of airborne microbes to other building spaces.

The height and shape of a building also will impact its exposure to wind. That exposure can in turn influence infiltration of outdoor air and microbes into the building, as well as amplify air pressure differences among building zones that affect indoor microbial migration.

Mechanical Ventilation

While HVAC systems meet important thermal and ventilation needs in buildings, these systems also provide routes for the entry of outdoor bioaerosols into buildings, as well as a means of circulating and dispersing indoor airborne contaminants. HVAC systems may themselves harbor microbial reservoirs, especially when water is involved (as described in the following section). These systems also affect temperature and moisture conditions throughout a building's interior spaces and within interior and exterior walls—conditions that affect the state of the indoor microbiome.

The delivery of outdoor air, or ventilation, through HVAC systems varies greatly among buildings and at different times within the same building, variations that are compounded by different approaches to HVAC control and operation. Most ventilation systems incorporate particle filtration, either of the outdoor air intake or of mixed airstreams of outdoor and recirculated air from occupied spaces. The efficiency with which particles are removed is a function of filter type, particle size, and airflow rate. The most common air filter effectiveness classification system in the United States is the MERV (Minimum Efficiency Reporting Value) rating,[8] with higher numbers corresponding to a higher fraction of particle removal. However, the MERV rating for filters commonly chosen for residential and commercial buildings indicates that they are not very efficient at removing submicron particles (<1.0 μm in mean aerodynamic diameter), which limits their effectiveness in reducing some airborne microbes.

In addition to the effectiveness of filters relative to microbial management, it is also important to gain a greater understanding of the effect of operating HVAC systems intermittently—for example, turning systems off over weekends and holidays—and the extent to which this intermittent operation enables microbes to collect on the filter medium, metabolize and multiply, and then be released when the system restarts (ASHRAE, 2009).

The design and placement of outdoor air intakes for mechanical ventilation systems can also influence the indoor microbial environment. The U.S. Environmental Protection Agency (EPA) guidance for schools suggests that "intakes should not be placed within 25 feet of any potential sources of air contaminants, including . . . mist from cooling towers"; that they should be screened to prevent birds and rodents from fouling and intro-

---

[8] American National Standards Institute (ANSI)/American Society of Heating, Refrigerating and Air-Conditioning Engineers (ASHRAE) Standard 52.2-2012. Method of Testing General Ventilation Air-Cleaning Devices for Removal Efficiency by Particle Size.

ducing microbes into HVAC systems; and that systems should be designed to cause "moisture to flow to the outside or to a drain if intake grilles are not designed to completely eliminate the intake of rain or snow" (EPA, 2017b). Similar requirements are contained in American National Standards Institute (ANSI)/American Society of Heating, Refrigerating and Air-Conditioning Engineers (ASHRAE) Standard 62.1-2016 (Ventilation for Acceptable Indoor Air Quality),[9] with the goal of keeping organic materials and moisture out of HVAC systems.

Studies have shown associations between increased ventilation rates and improved health outcomes, including reduced incidence of influenza and asthma and allergy symptoms[10] (Seppänen and Fisk, 2004; Sundell et al., 2011). Although higher prevalence of sick building syndrome symptoms is seen in air-conditioned buildings relative to naturally ventilated buildings (Finnegan et al., 1984), studies relating ventilation rates and health generally fail to describe how, when, and where ventilation rates were measured (Persily and Levin, 2011), and they often ignore buildings with strong indoor sources of air pollutants and locations with poor outdoor air quality (Sundell et al., 2011).

Existing Ventilation Standards and Measured Performance

Existing standards and building regulations include requirements for outdoor air ventilation rates and exhaust airflow rates for different building types and space uses intended to provide standards for model code requirements (ASHRAE, 2016a,b). For example, ASHRAE Standards 62.1, 62.2, and 170 contain minimum ventilation requirements for commercial and institutional buildings, residential buildings, and health care facilities, respectively.

ASHRAE's ventilation standard defines minimum values for acceptable HVAC system performance and is used in building design and construction when required by code. However, actual ventilation and exhaust airflow rates usually are quite different from those specified by codes or design documents as the result of a range of shortcomings in system installation, commissioning, operation, and maintenance. In some cases, building uses (and therefore, pollution sources) change after the system was designed,

---

[9]As defined by the standard, acceptable indoor air quality is "air in which there are no known contaminants at harmful concentrations as determined by cognizant authorities and with which a substantial majority (80% or more) of the people exposed do not express dissatisfaction" (ASHRAE, 2016a, pp. 6–7). Outside air used in ventilation also must meet National Ambient Air Quality Standards or be filtered, and local sources of concern may be identified. However, these evaluations are unlikely to specifically consider microorganisms.

[10]Factors other than ventilation also influence these respiratory health outcomes.

so that the design values are no longer relevant to the building as it exists (Persily et al., 2005).

Mechanical ventilation systems use a variety of approaches to control outdoor air entry and ventilation air distribution within buildings to achieve adequate minimum ventilation air delivery while saving energy. Minimum outdoor air intake rates are specified in the system design, based on standards and regulations. Ideally, these rates are verified during system commissioning and monitored occasionally during the life of the building, but this is rare in practice. Some buildings modulate the rate of outdoor air intake based on indicators of building occupancy (e.g., air temperature and carbon dioxide concentrations), providing less outdoor air during times of low occupancy as an energy-efficiency measure. This approach, referred to as "demand-controlled ventilation," may employ occupancy sensors, indoor carbon dioxide level detectors, or other strategies. Ventilation systems also may be configured to increase outdoor air intake when the outdoor air is cool and dry as an energy-efficient cooling mechanism, referred to as an "economizer cycle," replacing use of a mechanical system to lower the air temperature. Research to advance the development of real-time indoor and outdoor microbial sensors could be valuable, with data from such sensors being integrated into dynamic HVAC control systems.

Most studies of building ventilation performance to date have considered a small, nonrepresentative set of buildings, and authors often provide incomplete descriptions of the ventilation measurement methods employed (Persily, 2016). The most recent comprehensive studies yielded mixed results as to whether building ventilation systems actually meet current standards. The Building Assessment Survey and Evaluation (BASE) study, conducted in the 1990s by EPA, included ventilation and other indoor air quality measurements for 100 randomly selected large U.S. office buildings (Persily and Gorfain, 2008; Persily et al., 2005). The mean measured outdoor air ventilation rate was 49 L/s per person based on the occupant densities during the time of the ventilation measurements, a rate that exceeds the minimum requirements of ventilation standards. These high air change rates occurred in part because the systems often operated in economizer (energy-saving, free cooling) mode or because the actual space occupancies were, on average, 80 percent of the design values. Considering only minimum outdoor air intake operation and accounting for the lower occupancy levels, the mean ventilation rate was about 11 L/s per person at default occupancy values in ASHRAE Standard 62.1, which is based on achieving <20 percent dissatisfaction with perceived indoor air quality (ASHRAE, 2016a). Approximately one-half of the values were below the minimum requirements in Standard 62-1999 or the designed ventilation rates.

In low-rise residential buildings, uncontrolled air leakage across the building envelope can also be an important source of outdoor air ventilation, and these air change rates vary with geography and such factors as the age and size of the home (Koontz and Rector, 1995). A field study of 108 new single-family homes in California included measurements of outdoor air change rates (Offermann, 2009). The median 24-hour outdoor air change rate was 0.26 h^{-1}, with a range of 0.09 to 5.3 h^{-1}. Sixty-seven percent of the homes had outdoor air change rates below the minimum California Building Code requirement of 0.35 h^{-1}. The author notes that the combination of relatively tight envelope construction and "the fact that many people never open their windows for ventilation . . . resulted in many homes with low outdoor air exchange rates" (p. 210).

Mechanical Conditioning and Ventilation Systems and Indoor Microbiomes

Because building mechanical conditioning and ventilation systems are designed primarily for thermal environmental control (i.e., air temperature and relative humidity), conditions relevant to airborne microbes generally are affected by system status, design, and operation and other internal building conditions (including occupant numbers and activities). Where heating demand dominates design considerations, indoor air will be less humid than colder outdoor air because of the reduced relative humidity from warming of the air.[11] Where air conditioning is dominant, indoor air will be cooler than the air outdoors, and management of the air conditioning system will determine the moisture content of the indoor air and its potential effect on airborne microbes. Of special interest is the effect of air humidity on the survival and pathogenicity of infectious airborne agents, such as viruses and bacteria associated with seasonal flu, the common cold, and other diseases.

Mechanical systems may distribute microbes to occupants or disperse microbes from humans with bacterial or viral infections. HVAC design employs a range of ductwork configurations and components to distribute air to occupied zones and control the airflow as a function of thermal requirements in the zones and other factors. In many commercial buildings, zoned air distribution systems are designed to mix the ventilation air with the conditioned room air based on thermal comfort goals, a system known as "mixing ventilation." Displacement ventilation systems and task air systems that deliver air directly to occupants are also being implemented in some buildings, and research suggests that such systems could be more effective than mixing systems in creating healthier indoor environments (Kong et al., 2017). Also seeing increasing application are 100 percent outside air systems, also known as dedicated outdoor air systems (DOASs), with many such buildings being designed to separate ventilation from heating and

---

[11]The reduction in humidity from warming of the air will be offset by such indoor sources as cooking, showering, and occupant metabolism.

cooling. Such systems are more typically found in hospitals and are often combined with a means of heat recovery.

Natural Ventilation

Natural ventilation has been used for centuries to bring outdoor air into buildings and to circulate it within the building interior. It offers the benefits of resiliency in the face of power outages and the potential for higher levels of outside air to purge indoor pollutants when wind speeds or indoor–outdoor temperature differentials are sufficiently high. However, natural ventilation can lead to the entry of outdoor contaminants and moisture. It is limited by the driving pressures that induce airflows into the building, and the ventilating air entering the building may not be well distributed among occupied spaces or the rates of entry well controlled. As a result, to ensure a given continuous minimum rate of ventilation, natural ventilation systems increasingly are being designed and integrated with mechanical ventilation on a climate-specific basis. Critical design factors include such parameters as the ratio of window opening size to floor area; the ratio and locational relationships of openings to each other, as well as their relative positions in the room; and the influence of wind velocity and direction on air distribution in the room (Levin, 2010). Door openings also may have a significant influence on the amount of outside air that enters a building, especially in some types of commercial buildings. A Pacific Northwest National Laboratory report notes that "restaurants, strip-mall stores, retail stores, supermarkets, offices and hospitals are likely to have high door-opening frequency, either at certain time periods of day or in some cases throughout the occupied hours" (Cho et al., 2010, p. 1) that lead to high infiltration rates. Such airflows may raise or lower the concentration of indoor pollutants, including microbial agents, depending on their outdoor concentrations (Zaatari et al., 2014).

Design guidance and tools for natural ventilation include the Chartered Institution of Building Services Engineers' *Applications Manual AM10* (CIBSE, 2014) and the National Institute of Standards and Technology's *LoopDA* (NIST, 2017).

Indoor Air Sources and Reservoirs of Microbes: Summary of Findings

Air leakage through unintentional openings in the building envelope, internal migration of air between zones, the distribution of air by mechanical conditioning and ventilation systems, and natural ventilation affect the association between air and the indoor microbiome. These interrelationships involve bioaerosol and moisture generation, contaminant transport by natural and mechanical ventilation elements, and human behaviors that resuspend particles or modify ventilation. Building airflow, which constitutes an important transport mechanism for indoor air contaminants, is complex and often poorly controlled. Building ventilation systems—mechanical, natural, and hybrid—do not necessarily perform as intended or expected. These facts highlight the importance of considering microbial growth in relation to air-related transport in a building and the interplay of human behaviors and use patterns related to the indoor microbiome.

Ventilation and filtration have the potential to control the quality of airflow and indoor microbial conditions, but exercising this control requires sound maintenance practices and proper operation, especially the management of wet environments that may contribute to microbial growth. Generally speaking, over the past several decades commercial buildings in the United States have employed mechanical conditioning and ventilation systems along with tighter building envelopes without operable windows. This "sealed" approach to building ventilation is quite distinct from the earlier use of operable windows and natural ventilation (Banham, 1984). The implications of the shift to sealed buildings for indoor air quality, occupant satisfaction and performance, and indoor microbiomes have yet to be thoroughly researched.

INDOOR WATER SOURCES AND RESERVOIRS OF MICROBES

Water is essential to the viability (growth and survival) of microbes. The periodic or episodic presence of water can sustain some microbial life and even support reproduction and growth. As noted in Chapter 2, there is general agreement that dampness and mold within buildings are associated with unfavorable health outcomes, although many studies examining this issue have important methodological considerations that limit their generalizability. Water that enters a building from public or private supply sources can also contain pathogens, including *Legionella pneumophila* and *Mycobacterium avium*, that present health risks.

While buildings generally are intended to be and stay dry, water still can be present in four forms—liquid, vapor, adsorbed moisture, and ice—depending on air and surface temperatures, physical characteristics of surfaces and materials, and water concentration. Adsorbed moisture is water that is held on the surfaces of a material by intermolecular forces. It has inherently different properties from those of moisture in its liquid and frozen states, as well as from those of moisture that has been absorbed and chemically bound by a material, becoming part of its chemical structure. These different states influence how efficiently the water can play a role in biologic processes or chemical reactions; the more tightly bound it is, the less of it is available for such processes or reactions.

Buildings contain many different water sources and reservoirs. The most obvious is **municipal or well water** brought into the building deliberately through premise plumbing. **Unwanted liquid water** results from plumbing leaks and leakage through the building envelope (exterior walls, roofs, floor slabs, crawl space, or basement) or when capillary action draws groundwater through pores in building materials (so-called rising damp). **Water vapor** commonly migrates when air pressure differences between connected spaces result in transport of air and its contents from the region of higher pressure to that of lower pressure (although diffusion due to humidity differences can also be important). In addition to high outdoor vapor pressure, water vapor is also produced by such activities as showering, cooking, and washing clothes. Water vapor and **adsorbed moisture and ice** can enter the building enclosure when air pressures and temperatures and humidities inside and outside a building change, causing transfer of heat and moisture between the air and the building's materials. Adsorbed moisture in building materials, such as gypsum panels or rain-soaked construction lumber, can influence mold growth on surfaces, on interior walls, and in the concealed spaces within walls. **Recycled water**—which may be used indoors in such applications as toilet and urinal flushing under certain circumstances (Los Angeles County Department of Public Health, 2016)—can harbor microbial contaminants if not properly managed (Toze, 2006). Finally, green building practices such as rainwater harvesting or wastewater treatment and reuse can be a source of microbes in the indoor environment (Ghaitidak and Yadav, 2013).

The following sections address six examples of water-related sources of potentially problematic microbes:

- premise plumbing
- hot water heaters
- cooling towers, cooling coils, and drain pans
- leakage, flooding, and wet building materials
- indoor water sources and airborne moisture generators
- indoor humidity

Premise Plumbing

Water piped into buildings can come from municipal water treatment plants, wells, groundwater, or surface water sources. Water entering a building commonly contains bacteria, and some potable water supplies can contain fungi as well. Premise plumbing—a building's plumbing systems and equipment—can affect the concentrations of these microbes before they reach such outlets as faucets, spigots, showerheads, and other appliances, which often aerosolize microbes directly into the breathing zone of building occupants.

Showerheads, for example, have been identified as a source of exposure to aerosolized nontuberculous mycobacteria (Falkinham, 2011; Feazel et al., 2009). The type of showerhead, the frequency and duration of its use, and the duration of stagnancy affect the amount of aerosol generated and thus the total exposure to microbes, with those showerheads that produce a lower-pressure stream generally also producing less aerosolization.

Premise plumbing pathogens are responsible for a significant number of infections whose origins have been traced to drinking water (Beer et al., 2015). These opportunistic pathogens represent an emerging waterborne disease problem with a major economic cost of at least $1 billion annually (Falkinham, 2015). Water is the source of *Legionella pneumophila*, *Mycobacterium avium*, and *Pseudomonas aeruginosa*, which are ubiquitous in water systems and are estimated to be responsible for tens of thousands of infections each year (Collier et al., 2012; Falkinham, 2015). The design, operation, and maintenance of premise plumbing systems are critical to controlling exposures. Figure 3-1 illustrates how water chemistry and flow shape biofilms[12] in pipes and the resulting microbiome.

Premise water filters can themselves become sites for biofilms and microbial growth. Maintenance of filters and the frequency of their replacement are important factors in filter performance and the presence of microbes found in potable water at the point of use. Filters are used to remove inorganics (such as lead) or objectionable chemical agents (such as chlorine, benzene, or trichloroethylene) from the tap where water for cooking or drinking is drawn. Such filtering may leave little or no residual chlorine, which can result in growth of microbes within the premise plumbing system. The application of such filtration is an example of human behaviors that affect the microbial content (and its pathogenicity) of water in the premise plumbing system.

Building features and human behaviors that increase the likelihood of stagnant water are important to consider because stagnant water supports microbial growth. The frequency with which various components of the plumbing system are used and the design and materials of these systems determine both the locations of a building's stagnant water and the amount of time the water will remain stagnant. During the off season, for example, showers and water features of hotels and vacation resorts may see little use, thus allowing for more microbial growth. A similar problem can occur in vacation homes.

---

[12] Biofilm is defined as "a thin, normally resistant, layer of microorganisms such as bacteria that form on and coat various surfaces" (https://www.merriam-webster.com/dictionary/biofilm [accessed July 17, 2017]).

Water Chemistry and Flow Shapes Biofilm and Bulk Water Microbiome

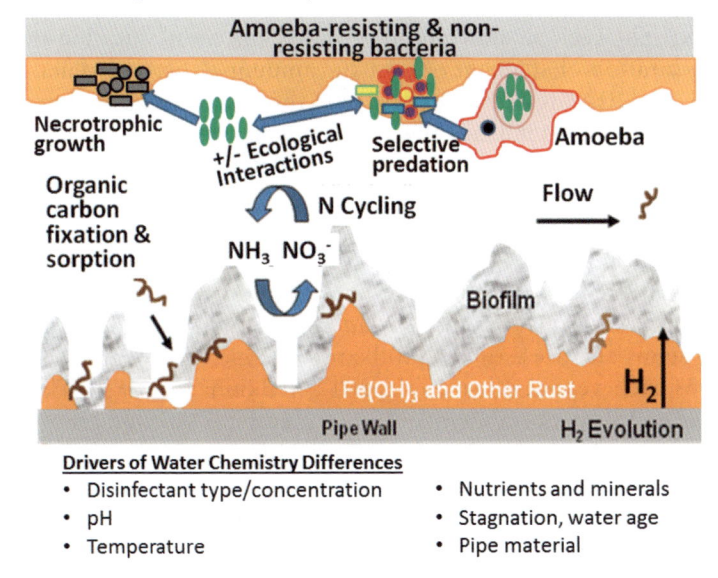

Drivers of Water Chemistry Differences
- Disinfectant type/concentration
- pH
- Temperature
- Nutrients and minerals
- Stagnation, water age
- Pipe material

FIGURE 3-1 The influence of water chemistry and flow on the microbiome of bulk water pipes.
SOURCE: Courtesy of Drs. Amy Pruden and Hong Wang, Virginia Polytechnic Institute and State University.

Hot Water Heaters

Researchers have long known that commercial and residential hot water heaters are reservoirs for thermophilic bacteria (Brock and Boylen, 1973) and are thus a primary source of waterborne pathogens in buildings in the United States (Brazeau and Edwards, 2013). As already noted, *Legionella pneumophila* is a particular concern in this regard, but Brazeau and Edwards (2013, pp. 617–618) note that "*Acanthamoeba, Mycobacterium avium* complex, and *Pseudomonas aeruginosa* can also grow within water heating systems and cause thousands of cases of infections annually."

Investigators have studied microbial communities in domestic hot water heated to different temperatures in systems with varying levels of use (Ji et al., 2017; Rhoads, 2017; Rhoads et al., 2015). The practice of lowering water heater temperatures to 120°F (49°C) to conserve energy and limit the possibility of scalding creates a circumstance whereby tap water is warm enough to support the growth of *L. pneumophila* but not to inactivate microbes in the biofilm normally found on the bottom of hot water tanks. And while 140°F (60°C) is hot enough to inactivate *L. pneumophila* in the tank, it may not eliminate the bacterium in the pipes of preheated water for infrequent hot water users. As noted earlier, the frequency of use of hot water fixtures (such as showers) determines the length of time the water remains stagnant and the corresponding potential for microbial growth. Furthermore, Rhoads and colleagues (2015) found that under experimental conditions, 124°F (51°C) may represent a "sweet spot" for *L. pneumophila* in conditions of low water use enriching the concentration at the tap.

Cooling Towers, Cooling Coils, and Drain Pans

Air conditioning system cooling towers are a documented source of community outbreaks of Legionnaires' disease, with higher rates of infection occurring closer to the source tower (Addiss et al., 1989; Weiss et al., 2017). Indeed, *Legionella pneumophila* have been identified as the main microbial risk in cooling towers (Torvinen et al., 2014), where they often are found growing in the warmer water associated with discharge from refrigeration and air conditioning systems. The cooling tower's role is heat rejection to the atmosphere, with the water being cooled by evaporation or exposure to air that is cooler than the water. Because of their rooftop location, which may be near or upwind of outdoor air intakes, cooling towers can release *L. pneumophila* into the airstream taken in by the ventilation systems, leading to illness in the buildings where they are located. The aerosol emitted into the air stream can also be carried to nearby and even distant locations, leading to community outbreaks (Weiss et al., 2017).

Testing of cooling tower water for *L. pneumophila* is not required. Reported concentrations vary quite widely. There is no official guidance on treatment requirements based on concentrations. The dominant form of treatment is the addition of chemicals, although cleaning of cooling towers to remove protozoans growing on surfaces is associated with reduced concentrations of *L. pneumophila* in the water (Pagnier et al., 2009).

Conventional HVAC system design in air-conditioned buildings—which involves frequently wet surfaces on cooling coils, drain pans, and sometimes humidifiers—may lead to as yet uncharacterized microbiologic exposures and consequent illness (Mendell et al., 2008; Menzies et al., 2003). Poor system condition and poor maintenance increase the risk of such problems. Accumulated dust and dirt and moisture in HVAC systems provide a nutrient source and growth medium for microorganisms (Morey et al., 2009; West and Hansen, 1989).

ASHRAE has a guideline—12-2000, Minimizing the Risk of Legionellosis Associated with Building Water Systems—and a standard—188-2015, Legionellosis: Risk Management for Building Water Systems—that address the minimization of Legionella contamination in building water systems. Separately, ANSI/ASHRAE Standard 62.1-2016, Ventilation for Acceptable

Indoor Air Quality, contains several general requirements related to moisture management in HVAC systems that include, for example, the cleanability of cooling coils, the ability of condensate drain pans to collect moisture effectively and for it to be removed from the system, the specification of duct surfaces to reduce microbial growth, and the provision of access to systems for inspection and maintenance. However, many HVAC systems still have drain pans with inadequate slope so that condensate remains stagnant and microbial growth, including biofilm formation, is facilitated. The standard also contains requirements designed to minimize the likelihood of envelope-related moisture problems via air leakage and building pressure control. ASHRAE/Illuminating Engineering Society (IES)/U.S. Green Building Council (USGBC) Standard 189.1 has analogous requirements, but they are more stringent than those in Standard 62.1 given that 189.1 is a high-performance standard. These ASHRAE standards are incorporated in some "model" codes, but widespread adoption of up-to-date standards is elusive given the thousands of local code agencies and shortage of trained inspectors or enforcement agents. The building codes are intended to govern design and issuance of building permits, but not operating conditions in occupied buildings.

Leakage, Flooding, and Wet Building Materials

Water can also enter a building from a host of unintentional sources, including enclosure leakage and flooding, plumbing leakage, rising damp, condensation, and human activities. Leakage through roofs, foundations, and walls during rain is common and often results in wetting of building materials and sometimes in the accumulation of moisture. During floods or other ground saturation conditions, hydrostatic pressure (the pressure exerted by a fluid at equilibrium at a given point within the fluid, due to the force of gravity[13]) can force groundwater in through cracks in the slab or foundation of a building. The increased frequency and intensity of floods and hurricanes have created extreme conditions of water intrusion that need to be followed by remediation to remove or limit future microbial growth (the installation of perimeter drains, for instance) and additional measures to reduce problems with hydrostatic pressure.

Building materials also may be damaged by rain or excessive moisture prior to or during construction, which changes their physical properties and potentially contaminates them with mold spores that grow when water is subsequently reintroduced. Andersen and colleagues (2017) found fungal species that had become embedded in gypsum wallboard during the manufacturing process, before the material reached retailers or construction

---

[13] See http://www.dictionary.com/browse/hydrostatic-pressure (accessed May 1, 2017).

sites. Keeping building materials dry prior to construction has long been recognized as a key strategy for limiting the likelihood of mold growth in buildings (EPA, 2013a).

Any excess liquid water or water vapor within a building will contribute to absorption by adjacent materials. The amount of absorption depends in part on the porosity of the material. Most building materials are hydrophilic, absorbing water readily (Straub, 2006). When water flooding a floor comes in contact with gypsum board, for example, it is slowly absorbed and by capillary action can saturate the board. The rate of drying of wet gypsum board[14] depends on several factors that interact dynamically, including the porosity, temperature, and moisture content of the gypsum board; its surface coating, if any; the absolute and relative humidity of the surrounding air; the velocity at which that drying air flows across the surface; and the temperature of the air compared with that of the wet board (Dedesko and Siegel, 2015).

Water in interstitial or concealed spaces in walls, crawl spaces, attics, and HVAC chases[15] supports the growth of mold. Mold and bacteria metabolism produces microbial volatile organic compounds (MVOCs) that can reach occupied spaces and be detected by the compounds' characteristic "earthy," "moldy," or "musty" odors. The presence of these odors is strong evidence of microbial growth resulting from persistent dampness in concealed spaces or inside HVAC systems (Mendell and Kumagai, 2017).

Standing water in crawl spaces beneath buildings—including single-family homes and portable classrooms—may come from rainwater, groundwater, water vapor in the soil, or soil gas (DOE, 2013). Such moisture can lead to high levels of mold growth within the crawl space from which mold spores and MVOCs can be carried into occupied spaces via airflow. There are also multiple pathways through which such moisture itself can enter buildings and contribute to mold growth within occupied spaces. These mechanisms include capillary action, bulk airflow through holes and other penetrations, and vapor diffusion (DOE, 2013).

Indoor Water Sources and Airborne Moisture Generators

In addition to rain penetration with the possibility of both direct wetting of materials and standing water, microbial growth can be supported

---

[14]Gypsum board is "the generic name for a family of panel products that consist of a noncombustible core, composed primarily of gypsum, and a paper surfacing on the face, back and long edges" (https://www.gypsum.org/technical/using-gypsum-board-for-walls-and-ceilings/using-gypsum-board-for-walls-and-ceilings-section-i [accessed July 17, 2017]). It may also be called drywall, wallboard, or plasterboard.

[15]A chase is a wall or a ceiling feature through which ducts, electrical wires, pipes, and the like are run.

by indoor and outdoor moisture migration and condensation. Microbial growth on surfaces and within building assemblies depends on moisture availability at the surface rather than on air humidity itself—high air humidity can support high surface water activity. Human activities that generate water vapor—cooking, showering, clothes and dish washing and drying, and occupant metabolism (which is an issue in densely occupied spaces), as well as use of humidifiers—are significant sources of indoor humidity. The frequency and duration of these activities determine the release of moisture into air, and the presence or absence of effective exhaust ventilation is critical.

The human activities that impact the water vapor emission intensity indoors include the

- frequency and duration of personal hygiene practices, including showers, baths, and toilet flushing[16];
- use of water for cleaning hard surfaces, such as flooring;
- frequency and duration of washing dishes and clothes;
- frequency and use of clothes dryers, with or without the exhaust of moisture to the outdoors;
- use of an exhaust hood to remove moisture emitted during food preparation; and
- use of exhaust fans or window opening during and after showering.

Hot tubs, whirlpool baths, and other spas and water features can aerosolize water and microbes and result in very high levels of exposure to aerosolized bacteria (e.g., *Legionella pneumophila* and *Mycobacteria* spp.). These water-based recreational and therapeutic systems often involve water temperatures in the range of 104–110°F (40–43°C), and they produce aerosolized bacteria in the surface water close to the head of the human occupant/user of the spa. In addition, increased humidity levels indoors can lead to higher water vapor on interior surfaces and within building materials.

Water vapor condenses more readily on cold surfaces, so interior surfaces of exterior walls and roofs in cold weather or the interior surfaces or layers of exterior walls and roofs of air-conditioned buildings in warm, humid climates increase the occurrence of unwanted condensation. Water supply pipes, wastewater pipes, and fire suppression system water pipes also tend to have colder temperatures than the air around them, which can cause condensation on these pipes. In addition, building materials that are

---

[16]The position of the toilet seat cover also affects the amount of aerosols released into the room air.

saturated with moisture themselves possess high water activity levels that can result in fungal growth.

The indoor moisture content of air and materials varies greatly depending on the season of the year. Inside walls, attics, and roofing assemblies, the moisture content of materials is affected by seasonal surface temperatures, with cooler internal surfaces collecting and retaining more moisture than warm surfaces. Air-conditioned interiors have increased moisture absorption and therefore higher surface water activity and more potential to support microbial growth.

Biowalls—also called green or living walls—have garnered attention as a possible means of promoting a healthy indoor environment through the introduction of plants that are intended to clean the air or create a "green" atmosphere. Very little research has been done on their effectiveness in this regard, however, and because they use water in liquid and aerosolized form as part of their operation and because plants loose water into the air as a result of metabolism (transpiration), they could elevate humidity and harbor pathogens (Girman et al., 2009; Waring, 2016).

While the relationships between moisture and microbial growth have been studied extensively—especially with respect to mold's association with allergy, asthma, and other health endpoints of interest (see Chapter 2)—the differences among geographic locations, building types, construction practices, and climates are too great to enable refinement of generalized design solutions or the development of broadly applicable guidelines or code requirements to ensure the elimination of surface and hidden mold in building assemblies.

Indoor Humidity

The moisture content of air is important to fungal viability and growth and to virus and bacteria survival and virulence. Mold growth on building materials depends on a variety of parameters, including moisture, temperature, time, and the presence of nutrients on the substrate and its pH level, but moisture is the primary driving factor (Haverinen, 2002). The relationships between air humidity and bacteria show different patterns of survival and pathogenicity depending on the species of concern.

Moisture in buildings can be characterized by various means. Relative humidity is perhaps used most commonly, both because it is comparatively easy to measure and because its levels are related to perceived occupant comfort. It is defined as the water vapor pressure of the air, expressed as a percentage of the saturated water vapor pressure at the same temperature, and thus reflects both the amount of water vapor in the air and the air temperature. However, airborne relative humidity by itself is not, in general, predictive of mold growth on indoor surfaces,[17] and it may vary considerably in an interior space depending on where it is measured.

Humidity level is thus only a part of the larger, more complex issue of how moisture affects the composition and viability of indoor microbiomes. Different humidity levels in combination with other parameters, such as ventilation and temperature, promote or suppress different viruses, bacteria, and fungi, and a microbe that thrives on one indoor surface may waste on another under the same humidity conditions. Therefore, generic advice about humidity levels needs to be viewed with skepticism.

Indoor Water Sources and Reservoirs of Microbes: Summary of Findings

Where there is water, there are likely to be microbial organisms. While few bacteria and molds have demonstrated adverse health effects, opportunistic pathogens and allergenic species are commonly found among indoor microbial contaminants. On the basis of its review of the literature and the prior findings of the Institute of Medicine (IOM) reports *Damp Indoor Spaces and Health* (IOM, 2004) and *Climate Change, the Indoor Environment, and Health* (IOM, 2011), the committee identified the need for more research on water quality supplied and delivered by premise plumbing; microbial management in building hot water heaters, cooling towers, cooling coils, and drain pans; leakage and flooding that results in moisture damage, especially in houses; indoor water fixtures, features, and airborne moisture generators; and the detection of mold and moisture inside building assemblies, especially walls and roof assemblies. More research is also needed on how to interpret moisture measurements in terms of the risk of mold and bacterial growth and the role of viruses in the evolution of the total indoor microbiome over time.

BUILDING SURFACES AND RESERVOIRS OF MICROBES

Microbes are introduced to and released from surfaces inside buildings through a number of mechanisms, including deposition of airborne microbes; transfer via occupants' direct contact with surfaces; the tracking and deposition of dirt, dust, pests, and water into buildings from the outdoors; resuspension of deposited microbes due to a variety of activities; and bodily emissions from exhalation, expectoration, skin shedding, cuts in the skin, and bladder and bowel waste. Occupants are exposed to these microbes when they touch the surfaces (via hand-to-mouth ingestion or

---

[17]The exception to this is high relative humidity conditions, which lead to damp indoor surfaces that are conducive to microbial growth.

direct dermal transmission) or when the surfaces are disturbed, aerosolizing the microbes and particles that may be attached to them, which are then inhaled. Indoor surfaces that can support microbial growth—including floor, wall, and ceiling materials, as well as plumbing and HVAC components—are important in designing and maintaining buildings to manage microbial communities to human advantage.

This section focuses on the relationship of building surfaces to microbes that impact human health. While the presence of fungi, bacteria, and viruses on surfaces or suspended in dust may be most medically serious in hospitals, important findings also have resulted from studies of homes, offices, fire stations, schools and kindergartens, gymnasiums, food service facilities, and dormitories. Because extensive interaction occurs among indoor air, water, and surfaces, some of the issues salient here relate to those introduced earlier in this chapter.

Direct Contact with Surfaces by Occupants

Bacterial, viral, and fungal communities are transferred to building surfaces by direct contact with occupants' skin, saliva, and mucosa. Recent advances in DNA sequencing analysis have facilitated research on bacterial and fungal communities on surfaces in classrooms, offices, homes, gymnasiums, and other building types (Barberán et al., 2015; Chase et al., 2016; Dunn et al., 2013; Flores et al., 2013; Kelley and Gilbert, 2013; Meadow et al., 2014a,b; Yamamoto et al., 2015). A study by Haleem and colleagues (2013) yielded information about numerous bacterial and fungal taxa from surface samples collected at a university, including *Bacillus* sp., *Candida albicans*, *E. coli*, *Fecal streptococcus*, *Pseudomonas aeruginosa*, *Klebsiella pneumonia*, *Staphylococcus* sp., *Streptococcus* sp., and *Trycophyton* sp.

Evidence indicates that contaminated surfaces also play a role in the spread of viral infections. Table 3-1 presents the results of a compilation by Boone and Gerba (2007) of buildings and surfaces where viruses have been detected or survived.

Understanding of human interactions with surfaces has advanced sufficiently to demonstrate that each human leaves a specific microbial signature on surfaces. It has been shown that bacterial communities on a surface can be traced back to an individual for forensic purposes (Fierer et al., 2010). When new occupants enter a home, their distinct microbes can be detected in the building's surface bacteria within days (Lax et al., 2014). This line of research demonstrates that bacterial communities on surfaces as distinct as floors, walls, chairs, tables, doorknobs, elevator buttons, keyboards, and other shared equipment contain human microbiota.

In addition to their ability to populate fomites, bacteria and viruses can remain infectious on surfaces for hours to days, and they have variable

TABLE 3-1 Buildings and Surfaces Where Viruses Have Been Detected or Survived

Virus	Location of Virus Buildings (reference[s])	Surfaces (reference[s])
Respiratory syncytial virus	Hospitals (23)	Countertops, cloth gowns, rubber gloves, paper facial tissue, hands (33)
Rhinovirus	Not found	Skin, hands (30), doorknob, faucet (52)
Influenza virus	Day care centers, homes, nursing home (51)	Towels, medical cart items (51)
Parainfluenza virus	Offices (data not published), hospitals (23)	Desks, phones, computer mouse (Boone and Gerba, submitted)
Coronavirus	Hospitals (23), apartments (62)	Phones, doorknobs, computer mouse, toilet handles (23), latex gloves, sponges (68)
Norovirus	Nursing home (6), hotels, hospital wards, cruise ships, recreational camps (22, 38, 61)	Carpets, curtains, lockers, bed covers, bed rails, drinking cup, water jug handle, lampshade (6, 38)
Rotavirus	Day care centers, pediatric ward (8)	Toys, phones, toilet handles, sinks, water fountains, door handles, play areas, refrigerator handles, water play tables, thermometers, play mats (8, 15, 38, 70), paper, china (2), cotton cloth, latex, glazed tile, polystyrene (1)
Hepatitis A virus	Hospitals, schools, institutions for mentally handicapped, animal care facilities, bars (72)	Drinking glasses (72), paper, china (2), cotton cloth, latex, glazed tile polystyrene (1)
Adenovirus	Bars, coffee shops (7, 24)	Drinking glasses (24), paper, china (2), cotton cloth, latex, glazed tile, polystyrene (1)
Astrovirus	Schools, pediatric wards, nursing homes (39)	Paper, china (2)

NOTE: References (in parentheses) listed in source document.
SOURCE: Boone and Gerba, 2007, Table 1.

die-off responses to disinfectants and other cleaning products. For example, one study of children's desks after cleaning found that fungal and bacterial communities recovered to precleaning loading levels after approximately 3 days (Kwan et al., 2016). In public restrooms, microbiota on floors that had been cleaned with bleach redeveloped within 5–8 hours and "showed remarkable stability over weeks to months" (Gibbons et al., 2015, p. 765). Other investigators have shown not only that school restrooms are dominated by microorganisms associated with the gastrointestinal and urogenital tracts (Flores et al., 2011) but also that microbes from these tracts are predominant on classroom chairs, along with those from the skin (Meadow et al., 2014b). Bean and colleagues (1982) found that viable influenza A and B virus could be transferred from such nonporous surfaces as steel and plastic to hands for 24 hours and from such porous surfaces as tissues to hands for 15 minutes. These and other data support the notion that viruses may be spread by indirect contact via fomites (Weinstein et al., 2003).

Identifying the surfaces of greatest significance for improved design or maintenance practices aimed at reducing bacterial and viral microbes continues to be a research challenge. In a study of intensive care units (ICUs) by Bures and colleagues (2000), the colonization rate for keyboards and faucet handles by "novel and unrecognized" taxa was greater than that for other surfaces in rooms with patients testing positive for methicillin-resistant *Staphylococcus aureus* (MRSA), revealing that a variety of surfaces may serve as reservoirs of pathogens and vectors for cross-transmission. A study of Boston-area homes found that the highest bacterial counts were associated with wet surfaces that are often touched, such as tubs, sinks, and faucet handles. These surfaces were found to be contaminated with bacteria, including *Enterobacteriaceae*, *Pseudomonas*, methicillin-sensitive *Staphylococcus aureus* (MSSA), and MRSA (Scott et al., 2009).

Of greatest concern are surfaces touched by multiple occupants—so-called high-touch surfaces—including both dry sites, such as door handles, elevator buttons, keyboards, light switches, and television remote controls, and wet sites, such as food preparation areas and kitchens, water fountains, bathroom faucets and counters, toilet seats and handles, and soap dispensers. Tables 3-2 and 3-3 illustrate the commonalities and differences among fomites of concern in residential buildings versus commercial buildings and hospitals. These tables also list surfaces of concern on which occupants sit and lie, from chairs to couches and bedding, where hand-to-mouth and mouth-to-mouth transfer is also possible. Exposure to bacterial, viral, or fungal fomites and their impact on human health depends on the sources and human susceptibility, as well as a host of physical and environmental conditions that need to be recorded in parallel with surface sampling to gain a more complete understanding of influences on microbial populations and communities (see Table 3-4). This topic is revisited in Chapter 4.

TABLE 3-2 Home High-Touch Surfaces and Bacterial Reservoirs

All rooms	Light switches, air, dust, floor, rugs, door knobs
Kitchen	Countertop, sink, faucet handles, drain, u-pipe, refrigerator handle, refrigerator shelves, microwave buttons, dish sponge, drying towels, drying rack
Bathroom	Countertop, sink, u-pipes, shower floor, shower curtain, showerhead, shower poufs, bar soap, toilet bowl, toilet water, toilet seat, toilet flush handle, hand towels
Bedroom	Pillows, sheets
Living room	Seats, arm rests, head rests, pillows, blankets, remote controls
Office, etc.	Keyboard, mouse, water from water heater, mop head, HVAC filters

SOURCE: Smith et al., 2013, Table A5-1.

TABLE 3-3 Hospital High-Touch Surfaces and Bacterial Reservoirs

Patient area	Bed rails, tray table, call boxes, telephone, bedside tables, patient chair, intravenous (IV) pole, floor, light switches, glove box, air, air exhaust filter
Patient restroom	Sink, faucet handles, inside faucet head, hot tap water, cold tap water, light switches, door knob, handrails, toilet seats, flush lever, bed pan cleaning equipment, floor, air, air exhaust filter
Additional equipment	IV pump control panel, monitor control panel, monitor touch screen, monitor cables, ventilator control panel, blood pressure cuff, janitorial equipment
Water	Cold tap water, hot tap water, water used to clean floors
Patient	Stool sample, nasal swab, hand
Staff	Nasal swab, bottom of shoe, dominant hand, cell phone, pager, iPad, computer mouse, work phone, shirt cuff, stethoscope
Travel areas	Corridor floor, corridor wall, steps, stairwell door knobs, stairwell door kick plates, elevator buttons, elevator floor, handrails, air
Lobby	Front desk surface, chairs, coffee tables, floor, air
Public restroom	Floor, door handles, sink controls, sink bowl, soap dispenser, towel dispenser, toilet seats, toilet lever, stall door lock, stall door handle, urinal flush lever, air, air exhaust filter

SOURCE: Smith et al., 2013, Table A5-2.

A range of solutions are used to reduce the opportunity for infection through fomites from human contact with surfaces, including

- hand hygiene—washing with soap and water or hand sanitizers;
- surface washing with disinfectants;

TABLE 3-4 Environmental, Location, and Surface Parameters That May Influence Microbial Populations and Communities

Building	Room	Surface
• Latitude, longitude, altitude • Foundation type • HVAC sterility • Surrounding flora • Construction materials	• Window closed vs. open • Window direction (N, S, E, W) • Light bulb type • Hours per day occupied • Barefoot versus shoe traffic • Exposure to pets, vermin, etc. • Plants or water features • Number of occupants • Connections to other rooms • Air temperature • Relative humidity • Percentage recirculated air • Air change rate	• Material (carpet, granite, etc.) • Water activity • Time since last cleaning • Type of cleaning • Light exposure • Surface temperature • History of moisture events • Occupant proximity and interaction

SOURCE: Smith et al., 2013, Table A5-3.

- surface sterilization with antimicrobial agents, such as bleach, ethanol, and peroxide;
- hands-free lights, doors, and elevators;
- protective covers, easy-to-disinfect surfaces, and built-in periodic cleaning reminders for keyboards and computer mouses in medical environments;
- ultraviolet (UV) light irradiation via lamps or sunlight; and
- architectural design that is mindful of the potential for microbial contamination, such as separate wet and dry walking areas in gyms and pools or not placing restrooms next to areas where food is prepared.

While some of these approaches are marketed as effective tools for infection control, evidence of their efficacy often is sparse or nonexistent, and improper use of some interventions can result in the development or promotion of disinfectant-resistant microbes. Research thus is needed to better understand the determinants of effective infection control. Such research might include, for example, examining whether biofilm-resistant or antimicrobial compounds and materials actually reduce the accumulation of bacterial, viral, or fungal microorganisms on a surface and further experimentation with articulated surface topography as an alternative to chemistry to reduce biofilm development (Xu and Siedlecki, 2012).

Surfaces That Support Microbes Tracked in from Outdoor Sources

While deposition of airborne microbes via settling is a primary means of populating indoor surfaces, occupants also affect the indoor microbiome by tracking dirt, dust, pests, and water into buildings from the outdoors. Fungal and bacterial microorganisms are brought indoors on shoes and clothing, along with pests and flora that carry microbes of concern. Clothing and shoes are a source of moisture, dirt, pollen, and mold from soil, plants, pests, and animals outdoors, along with fungi and bacteria that can be aerosolized into the breathing zone or contacted directly from surfaces. In a chamber bioaerosol study, Adams and colleagues (2015) compared the relative abundance of bacterial and fungal taxa in indoor air, outdoor air, and dust. The authors concluded that "the microbial communities observed in the indoor air samples largely tracked those simultaneously measured outdoors, and taxa known to be associated with the human body played a secondary although important role" (p. 5 of 18). Meadow and colleagues (2014b) found that the floors and walls of a university building were dominated by species from the outdoors that may have been introduced via foot traffic or the HVAC system.

Simple modifications may be effective at reducing the transport of outdoor microbes indoors, although the support for such measures is derived largely from common sense rather than data. For example, Simcox and colleagues (2012) identify several steps they deem "prudent" for reducing MRSA contamination in Washington State firehouses from sources external to the building, including placing multilevel scraper walk-off mats at entrances,[18] leaving work boots outside of living quarters, vacuuming frequently, and replacing cloth surfaces with hard surfaces wherever possible. However, the investigators acknowledge that one research study on the issue[19] found that the use of walk-off mats was statistically significantly associated with the *presence* of MRSA on surfaces ($p = 0.02$), a counterintuitive result they suggest could be due to a statistical issue known as the multiple comparisons or multiplicity problem. They do, however, deem that finding worthy of further investigation. Research on such related issues as the effect of removing shoes upon entering a home on indoor microbial communities would be valuable as well.

---

[18]Such mats are required by green building standards such as Leadership in Energy and Environmental Design (LEED).

[19]Later expanded on by Roberts and No (2014).

Surfaces That Suspend Dust and Enable Resuspension

Humans also alter the indoor microbiome by shedding microbes onto receptive surfaces and by engaging in activities that resuspend existing surface microbes, which expose occupants to bacteria or fungi on dust particles. Depending on their aerodynamic diameters, aerosolized microbes and particles can be deposited on surfaces as a result of Brownian motion (below 0.1 μm); accumulate in the air and be inhaled directly (0.1–1 μm); or settle on surfaces as a result of gravity, impaction, and interception (above 1 μm).

Interest in human shedding of microbes from skin (Noble et al., 1976), as well as through coughing, sneezing, and talking (Morawska, 2006), is long-standing. Researchers have found shed or flaked (desquamated) human skin cells in aerosols in occupied indoor environments (Clark, 1974; Fox et al., 2008) and routinely have noted the presence of human microbiota such as *Staphylococcus*, *Propionibacteria*, *Corynebacteria*, and enteric bacteria (Täubel et al., 2009). A 2015 study found that human "bacterial clouds" are distinct with respect to their community structure and that humans carry personalized microbial clouds (Meadow et al., 2015). Findings of statistical associations and material balance–based studies suggest that resuspension can be a significant source of airborne bacteria and fungi; in densely occupied settings, it is the primary source (Hospodsky et al., 2012; Qian et al., 2012).

While shed skin can be inhaled directly by other building occupants, it is more typically shed into dust on floors and building surfaces and then resuspended through walking, vacuuming, dusting, and other activities (Meadow et al., 2015). Resuspension most critically exposes those crawling or seated, who are closer to the highest concentrations (Täubel, 2016). The emission and aerosolization of dust during vacuuming can potentially spread bacteria, including *Salmonella* spp. and *Clostridium botulinum* (Veillette et al., 2013). In the absence of combustion, cooking, and smoking, resuspension is a major source of total airborne particulate matter in occupied indoor environments, suggesting it is an important mechanism for the aerosolization of microbes (Qian et al., 2014). Yamamoto and colleagues (2015, p. 5104) indicate that research on this topic highlights "the importance of reducing indoor emissions associated with occupancy, potentially through more regular and effective floor cleaning and through the choice of flooring materials that limit particle resuspension."

There is strong evidence that human occupancy increases fungal and bacterial concentrations in indoor air through resuspension. Dust is hygroscopic and contains a substantial portion of microbes. Luoma and Batterman (2001) found that resuspension due to walking explained 24–55 percent of variation in 1–25 μm diameter particle concentrations in homes. Similarly, a study of six school classrooms found that human occupancy resulted in significantly elevated airborne bacterial (by a factor of 81, on average) and fungal (factor of 15) concentrations (Hospodsky et al., 2015). Microbial community analyses conducted in schools demonstrate that resuspended floor dust is enriched in bacteria and fungi associated with human skin and that surface-based resuspension may be a major source of airborne fungal allergens, rather than infiltration of these microbes into buildings from outdoor air (Hospodsky et al., 2015; Yamamoto et al., 2015). Other research revealed that classroom emission rates from shedding and resuspension were on the order of 10 million bacteria or fungi per person per hour (Hospodsky et al., 2015). If total particles are used as a measure of human impact on the indoor microbiome, the effect of resuspension appears to be much stronger than the effect of direct human emissions (Hospodsky et al., 2012). Importantly, it has been demonstrated that particles larger than 5 µm are suspended more easily than smaller ones, indicating a differential impact of human occupancy on this aspect of indoor microbial exposure (Thatcher and Layton, 1995).

Beds represent a significant but underrecognized exposure microenvironment because they are a place where people spend large amounts of time in close proximity to sources. Boor and colleagues (2015, p. 442) note that "mattresses are possible sources of a myriad of chemical species, such as volatile organic compounds (VOCs), plasticizers, flame retardants, and unreacted isocyanates (Boor et al., 2014; Stapleton et al., 2011) [and] mattresses, pillows, and bedding serve as an accumulation zone for a diverse spectrum of particles, many of which are of biological origin." The biologic matter includes multiple allergens, fungi, and bacteria. A chamber study conducted by the authors of that study revealed that resuspension from typical sleep activities was an important source mechanism.

It is not clear how best to address shedding and resuspension as sources of microbes that may be harmful to human health. Studies suggest that rates of resuspension of microbes are influenced by flooring material type (Qian and Ferro, 2008), floor dust loading (Qian et al., 2014), human activity level (Ferro et al., 2004), and particle size (Qian and Ferro, 2008). Other studies reveal the benefits of hard surfaces over carpet and upholstery in reducing microbial levels in resuspended dust (Buttner et al., 2002) and of UV-C equipped vacuum cleaners in reducing the viable microbial load in carpets (Lutz et al., 2010). However, substantial gaps remain in research on the determinants of and the most effective way of reducing the resuspension of microbes in the full range of building types and indoor environmental conditions.

As previously discussed in this chapter, humans also affect the indoor microbiome through bodily emissions from exhalation, expectoration, skin shedding, cuts in the skin, and bladder and bowel waste. Through such actions as coughing and sneezing, the body emits aerosols that become incorporated into a building's microbiome. Such factors as the diameter of the viral-, bacterial-, or fungal-bearing particles; occupant density; occupant proximity (e.g., face-to-face exchange); and ventilation system configurations affect how these particles are distributed once they are emitted from the body (Li et al., 2007; Liu et al., 2017).

Surfaces That Sustain Dampness and Mold

The final surface design and maintenance issue of concern is surfaces that support growth of mold because they are or may become damp or wet. These include surfaces found in the living spaces of buildings—particularly the floor, wall, and ceiling materials—and those that are in concealed spaces, such as HVAC components and insulation within framed walls (Mensah-Attipoe et al., 2015).

Building materials, furnishings, and other surfaces differ in their ability to sustain damp conditions and their susceptibility to deterioration resulting from the presence of those conditions. A key factor in this regard is water activity (a_w) or equilibrium relative humidity (ERH), which indicates whether a particular material is damp enough to support microbial growth. ERH characterizes the relative humidity of the atmosphere in equilibrium with a material that has a particular moisture content. Different materials used indoors, such as brick, concrete, drywall, textiles, and wood, may have very different ERHs in the presence of the same airborne relative humidity level. ERH is difficult to measure accurately in situ, however, with results varying depending on how the sample is taken (Dedesko and Siegel, 2015).

The implications of damp surfaces are discussed in detail in the IOM report *Damp Indoor Spaces and Health* (IOM, 2004), which addresses the interactions among moisture, materials, and environmental conditions within and outside a building that determine whether the building may become a source of potentially harmful dampness-related microbial and chemical exposures. In brief, that report notes that mold spores are found regularly on indoor surfaces and materials, and their growth, which usually is accompanied by bacterial growth, is determined primarily by the availability of moisture since the nutrients and temperature range they need to grow are usually present. The primary risk factors for the dampness that supports microbial growth differ across climates, geographic areas, and building types. This dampness can also damage building materials and furnishings, causing or exacerbating their release of chemicals and nonbiologic particles.

More recently, Adams and colleagues (2015) examined how the composition of building materials influences the indoor microbiome. Their overview of the literature finds that "different building materials and environmental conditions (e.g., temperature, available water, cleaning chemicals and frequency, light intensity at certain wavelengths, and carbon sources) can create different selective pressures for microorganisms if varied over wide ranges, which can result in differential survival and persistence rates" (p. 226). They note, for example, that "wooden materials show greater fungal diversity than plasterboard or ceramics, and cellulose-based materials are more sensitive to contamination by fungal growth than inorganic materials such as gypsum, mortar, and concrete" (p. 228). Verdier and colleagues' (2014) review of the literature on indoor microbial growth across building materials and sampling and analysis methods finds that the bioreceptivity of materials is determined most strongly by their water activity, chemical composition (in particular, nutrient sources), pH, and surface physical properties (porosity, surface roughness, and the like). The presence of fungi, bacteria, and viruses and their by-products in materials has led some manufacturers to introduce antimicrobial agents into their products. However, a white paper by the design firm Perkins+Will (2017) asserts that there is no evidence that these additives result in a healthier indoor environment, and they may in fact have negative effects on occupants and the environment through, for instance, the promotion of antibiotic-resistant species.

Several factors have led to increases in indoor dampness problems in the United States in the past several years. These include the construction of air-conditioned buildings in hot, humid regions and in areas previously considered to be wetlands, and the greater use of moisture-retaining gypsum board (Weschler, 2009). The health implications of damp indoor spaces—which are addressed in Chapter 2 of this report—are illustrated by a study by Sordillo and colleagues (2013) in 376 Boston area homes, in which visible water damage and mold or mildew were associated with a 20–66 percent increase in levels of Gram-negative bacteria associated with childhood asthma. This and other studies reveal the importance of designing building surfaces and systems so as to limit the opportunity for water damage.

Such interventions are especially important at a time when increases in the frequency and geographic spread of deluge rain, high temperatures, and high humidity conspire to speed the growth and migration of mold spores. A number of IOM and National Research Council reports have identified the importance of managing dampness and mold to improve human health and have listed critically needed research (IOM, 1993, 2000, 2004, 2011; NRC, 2006).

Numerous means exist for controlling moisture in buildings and thus limiting the growth of mold and reducing the spread of mold spores. EPA (2013a, p. 1) identifies the two key principles as

- preventing water intrusion and condensation in areas of a building that must remain dry; and

- limiting the areas of a building that are routinely wet because of their use (e.g., bathrooms, spas, kitchens, and janitorial closets) and drying them out when they do get wet.

Voluntary standards specify actions that can achieve these goals, but few of these standards have made their way into building regulations and other enforceable instruments that would lead to their widespread implementation. And research to evaluate the effectiveness of these interventions—especially in the long term—is lacking.

Building Surfaces and Reservoirs of Microbes: Summary of Findings

Surfaces constitute a critical reservoir for microbial growth and transfer in the built environment. The specification, design, and maintenance of building surfaces need to be critically evaluated to reduce microbial sources and transfers that occur by (1) direct contact with building surfaces; (2) tracking of dirt, dust, pests, and water into buildings from the outdoors; (3) activities that resuspend existing, surface-bound microbes; (4) bodily emissions from exhalation, expectoration, skin shedding, cuts in the skin, and bladder and bowel waste; and (5) surfaces that engender dampness and mold. With advances in microbial field research, the solution sets and priorities for the design community will become more evident.

IMPACTS OF MICROBES ON DEGRADATION OF BUILDING MATERIALS AND ENERGY USAGE

In addition to the important connections between building air, water, and surfaces and human exposures to indoor microorganisms, microorganisms within the built environment can impact building systems and materials in ways that can have associated sustainability, energy usage, and economic effects. For example, the formation of biofilms in water systems can lead to corrosion in pipes and holding tanks (Liu et al., 2016), which carries economic costs when the corrosion requires remediation or the infrastructure must be replaced. Also, biofilms and fungi frequently collect on damp cooling coils (Hugenholtz and Fuerst, 1992; Levetin et al., 2001), which can impair effective heat transfer (Wang et al., 2016). This reduction in heat transfer efficiency means that the cooling coils will consume more energy to accomplish their desired effects, with both economic and sustainability consequences. Microbial decay of building materials, which can include fungal decay of wood, microbial decay of paint, and more general processes known as biodeterioration, can significantly shorten the life of building materials, which also has direct economic and sustainability impacts (Viitanen et al., 2010). Biodeterioration is linked most commonly to moisture in a building and can therefore be managed, but other microbial impacts on building materials and structures are less well understood. Moisture within a building also frequently leads to the growth of molds or fungi that can decay wood and is linked to the failure or rapid deterioration of interior finishes and coatings (EPA, 2013a). Improved frameworks are needed for assessing the effects of microorganisms on and in the built environment and the benefits and costs of potential interventions to manage microbial communities.

BUILDING CODES AND STANDARDS THAT MAY AFFECT THE MICROBIOME

Regulations and Guidance

The regulatory environment for indoor environmental conditions in the United States is defined by several different federal agencies, as well as by state and local authorities. The Occupational Safety and Health Administration has purview over worker health and safety in all workplaces, but it has no specific requirements related to indoor air quality in nonindustrial workplaces such as offices and schools. The General Services Administration, the U.S. Department of Defense, and other federal agencies maintain requirements for the design and operation of their own facilities, some of which are related to indoor environmental conditions. U.S. Department of Housing and Urban Development regulations cover public housing, as well as manufactured homes, some of which again affect the indoor environment. EPA does not have the authority to regulate indoor air quality, although it has the authority to issue reporting, record-keeping, and testing requirements and restrictions related to some chemicals and materials used indoors, such as pesticides and biocides. EPA does, however, produce a wide range of guidance documents[20] for homeowners and commercial building designers, owners, and operators on a range of indoor air issues, including radon, asthma, moisture, and exposure to particulate matter.

Many states and local governments have their own health, safety, and environmental quality agencies and building regulations or codes that contain minimum requirements designed to protect occupant health and safety and that are enforced by local building officials as part of the process of obtaining a permit to construct a new building or make significant renovations. These local regulations historically have been focused on structural, electrical, and fire safety issues, although more recently they have also en-

---

[20]EPA guidance documents on indoor environments are available via links found at https://www.epa.gov/indoor-air-quality-iaq (accessed July 25, 2017).

compassed building energy-efficiency requirements and, to a lesser degree, indoor air quality. These local building regulations are a key mechanism for effecting change in how buildings are designed and built. ASHRAE also offers guidelines—*0-2013, The Commissioning Process*; *1.1-2007, HVAC&R Technical Requirements for the Commissioning Process*—and standards—*202-2013, Commissioning Process for Buildings and Systems*—that could influence the indoor microbiome through their intent and focus on ensuring that building systems are performing as designed and in a manner that promotes occupant comfort and health.

Sustainable, Green, and Healthy Building Standards and Certifications

A number of sustainable, green, and healthy building standards and certification programs address indoor environmental quality, as well as other aspects of building performance. EPA (2016) defines green or sustainable design as "the practice of creating and using healthier and more resource-efficient models of construction, renovation, operation, maintenance and demolition," and the agency's *Sustainable Design and Green Building Toolkit for Local Governments* offers advice on how to achieve such design within the context of permitting processes (EPA, 2013b). The Office of the Federal Environmental Executive[21] defines green building as the practice of (1) increasing the efficiency with which buildings and their sites use energy, water, and materials, and (2) reducing building impacts on human health and the environment through better siting, design, construction, operation, maintenance, and removal—the complete building life cycle (OFEE, 2003).

The relationship of green design standards to both healthy and harmful microbial communities within and near buildings is a subject of significant importance for ongoing research and the identification of opportunities to improve practice based on new research findings. Table 3-5 lists some of the green design standards, guidelines, and certifications used in the United States that are aimed at improving site, energy, water, materials, and indoor environmental quality and address microbial communities.

These standards and guidelines promote building features that reduce adverse exposures to microbes, including increased ventilation rates with better filtration of outside air to remove particulate matter, commissioning and continuous commissioning of building mechanical systems, design for cleanability and quality cleaning practices, walk-off mats for reducing particulate matter and pest intrusion, cooling coil and cooling tower management, and increased access to the outdoors for occupants. These standards

---

[21]Currently called the Office of Federal Sustainability (https://sustainability.gov/home.html [accessed September 22, 2017]).

TABLE 3-5 Sustainable, Green, and Healthy Codes, Standards, Guidelines, and Certifications That Address Microbiome-Related Issues

Code, Standard, Guideline, or Certification	Source
ANSI/ASHRAE/IES/USGBC Standard 189.1-2014, Standard for the Design of High-Performance Green Buildings	ASHRAE, 2014
Federal Green Construction Guide for Specifiers	EPA, 2010
Fitwel System	HHS, 2017
Green Guide for Healthcare	CMPBS and HCWH, 2007
International Green Construction Code (IgCC)	ICC, 2012
LEED (Leadership in Energy and Environmental Design)	USGBC, 2017
Living Building Challenge Standard	ILFI, 2016
WELL Building Standard	IWBI, 2014

NOTE: ANSI = American National Standards Institute; ASHRAE = American Society of Heating, Refrigerating and Air-Conditioning Engineers; IES = Illuminating Engineering Society; USGBC = U.S. Green Building Council.

and guidelines also, in some cases, raise issues related to microbial communities indoors, such as the use of natural materials and indoor landscape features; prohibition of the use of selected cleaning, pesticide, and disinfectant agents; the use of antimicrobials; and rain capture and grey and black water[22] systems. However, there are opportunities to improve how some of these standards and guidelines address the building envelope and plumbing or mechanical design as they relate to mold and moisture management. For example, one of the newer efforts, the WELL Building Standard, includes several design "features" that address microbiome-related issues (see Table 3-6). It is essential for professionals in sustainable, green, and healthy design to update these standards regularly in accordance with changes in the knowledge base on good and bad microbial communities and to promote making such voluntary standards mandatory when appropriate.

THE INFLUENCE OF CLIMATE AND CLIMATE CHANGE ON THE BUILT ENVIRONMENT AND MICROBIAL COMMUNITIES

The indoor microbiome depends strongly on climate, and a better understanding of how buildings are designed and used in different climates is essential for improving understanding of the relevant issues. Different

---

[22] "Grey water" and "black water" are both forms of wastewater generated from human activities. Black water (toilet water, for instance) is presumed to be contaminated with fecal and organic matter that could carry or promote disease, while grey water (drainage water from a sink and the like) is presumed not to contain such contaminants.

TABLE 3-6 WELL Building Standard Features That Address Microbiome-Related Issues

Section/Feature	No.	Title
Air	06	Microbe and Mold Control
	08	Healthy Entrance
	09	Cleaning Protocol
	12	Moisture Management
	16	Humidity Control
	27	Antimicrobial Surfaces
	28	Cleanable Environment
	29	Cleaning Equipment
Water	30	Fundamental Water Quality
	34	Public Water Additives
	35	Periodic Water Quality Testing
	36	Water Treatment
Nourishment	41	Hand Washing
	42	Food Contamination

NOTE: There are also a number of features that address ventilation and air cleaning: 03, 05, 14, 15, 17, 19, 21, 23.
SOURCE: Table created using data from IWBI (2014).

climates have diverse impacts on how water behaves in buildings. In cold winter climates, for example, the cold outdoor air contains relatively little water vapor. When this air infiltrates a heated building, the relative humidity declines, and the indoor air is perceived to be very dry. Although mold growth may be limited by the low moisture content, occupants can experience discomfort from dry skin, eyes, and mucous membranes and rapid evaporation of moisture from the skin. These conditions may lead to the use of humidifiers, which have been shown to be sources of microbial exposure.

Climatic conditions and their changes also have an effect on building ventilation and space conditioning. Outdoor weather conditions are major determinants of envelope infiltration and natural ventilation rates, given their strong dependence on indoor–outdoor temperature differences, as well as on wind speed and direction. In the United States, these conditions often lead to sealing the building windows, eliminating the possibility of natural ventilation and demanding effective design and maintenance of the mechanical system.

Local climate is a major factor in the design and operation of HVAC systems, as their capacity is based on expected heating and cooling loads—

both the so-called sensible load, which is related to temperature control, and the latent load, which is related to humidity control. Many systems modulate the rate of outdoor air intake as a function of outdoor air temperature and humidity to minimize energy use when a building is being cooled mechanically. As noted in the discussion of interventions in Chapter 5, some systems are designed to increase the outdoor air intake rate to cool the building when the conditions of the outdoor air are conducive to cooling without air tempering by mechanical equipment (air conditioning). Changes in outdoor air temperature and humidity over time impact the ability of the system to maintain desirable indoor air temperatures and to control indoor humidity, and if these changes are large enough, the system may not be able to provide the desired indoor conditions.

Seasonal differences in climatic conditions and occupant behavior also have an effect on the indoor microbiome. The water inside buildings in areas with hot, humid summers and cold, dry winters will behave differently at different times of the year. In these areas, the indoor moisture content of air and materials varies greatly depending on the season. Inside walls, attics, and roofing assemblies, the moisture content of materials is affected by seasonal surface temperatures. In these spaces, cooler internal surfaces collect and retain more moisture than when the same surfaces are warm. In cold climates or at colder times of the year, moisture levels and relative humidity due to condensation on cold surfaces close to the exterior walls tend to increase. In hot and humid climates, moisture inside walls tends to be higher in the summer as a result of infiltration of humid air into the walls from outside the structure if indoor spaces are being cooled mechanically.

It is challenging to design and maintain buildings in climates that feature significant seasonal variation because air (along with the moisture it contains) is influenced by temperature differences among spaces. Thus, attic air can become very hot and humid in the summer, and that air can flow into the occupied zone, carrying with it microbes and their metabolites that may have grown therein. Microbes and their metabolic products in crawl spaces can then enter the living area as a result of the stack effect.

Climate change[23] has the potential to affect the indoor environment and thus the indoor microbiome. Outdoor air temperature, humidity, air quality, precipitation, wind direction and velocity, land surface wetness, and catastrophic weather events all can influence the indoor environment, depending on such factors as the integrity of a building's envelope, the design and conditions of its HVAC systems, the microbial composition of the outdoor ecosystem, and the characteristics of the surrounding buildings. If climatic conditions in a particular area change—for example, if the climate

---

[23]The text regarding climate change in this section is excerpted or derived from the IOM report *Climate Change, the Indoor Environment, and Health* (IOM, 2011).

becomes warmer or if there are more severe or more frequent episodes of high heat or intense precipitation—buildings (and other infrastructure) that were designed to operate under the previous conditions may not function well under the new ones. Furthermore, in responding to climate change, people and societies will seek to mitigate undesirable changes and adapt to changes that cannot be mitigated. Some of their responses will play out in how buildings are designed, constructed, used, maintained, and in some cases retrofitted, and the actions taken may well have consequences for indoor environmental quality and public health.

The IOM report *Climate Change, the Indoor Environment, and Health* (IOM, 2011) addresses this topic in detail. The present committee did not attempt to review the literature in this area as many of details are outside the statement of task for this study (see Box 1-1 in Chapter 1). Instead, it draws on the research and conclusions contained in the 2011 report, which include the following:

> There is inadequate evidence to determine whether an association exists between climate-change–induced alterations in the indoor environment and any specific adverse health outcomes. However, available research indicates that climate change may make existing indoor environmental problems worse and introduce new problems by
>
> - Altering the frequency or severity of adverse outdoor conditions that affect the indoor environment.
> - Creating outdoor conditions that are more hospitable to pests, infectious agents, and disease vectors that can penetrate the indoor environment.
> - Leading to mitigation or adaptation measures and changes in occupant behavior that cause or exacerbate harmful indoor environmental conditions. (p. 241)
>
> Opportunities exist to improve public health while mitigating or adapting to alterations in indoor environmental quality induced by climate change. (IOM, 2011, p. 243)

Supporting literature and additional details may be found in the 2011 report.

SUMMARY OBSERVATIONS AND KNOWLEDGE GAPS

Summary Observations

The composition and viability of indoor microbial communities are dependent on the physical attributes and environmental conditions of the buildings in which they are located. Understanding the conditions in which microbial communities form and are maintained requires consideration of the variability of building systems, their management, and the surrounding climates.

The building characteristics that affect microbial sources in the air, in water, and on surfaces are an interconnected system that also interacts with occupants. Viewing indoor microbiomes—and by extension, the buildings they inhabit—as part of a dynamic system rather than a collection of individual components will help better define a research agenda that can identify the key drivers of harmful indoor microbial communities. At the most fundamental level, it will be important to recognize that steps taken to address one source of these communities may affect others in ways that are not necessarily easy to anticipate. To take one simple example, limiting outdoor air sources by tightening the building envelope without otherwise providing adequate ventilation could increase indoor moisture and airborne microbial levels, resulting in enhanced microbial growth on damp interior surfaces.

Little is known about what constitutes a "good" indoor microbiome and even less about which building characteristics might foster one. As the literature reviewed in this chapter makes clear, advances have been achieved in the understanding of how building and environmental characteristics influence the presence, abundance, and transmission indoors of microbes known to have adverse health effects. While much remains to be learned, there is an information base on which interventions can be built. Evidence is also starting to emerge for those microbes that either are benign in isolation[24] or may have beneficial effects, but this knowledge remains preliminary and mostly speculative. Research addressing good versus bad microbial communities will need to include an examination of the building factors that support each.

Building operations and maintenance are critical contributors to the condition of indoor microbiomes. The roles of building operations and maintenance in determining the health-supporting aspects of the indoor environment often are overlooked. Yet such easily neglected elements as failing to replace air filters regularly can have a large effect. Building operations and maintenance are affected not only by the funds devoted to these activities but also by myriad factors related to the design, age, and use of the building's enclosure and systems, as well as the actions of the building's staff and occupants. Tracking the performance of buildings over time compared with design intent, investigating the effectiveness of various commissioning strategies, and examining the payback generated by various maintenance investments is essential when formulating research strategies.

Climatic conditions—which influence such factors as indoor water,

---

[24]This may have indirect beneficial effects if they displace or hinder microbes that have adverse effects.

relative humidity, and the use of natural ventilation and HVAC systems—strongly affect the survival of bacteria, fungi, and viruses. The building is a mediator of these effects. The literature on the myriad ways in which changes in the outdoor environment affect the conditions inside buildings is summarized and reviewed in the IOM report *Climate Change, the Indoor Environment, and Health* (IOM, 2011).

In the future, it may be possible to design buildings that sustain healthy microbiomes. While research has not advanced nearly far enough to inform the intentional design of buildings that can maintain a healthy indoor microbiome, the knowledge base needed to accomplish this goal is becoming progressively broader. At this point in time (mid-2017), it is possible to formulate and test healthy microbial designs that include such features as moisture and dirt management, microbial and pest management, and reduced exposure to "bad" microbes found in the air and water and on surfaces. Such research will need to reflect cognizance of how differences in building type, location, use, and occupants influence microbial communities and health outcomes. The information gained by evaluating what does and does not work can then be used to develop design strategies for built environments that sustain healthy microbiomes.

Knowledge Gaps

On the basis of the above summary observations and the information developed in this chapter connecting the built environment and microbial communities, the committee identified the following goals for research to address knowledge gaps and advance the field:

1. **Improve understanding of how building attributes are associated with microbial communities, and establish a common set of building and environmental data for collection in future research efforts.** The building attributes that are associated with various microbial communities need to be gathered for the full range of building types in different climates to foster a better understanding of the differences between how buildings are designed and how they actually function—an important element of determining how to achieve healthy indoor microbiomes. For example, information is lacking on how to interpret test results for water samples from premise plumbing or cooling towers for the purpose of improving building management practices. Such research could be employed to develop guidelines that are test- and climate-specific. There is also a lack of guidance on how to interpret results of microbial air or surface samples. Future research on the indoor microbiome would benefit greatly from the systematic collection of a common set of data on

building attributes and indoor environmental conditions so that these factors can be taken into account and examined across studies. This could be achieved by refining the information covered by such existing instruments as EPA's Building Assessment Survey and Evaluation (BASE) study (EPA, 2017a), the U.S. Centers for Disease Control and Prevention's MicrobeNet[25] survey (CDC, 2017), and the U.S. Department of Energy's Commercial Buildings Energy Consumption Survey (CBECS) (DOE, 2017a) and Residential Energy Consumption Survey (RECS) (DOE, 2017b), as well as such efforts as MIxS-BE (see Chapter 4). Data on air source–related features, such as HVAC system design information, building ventilation rates, filter efficiencies, and filter replacement practices, are important, as is documentation concerning such water sources and features as premise plumbing material, filtration, water treatment, water use patterns, cooling towers and coils, hot water heater types and temperature settings, building enclosure materials, and moisture damage.

2. **Collect better information on air, water, and surface microbiome sources and reservoirs in the built environment.** The committee's literature review identified a number of knowledge gaps associated with indoor microbial sources and reservoirs. For air sources, the implications of the shift to more completely sealed buildings for indoor air quality, occupant satisfaction and performance, and indoor microbiomes have yet to be thoroughly researched. For water sources, better means of detecting mold and moisture inside building assemblies, interpreting water activity and moisture measurements in terms of the risk of fungal and bacterial growth, and responding to water damage and subsequent mold growth are needed, as well as more complete understanding of the role of viruses in the evolution of the total indoor microbiome over time. And for surfaces, research is required to determine the relative importance of surfaces occupants touch, sit on, and lie on; surfaces that store and suspend dust and enable resuspension; and surfaces that support dampness and mold.

3. **Clarify the association of building attributes and conditions with the presence of indoor microorganisms that have beneficial effects.** There is a need to understand when the outdoor environment and its microbial communities may have beneficial effects indoors. Examples include examining both whether natural ventilation reduces or

---

[25] MicrobeNet is an online database containing genetic sequencing, biochemical and morphological characterization, and antibiotic resistance profile information on more than 2,400 rare disease-causing microbes (https://www.cdc.gov/microbenet [accessed May 11, 2017]).

increases indoor harmful microbes compared with mechanical ventilation or air conditioning and whether sunlight entering through windows reduces indoor harmful microbes.
4. **Develop means to better monitor and maintain the built environment, including for concealed spaces, to promote a healthy microbiome.** Concealed spaces in buildings play roles in the growth and transport of indoor microorganisms, but they are not easily accessed and thus typically are left unmonitored or unmaintained. Design advances are needed to address monitoring and long-term maintenance in concealed spaces. Maintenance practices in the built environment also will need to avoid investment in unsubstantiated remedies or preventive measures and in interventions that cause, however unintentionally, more problems than they address.
5. **Deepen knowledge on the impact of climate and climate variations on the indoor environment.** Further studies to explore the impact of climate on the survival of bacteria, fungi, and viruses will be useful, especially with the potential of climate change to affect such factors as relative humidity, outdoor microbial communities, the frequency of water penetration into buildings, whether and when windows are opened for ventilation, and how often air conditioning is used. Research on human responses to climate change—from "tight" buildings to the use of biocides, humidifiers, and dehumidifiers—will also need to be pursued to examine the potential for unintended adverse health effects and changes to the microbiome.

REFERENCES

Adams, R. I., S. Bhangar, W. Pasut, E. A. Arens, J. W. Taylor, S. E. Lindow, W. W. Nazaroff, and T. D. Bruns. 2015. Chamber bioaerosol study: Outdoor air and human occupants as sources of indoor airborne microbes. *PLOS ONE* 10(5):e0128022.

Adams, R. I., S. Bhangar, K. C. Dannemiller, J. A. Eisen, N. Fierer, J. A. Gilbert, J. L. Green, L. C. Marr, S. L. Miller, J. A. Siegel, B. Stephens, M. S. Waring, and K. Bibby. 2016. Ten questions concerning the microbiomes of buildings. *Building and Environment* 109:224-234.

Addiss, D. G., J. P. Davis, M. LaVenture, P. J. Wand, M. A. Hutchinson, and R. M. McKinney. 1989. Community-acquired Legionnaires' disease associated with a cooling tower: Evidence for longer-distance transport of Legionella pneumophila. *American Journal of Epidemiology* 130(3):557-568.

Andersen, B., I. Dosen, A. M. Lewinska, and K. F. Nielsen. 2017. Pre-contamination of new gypsum wallboard with potentially harmful fungal species. *Indoor Air* 27(1):6-12.

ASHRAE (American Society of Heating, Refrigerating and Air-Conditioning Engineers). 2009. *Indoor air quality guide: Best practices for design, construction, and commissioning.* Atlanta, GA: ASHRAE.

ASHRAE. 2014. *Standard for the design of high-performance green buildings.* ANSI/ASHRAE/USGBC/IES Standard 189.1-2014. https://www.ashrae.org/resources--publications/bookstore/standard-189-1 (accessed April 18, 2017).

ASHRAE. 2016a. *ANSI/ASHRAE Standard 62.1: Ventilation for acceptable indoor air quality*. Atlanta, GA: ASHRAE.

ASHRAE. 2016b. *ANSI/ASHRAE Standard 62.2: Ventilation and acceptable indoor air quality in low-rise residential buildings*. Atlanta, GA: ASHRAE.

Banham, R. A. 1984. *The architecture of the well-tempered environment*. Chicago, IL: University of Chicago Press.

Barberán, A., R. R. Dunn, B. J. Reich, K. Pacifici, E. B. Laber, H. L. Menninger, J. M. Morton, J. B. Henley, J. W. Leff, S. L. Miller, and N. Fierer. 2015. The ecology of microscopic life in household dust. *Proceedings of the Royal Society B: Biological Sciences* 282(1814):20151139. doi:10.1098/rspb.2015.1139.

Bean, B., B. M. Moore, B. Sterner, L. R. Peterson, D. N. Gerding, and H. H. Balfour, Jr. 1982. Survival of influenza viruses on environmental surfaces. *The Journal of Infectious Diseases* 146(1):47-51.

Beer, K. D., J. W. Gargano, V. A. Roberts, H. E. Reses, V. R. Hill, L. E. Garrison, P. K. Kutty, E. D. Hilborn, T. J. Wade, K. E. Fullerton, and J. S. Yoder. 2015 Outbreaks associated with environmental and undetermined water exposures—United States, 2011–2012. *Morbidity and Mortality Weekly Report* 64(31):849-851.

Bennett, D., M. Apte, X. Wu, A. Trout, D. Faulkner, R. Maddalena, and D. Sullivan. 2011. *Indoor environmental quality and heating, ventilating, and air conditioning survey of small and medium size commercial buildings: Field study*. CEC-500-2011-043. Sacramento, CA: California Energy Commission.

Boone, S. A., and C. P. Gerba. 2007. Significance of fomites in the spread of respiratory and enteric viral disease. *Applied and Environmental Microbiology* 73(6):1687-1696. http://aem.asm.org/content/73/6/1687.full (accessed May 1, 2017).

Boor, B. E., H. Järnström, A. Novoselac, and Y. Xu. 2014. Infant exposure to emissions of volatile organic compounds from crib mattresses. *Environmental Science & Technology* 48(6):3541-3549.

Boor, B. E., M. P. Spilak, R. L. Corsi, and A. Novoselac. 2015. Characterizing particle resuspension from mattresses: Chamber study. *Indoor Air* 25(4):441-456.

Brazeau, R. H., and M. A. Edwards. 2013. Role of hot water system design on factors influential to pathogen regrowth: Temperature, chlorine residual, hydrogen evolution, and sediment. *Environmental Engineering Science* 30(10):617-627.

Brock, T. D., and K. L. Boylen. 1973. Presence of thermophilic bacteria in laundry and domestic hot-water heaters. *Applied Microbiology* 25(1):72-76.

Bures, S., J. T. Fishbain, C. F. Uyehara, J. M. Parker, and B. W. Berg. 2000. Computer keyboards and faucet handles as reservoirs of nosocomial pathogens in the intensive care unit. *American Journal of Infection Control* 28(6):465-471. http://www.sciencedirect.com/science/article/pii/S0196655300906552 (accessed May 1, 2017).

Buttner, M. P., P. Cruz-Perez, L. D. Stetzenbach, P. J. Garrett, and A. E. Luedtke. 2002. Measurement of airborne fungal spore dispersal from three types of flooring materials. *Aerobiologia* 18(1):1.

CDC (U.S. Centers for Disease Control and Prevention). 2017. *MicrobeNet*. https://www.cdc.gov/microbenet (accessed March 16, 2017).

Chase, J., J. Fouquier, M. Zare, D. L. Sonderegger, R. Knight, S. T. Kelley, J. Siegel, and J. G. Caporaso. 2016. Geography and location are the primary drivers of office microbiome composition. *mSystems* 1(2):e00022-16.

Checinska, A., A. J. Probst, P. Vaishampayan, J. R. White, D. Kumar, V. G. Stepanov, G. E. Fox, H. R. Nilsson, D. L. Pierson, J. Perry, and K. Venkateswaran. 2015. Microbiomes of the dust particles collected from the International Space Station and Spacecraft Assembly Facilities. *Microbiome* 3:50.

Chenari, B., J. D. Carrilho, and M. G. da Silva. 2016. Towards sustainable, energy-efficient and healthy ventilation strategies in buildings: A review. *Renewable and Sustainable Energy Reviews* 59:1426-1447.

Cho, H., Liu, B., and K. Gowri. 2010. *Energy saving impact of ASHRAE 90.1 vestibule requirements: Modeling of air infiltration through door openings*. No. PNNL-20026. Richland, WA: Pacific Northwest National Laboratory.

CIBSE (Chartered Institution of Building Services Engineers). 2014. *Natural ventilation in non-domestic buildings—CIBSE applications manual AM10*. http://www.cibse.org (accessed May 1, 2017).

Clark, R. P. 1974. Skin scales among airborne particles. *Journal of Hygiene* 72(1):47-51.

CMPBS (Center for Maximum Potential Building Systems) and HCWH (Health Care Without Harm). 2017. *Green guide for healthcare*, version 2.2. http://www.gghc.org (accessed August 7, 2017).

Collier, S. A., L. J. Stockman, L. A. Hicks, L. E. Garrison, F. J. Zhou, and M. J. Beach. 2012. Direct healthcare costs of selected diseases primarily or partially transmitted by water. *Epidemiology and Infection* 140(11):2003-2013.

Dedesko, S., and J. A. Siegel. 2015. Moisture parameters and fungal communities associated with gypsum drywall in buildings. *Microbiome* 3(1):71.

DOE (U.S. Department of Energy). 1999. *Building commissioning. The key to quality assurance*. https://www.michigan.gov/documents/CIS_EO_commissioningguide_75698_7.pdf (accessed March 18, 2017).

DOE. 2007. Chapter 5. Retrocommissioning. In *ENERGY STAR building upgrade manual*. https://www.energystar.gov/sites/default/files/buildings/tools/EPA_BUM_CH5_RetroComm.pdf (accessed March 19, 2017).

DOE. 2013. *Guide to closing and conditioning ventilated crawlspaces*. http://www.nrel.gov/docs/fy13osti/54859.pdf (accessed June 14, 2017).

DOE. 2017a. *Commercial Buildings Energy Consumption Survey (CBES)*. https://www.eia.gov/consumption/commercial (accessed March 18, 2017).

DOE. 2017b. *Residential Energy Consumption Survey (RECS)*. https://www.eia.gov/consumption/residential (accessed March 18, 2017).

Dunn, R. R., N. Fierer, J. B. Henley, J. W. Leff, and H. L. Menninger. 2013. Home life: Factors structuring the bacterial diversity found within and between homes. *PLOS ONE* 8(5):e64133.

ED (U.S. Department of Education). 2008. National Center for Education Statistics, Schools and Staffing Survey (SASS), "Public School Data File," 2007–08. "Average number of hours in the school day and average number of days in the school year for public schools, by state: 2007–08." https://nces.ed.gov/surveys/sass/tables/sass0708_035_s1s.asp (accessed June 7, 2017).

EPA (U.S. Environmental Protection Agency). 2010. *Federal green construction guide for specifiers*. https://www.wbdg.org/ffc/epa/federal-green-construction-guide-specifiers (accessed August 8, 2017).

EPA. 2013a. *Moisture control guidance for building design, construction and maintenance*. https://www.epa.gov/sites/production/files/2014-08/documents/moisture-control.pdf (accessed March 16, 2017).

EPA. 2013b. *Sustainable design and green building toolkit for local governments*. EPA 904B10001. https://archive.epa.gov/greenbuilding/web/pdf/sustainable-design-permitting-toolkit-06_27_13_formatted.pdf (accessed April 18, 2017).

EPA. 2016. *Green building*. https://archive.epa.gov/greenbuilding/web/html (accessed April 18, 2017).

EPA. 2017a. *Building Assessment Survey and Evaluation Study*. https://www.epa.gov/indoor-air-quality-iaq/building-assessment-survey-and-evaluation-study (accessed March 16, 2017).

EPA. 2017b. Location of outdoor air intakes and exhaust. In *Heating, ventilation and air-conditioning systems, part of indoor air quality design tools for schools*. https://www.epa.gov/iaq-schools/heating-ventilation-and-air-conditioning-systems-part-indoor-air-quality-design-tools#Location (accessed June 15, 2017).

Falkinham, J. O. 2011. Nontuberculous mycobacteria from household plumbing of patients with nontuberculous mycobacteria disease. *Emerging Infectious Disease* 17(3):419-424.

Falkinham, J. O. 2015. Common features of opportunistic premise plumbing pathogens. *International Journal of Environmental Research and Public Health* 12(5):4533-4545.

Feazel, L. M., L. K. Baumgartner, K. L. Peterson, D. N. Frank, J. K. Harris, and N. R. Pace. 2009. Opportunistic pathogens enriched in showerhead biofilms. *Proceedings of the National Academy of Sciences of the United States of America* 106(38):16393-16399.

Ferro, A. R., R. J. Kopperud, and L. M. Hildemann. 2004. Source strengths for indoor human activities that resuspend particulate matter. *Environmental Science & Technology* 38(6):1759-1764.

Fierer, N., C. L. Lauber, N. Zhou, D. McDonald, E. K. Costello, and R. Knight. 2010. Forensic identification using skin bacterial communities. *Proceedings of the National Academy of Sciences of the United States of America* 107(14):6477-6481.

Finnegan, M. J., C. A. Pickering, and P. S. Burge. 1984. The sick building syndrome: Prevalence studies. *British Medical Journal (Clinical Research Edition)* 289(6458):1573-1575.

Flores, G. E., S. T. Bates, D. Knights, C. L. Lauber, J. Stombaugh, R. Knight, and N. Fierer. 2011. Microbial biogeography of public restroom surfaces. *PLOS ONE* 6(11):e28132.

Flores, G. E., S. T. Bates, J. G. Caporaso, C. L. Lauber, J. W. Leff, R. Knight, and N. Fierer. 2013. Diversity, distribution and sources of bacteria in residential kitchens. *Environmental Microbiology* 15(2):588-596.

Fox, K., E. Castanha, A. Fox, C. Feigley, and D. Salzberg. 2008. Human K10 epithelial keratin is the most abundant protein in airborne dust of both occupied and unoccupied school rooms. *Journal of Environmental Monitoring* 10(1):55-59.

Ghaitidak, D. M., and K. D. Yadav. 2013. Characteristics and treatment of greywater—A review. *Environmental Science and Pollution Research* 20(5):2795-2809.

Gibbons, S. M., T. Schwartz, J. Fouquier, M. Mitchell, N. Sangwan, J. A. Gilbert, and S. T. Kelley. 2015. Ecological succession and viability of human-associated microbiota on restroom surfaces. *Applied and Environmental Microbiology* 81(2):765-773.

Girman, J. R., T. Phillips, and H. Levin. 2009. Critical review: How well do house plants perform as indoor air cleaners? *Proceedings: Healthy Buildings 2009* 667-671.

Grot, R. A., A. Persily, A. T. Hodgson, and J. M. Daisey. 1989. *Environmental evaluation of the Portland East federal office building preoccupancy and early occupancy results*. NISTIR 89-4066. Gaithersburg, MD: National Institute of Standards and Technology.

Haleem, A. M., D. M. Hassan, and S. A. Al-Hiyaly. 2013. Comparative assessment of microbial contamination from swabs collected within university facilities. *Journal of Health Science* 3(2):25-28. http://www.sapub.org/global/showpaperpdf.aspx?doi=10.5923/j.health.20130302.04 (accessed May 1, 2017).

Haverinen U. 2002. *Modeling moisture damage observations and their association with health symptoms*. Ph.D. dissertation. Kuipio, Finland: National Public Health Institute, Department of Environmental Health.

Heiselberg, P. 2006. Hybrid ventilation in non-residential buildings. In *Building ventilation: The state of the art*, edited by M. Santamouris and P. Wouters. London, UK: Earthscan Publications, Ltd. Pp. 191-216.

HHS (U.S. Department of Health and Human Services). 2017. *The Fitwel System*. https://fitwel.org/system (accessed April 18, 2017).

Hoisington, A., J. P. Maestre, K. A. Kinney, and J. A. Siegel. 2016. Characterizing the bacterial communities in retail stores in the United States. *Indoor Air* 26(6):857-868.

Hospodsky, D., J. Qian, W. W. Nazaroff, N. Yamamoto, K. Bibby, H. Rismani-Yazdi, and J. Peccia. 2012. Human occupancy as a source of indoor airborne bacteria. *PLOS ONE* 7(4):e34867.

Hospodsky, D., N. Yamamoto, W. W. Nazaroff, D. Miller, S. Gorthala, and J. Peccia. 2015. Characterizing airborne fungal and bacterial concentrations and emission rates in six occupied children's classrooms. *Indoor Air* 25(6):641-652.

Hugenholtz, P., and J. A. Fuerst. 1992. Heterotrophic bacteria in an air-handling system. *Applied and Environmental Microbiology* 58(12):3914-3920.

ICC (International Code Council). 2012. *2012 International green construction code.* https://www.iccsafe.org/codes-tech-support/international-green-construction-code-igcc/international-green-construction-code (accessed April 18, 2017).

ILFI (International Living Future Institute). 2016. *Living Building Challenge 3.1.* Seattle, WA: ILFI. https://living-future.org/lbc (accessed August 7, 2017).

IOM (Institute of Medicine). 1993. *Indoor allergens: Assessing and controlling adverse health effects.* Washington, DC: National Academy Press.

IOM. 2000. *Clearing the air: Asthma and indoor air exposures.* Washington, DC: National Academy Press.

IOM. 2004. *Damp indoor spaces and health.* Washington, DC: The National Academies Press.

IOM. 2011. *Climate change, the indoor environment, and health.* Washington, DC: The National Academies Press.

IWBI (International WELL Building Institute). 2014. The WELL Building Standard, version 1.0. New York: Delos Living, LLC.

Ji, P., W. J. Rhoads, M. A. Edwards, and A. Pruden. 2017. Impact of water heater temperature setting and water use frequency on the building plumbing microbiome. *The ISME Journal* 11(6):1318.

Kelley, S. T., and J. A. Gilbert. 2013. Studying the microbiology of the indoor environment. *Genome Biology* 14(2):202.

Kong, M., T. Q. Dang, J. Zhang, and H. E. Khalifa. 2017. Micro-environmental control for efficient local cooling. *Building and Environment* 118:300-312.

Koontz, M. D., and H. E. Rector. 1995. *Estimation of distributions for residential air exchange rates: Final report.* Document No. 600R95180. Washington, DC: EPA.

Kwan, S., R. Shaughnessy, U. Haverinen-Shaughnessy, and J. Peccia. 2016. *Exploring the rate of surface microbial redevelopment after cleaning in schools; DNA-sequencing and ATP analysis.* Presented at Indoor Air 2016, The 14th International Conference of Indoor Air Quality and Climate, Ghent, Belgium, July 3-8.

Lax, S., D. P. Smith, J. Hampton-Marcell, S. M. Owens, K. M. Handley, N. M. Scott, S. M. Gibbons, P. Larsen, B. D. Shogan, S. Weiss, and J. L. Metcalf. 2014. Longitudinal analysis of microbial interaction between humans and the indoor environment. *Science* 345(6200):1048-1052.

Levetin, E., R. Shaughnessy, C. A. Rogers, and R. Scheir. 2001. Effectiveness of germicidal UV radiation for reducing fungal contamination within air-handling units. *Applied Environmental Microbiology* 67(8):3712-3715.

Levin, H. 2010. Natural ventilation: A sustainable solution to infection control in healthcare settings? In *Proceedings, ASHRAE IAQ 2010.* Atlanta, GA: ASHRAE. http://www.buildingecology.com/articles/natural-ventilation-a-sustainable-solution-to-infection-control-in-healthcare-settings/at_download/file (accessed April 14, 2017).

Li, Y., G. M. Leung, J. W. Tang, X. Yang, C. Y. Chao, J. Z. Lin, J. W. Lu, P. V. Nielsen, J. Niu, H. Qian, and A. C. Sleigh. 2007. Role of ventilation in airborne transmission of infectious agents in the built environment—a multidisciplinary systematic review. *Indoor Air* 17(1):2-18.

Liu, D. L., and W. W. Nazaroff. 2001. Modeling pollutant penetration across building envelopes. *Atmospheric Environment* 35(26):4451-4462.
Liu, L., Y. Li, P. V. Nielsen, J. Wei, and R. L. Jensen. 2017. Short-range airborne transmission of expiratory droplets between two people. *Indoor Air* 27(2):452-462.
Liu, S., C. Gunawan, N. Barraud, S. A. Rice, E. J. Harry, and R. Amal. 2016. Understanding, monitoring, and controlling biofilm growth in drinking water distribution systems. *Environmental Science & Technology* 50(17):8954-8976.
Los Angeles County Department of Public Health. 2016. *Guidelines for alternate water sources: Indoor and outdoor non-potable uses.* http://publichealth.lacounty.gov/eh/docs/ep_cross_con_AltWaterSourcesGuideline.pdf (accessed March 20, 2017).
Luoma, M., and S. A. Batterman. 2001. Characterization of particulate emissions from occupant activities in offices. *Indoor Air* 11(1):35-48. http://www.aivc.org/sites/default/files/airbase_13651.pdf (accessed May 1, 2017).
Lutz, E. A., S. Sharma, B. Casto, G. Needham, and T. J. Buckley. 2010. Effectiveness of UV-C equipped vacuum at reducing culturable surface-bound microorganisms on carpets. *Environmental Science & Technology* 44(24):9451-9455.
Mayer, T., A. Blachowicz, A. J. Probst, P. Vaishampayan, A. Checinska, T. Swarmer, P. de Leon, and K. Venkateswaran. 2016. Microbial succession in an inflated lunar/Mars analog habitat during a 30-day human occupation. *Microbiome* 4(1):22.
Meadow, J. F., A. E. Altrichter, S. W. Kembel, J. Kline, G. Mhuireach, M. Moriyama, D. Northcutt, T. K. O'Connor, A. M. Womack, G. Z. Brown, and J. L. Green. 2014a. Indoor airborne bacterial communities are influenced by ventilation, occupancy, and outdoor air source. *Indoor Air* 24(1):41-48.
Meadow, J. F., A. E. Altrichter, S. W. Kembel, M. Moriyama, T. K. O'Connor, A. M. Womack, G. Z. Brown, J. L. Green, and B. J. Bohannan. 2014b. Bacterial communities on classroom surfaces vary with human contact. *Microbiome* 2(1):7.
Meadow, J. F., A. E. Altrichter, A. C. Bateman, J. Stenson, G. Z. Brown, J. L. Green, and B. J. Bohannan. 2015. Humans differ in their personal microbial cloud. *PeerJ* 3:e1258.
MedicineNet.com. 2017. *Medical definition of reservoir of infection.* http://www.medicinenet.com/script/main/art.asp?articlekey=14969 (accessed June 7, 2017).
Mendell, M. J., and K. Kumagai. 2017. Observation-based metrics for residential dampness and mold with dose-response relationships to health: A review. *Indoor Air* 27(3):506-517.
Mendell, M. J., Q. Lei-Gomez, A. G. Mirer, O. Seppänen, and G. Brunner. 2008. Risk factors in heating, ventilating, and air-conditioning systems for occupant symptoms in U.S. office buildings: The U.S. EPA BASE study. *Indoor Air* 18(4):301-316.
Mensah-Attipoe, J., T. Reponen, A. Salmela, A. M. Veijalainen, and P. Pasanen. 2015. Susceptibility of green and conventional building materials to microbial growth. *Indoor Air* 25(3):273-284.
Menzies, D., J. Popa, J. A. Hanley, T. Rand, and D. K. Milton. 2003. Effect of ultraviolet germicidal lights installed in office ventilation systems on workers' health and wellbeing: Double-blind multiple crossover trial. *The Lancet* 362(9398):1785-1791.
Morawska, L. 2006. Droplet fate in indoor environments, or can we prevent the spread of infection? *Indoor Air* 16(5):335-347.
Morey, P. R., T. Rand, and T. Phoenix. 2009. *On the penetration of mold into the fiberboard used in HVAC ductwork.* Presented at Healthy Buildings 2009—9th International Conference and Exhibition, Syracuse, NY.
Morrison, G. 2015. Recent advances in indoor chemistry. *Current Sustainable/Renewable Energy Reports* 2(2):33-40.
Nazaroff, W. W. 2016. Indoor bioaerosol dynamics. *Indoor Air* 26(1):61-78.
Nazaroff, W. W., and A. H. Goldstein. 2015. Indoor chemistry: Research opportunities and challenges. *Indoor Air* 25(4):357-361.

Ng, L., A. Persily, and S. Emmerich. 2015. *Infiltration and ventilation in a very tight, high performance home*. Presented at 36th AIVC Conference Effective Ventilation in High Performance Buildings, Madrid, Spain, Air Infiltration and Ventilation Centre, Madrid, Spain.

NIST (National Institute of Standards and Technology). 2017. *LoopDA*. https://www.nist.gov/services-resources/software/loopda (accessed March 20, 2017).

Noble, W. C., J. D. Habbema, R. Van Furth, I. Smith, and C. A. De Raay. 1976. Quantitative studies on the dispersal of skin bacteria into the air. *Journal of Medical Microbiology* 9(1):53-61.

NRC (National Research Council). 2006. *Green schools: Attributes for health and learning*. Washington, DC: The National Academies Press.

OFEE (Office of Federal Environmental Executive). 2003. *The federal commitment to green building: Experiences and expectations*. https://archive.epa.gov/greenbuilding/web/pdf/fedcomm_greenbuild.pdf (accessed April 18, 2017).

Offermann, F. J. 2009. *Ventilation and indoor air quality in new homes*. CEC-500-2009-085. https://www.arb.ca.gov/research/apr/past/04-310.pdf (accessed July 17, 2017).

Pagnier, I., M. Merchat, and B. La Scola. 2009. Potentially pathogenic amoeba-associated microorganisms in cooling towers and their control. *Future Microbiology* 4(5):615-629.

Perkins+Will. 2017. *Healthy environments: Understanding antimicrobial ingredients in building materials*. http://perkinswill.com/sites/default/files/Antimicrobial_WhitePaper_PerkinsWill.pdf (accessed June 19, 2017).

Persily, A. K. 2016. Field measurement of ventilation rates. *Indoor Air* 26(1):97-111.

Persily, A., and J. Gorfain. 2008. *Analysis of ventilation data from the U.S. Environmental Protection Agency Building Assessment Survey and Evaluation (BASE) Study*. NISTIR 7145-Revised. http://ws680.nist.gov/publication/get_pdf.cfm?pub_id=916664 (accessed May 1, 2017).

Persily, A., and H. Levin. 2011. Ventilation measurements in IAQ studies: Problems and opportunities. In *Proceedings of Indoor Air 2011, 12th International Conference on Indoor Air Quality and Climate*. http://ws680.nist.gov/publication/get_pdf.cfm?pub_id=907718 (accessed September 22, 2017).

Persily, A., J. Gorfain, and G. Brunner. 2005. Ventilation rates in U.S. office buildings from the EPA Base Study. *Indoor Air* 15(11):917-922.

Qian, J., and A. R. Ferro. 2008. Resuspension of dust particles in a chamber and associated environmental factors. *Aerosol Science and Technology* 42(7):566-578.

Qian, J., D. Hospodsky, N. Yamamoto, W. W. Nazaroff, and J. Peccia. 2012. Size-resolved emission rates of airborne bacteria and fungi in an occupied classroom. *Indoor Air* 22(4):339-351.

Qian, J., J. Peccia, and A. R. Ferro. 2014. Walking-induced particle resuspension in indoor environments. *Atmospheric Environment* 89:464-481.

Quimby, S. C. 2016. Concealed spaces. *Insurance Advocate*, October 17. p. 13. http://www.msonet.com/wp-content/uploads/2016/12/Concealed-Spaces.pdf (accessed July 14, 2017).

Rhoads, W. J. 2017. *Growth of opportunistic pathogens in domestic plumbing: Building standards, system operation, and design*. Ph.D. dissertation. Blacksburg, VA: Virginia Tech. https://vtechworks.lib.vt.edu/bitstream/handle/10919/76653/Rhoads_WJ_D_2017.pdf (accessed May 1, 2017).

Rhoads, W. J., P. Ji, A. Pruden, and M. A. Edwards. 2015. Water heater temperature set point and water use patterns influence *Legionella pneumophila* and associated microorganisms at the tap. *Microbiome* 3(1):67. https://microbiomejournal.biomedcentral.com/articles/10.1186/s40168-015-0134-1 (accessed May 1, 2017).

Roberts, M. C., and D. B. No. 2014. Environment surface sampling in 33 Washington State fire stations for methicillin-resistant and methicillin-susceptible *Staphylococcus aureus*. *American Journal of Infection Control* 42(6):591-596.

Scott, E., S. Duty, and K. McCue. 2009. A critical evaluation of methicillin-resistant *Staphylococcus aureus* and other bacteria of medical interest on commonly touched household surfaces in relation to household demographics. *American Journal of Infection Control* 37(6):447-453.

Seppänen, O. A., and W. J. Fisk. 2004. Summary of human responses to ventilation. *Indoor Air* 14(S7):102-118.

Simcox, N. J., J. Camp, and M. C. Roberts. 2012. *Environmental surface sampling for MRSA in Washington State fire stations. Final report.* http://deohs.washington.edu/sites/default/files/images/MRSA_report_8-16-12.pdf (accessed March 16, 2017).

Smith, D. P., J. C. Alverdy, J. A. Siegel, and J. A. Gilbert. 2013. Design considerations for home and hospital microbiome studies. In *The science and applications of microbial genomics: Workshop summary*. Washington, DC: The National Academies Press.

Sordillo, J. E., U. K. Alwis, E. Hoffman, D. R. Gold, and D. K. Milton. 2013. Bacterial and fungal microbial biomarkers in house dust. In *Environmental health: Indoor exposures, assessments and interventions*. Oakville, ON: Apple Academic Press. Pp. 63-86.

Stapleton, H. M., S. Klosterhaus, A. Keller, P. L. Ferguson, S. van Bergen, E. Cooper, T. F. Webster, and A. Blum. 2011. Identification of flame retardants in polyurethane foam collected from baby products. *Environmental Science & Technology* 45(12):5323–5331.

Straub, J. 2006. *Building Science Digests. BSD-138: Moisture and materials.* https://buildingscience.com/documents/digests/bsd-138-moisture-and-materials (accessed March 5, 2017).

Sundell, J., H. Levin, W. W. Nazaroff, W. S. Cain, W. J. Fisk, D. T. Grimsrud, F. Gyntelberg, Y. Li, A. K. Persily, A. C. Pickering, J. M. Samet, J. D. Spengler, S. T. Taylor, and C. J. Weschler. 2011. Ventilation rates and health: Multidisciplinary review of the scientific literature. *Indoor Air* 21(3):191-204.

Täubel, M. 2016. *Of house dust and a crawling baby robot—indoor microbial exposure assessment.* Presented at the 5th Conference—Microbiology of the Built Environment, Boulder, CO, June 3.

Täubel, M., H. Rintala, M. Pitkäranta, L. Paulin, S. Laitinen, J. Pekkanen, A. Hyvärinen, and A. Nevalainen. 2009. The occupant as a source of house dust bacteria. *Journal of Allergy and Clinical Immunology* 124(4):834-840.

TEES (Texas A&M Experimental Energy Station). 2017. *Continuous commissioning.* College Station, TX: Energy Systems Laboratory. http://esl.tamu.edu/CC (accessed March 19, 2017).

Thatcher, T. L., and D. W. Layton. 1995. Deposition, resuspension, and penetration of particles within a residence. *Atmospheric Environment* 29(13):1487-1497.

Torvinen, E., S. Suomalainen, L. Paulin, and J. Kusnetsov. 2014. Mycobacteria in Finnish cooling tower waters. *APMIS* 122(4):353-358.

Toze, S. 2006. Water reuse and health risks—real vs. perceived. *Desalination* 187(1):41-51.

TRB (Transportation Research Board). 2015. *Optimizing airport building operations and maintenance through retrocommissioning: A whole-systems approach.* Washington, DC: The National Academies Press.

USGBC (U.S. Green Building Council). 2017. *LEED v4 for building design and construction.* http://www.usgbc.org/resources/leed-v4-building-design-and-construction-current-version (accessed April 18, 2017).

Veillette, M., L. D. Knibbs, A. Pelletier, R. Charlebois, P. B. Lecours, C. He, L. Morawska, and C. Duchaine. 2013. Microbial contents of vacuum cleaner bag dust and emitted bioaerosols and their implications for human exposure indoors. *Applied and Environmental Microbiology* 79(20):6331-6336.

Venkateswaran, K. 2016. Environmental *"omics"* of ISS. Presentation to the Committee on Microbiomes of the Built Environment: From Research to Application, Irvine, CA, October 17.

Verdier, T., M. Coutand, A. Bertron, and C. Roques. 2014. A review of indoor microbial growth across building materials and sampling and analysis methods. *Building and Environment* 80:136-149.

Viitanen, H., J. Vinha, K. Salminen, T. Ojanen, R. Peuhkuri, L. Paajanen, and K. Lähdesmäki. 2010. Moisture and bio-deterioration risk of build materials and structures. *Journal of Building Physics* 33(3):201-224.

Wang, Y., C. Sekhar, W. P. Bahnfleth, K. W. Cheong, and J. Firantello. 2016. Effectiveness of an ultraviolet germicidal irradiation system in enhancing cooling coil energy performance in a hot and humid climate. *Energy and Buildings* 130:321-329.

Waring, M. S. 2016. *Bio-walls and indoor houseplants: Facts and fictions*. Presentation before the Committee on Microbiomes of the Built Environment: From Research to Application, October 17. http://nas-sites.org/builtmicrobiome/files/2016/07/Michael-Waring-FOR-POSTING.pdf (accessed March 20, 2017).

Weinstein, R. A., C. B. Bridges, M. J. Kuehnert, and C. B. Hall. 2003. Transmission of influenza: Implications for control in health care settings. *Clinical Infectious Diseases* 37(8):1094-1101. https://academic.oup.com/cid/article/37/8/1094/2013282/Transmission-of-Influenza-Implications-for-Control (accessed May 1, 2017).

Weiss, D., C. Boyd, J. L. Rakeman, S. K. Greene, R. Fitzhenry, T. McProud, K. Musser, L. Huang, J. Kornblum, E. J. Nazarian, A. D. Fine, S. L. Braunstein, D. Kass, K. Landman, P. Lapierre, S. Hughes, A. Tran, J. Taylor, D. Baker, L. Jones, L. Kornstein, B. Liu, R. Perez, D. E. Lucero, E. Peterson, I. Benowitz, K. F. Lee, S. Ngai, M. Stripling, and J. K. Varma. 2017. A large community outbreak of Legionnaires' disease associated with a cooling tower in New York City, 2015. *Public Health Reports* 132(2):241-250.

Weschler, C. J. 2009. Changes in indoor pollutants since the 1950s. *Atmospheric Environment* 43:153-169.

Weschler, C. J. 2011. Chemistry in indoor environments: 20 years of research. *Indoor Air* 21(3):205-218.

Weschler, C. J. 2016. Roles of the human occupant in indoor chemistry. *Indoor Air* 26(1):6-24.

West, M., and E. Hansen. 1989. Determination of material hygroscopic properties that affect indoor air quality. In *Proceedings of the IAQ 89, the human equation: Health and comfort*. Atlanta, GA: ASHRAE.

Xu, L.-C., and C. A. Siedlecki. 2012. Submicron-textured biomaterial surface reduces staphylococcal bacterial adhesion and biofilm formation. *Acta Biomaterials* 8(1):72-81.

Yamamoto, N., D. Hospodsky, K. C. Dannemiller, W. W. Nazaroff, and J. Peccia. 2015. Indoor emissions as a primary source of airborne allergenic fungal particles in classrooms. *Environmental Science & Technology* 49(8):5098-5106.

Zaatari, M., E. Nirlo, D. Jareemit, N. Crain, J. Srebric, and J. Siegel. 2014. Ventilation and indoor air quality in retail stores: A critical review (RP-1596). *HVAC&R Research* 20(2):276-294.

4

Tools for Characterizing Microbiome–Built Environment Interactions

Chapter Highlights

- Approaches for characterizing and managing buildings include sensors to measure and monitor physical and chemical characteristics of indoor environments and building modeling tools to analyze measured data and support building design. Understanding human activities is also important.
- Approaches for characterizing the microbial communities in built environments include direct culturing, use of a variety of "omics" techniques, and other molecular measurements; the goal is to identify microorganisms and characterize their functional activities.
- Key tool and infrastructure gaps include furthering agreement on sampling and on the building and occupant data to collect in parallel with microbial samples; developing reference materials, standards, and assessment approaches; and improving the ability to share and access data effectively and to compare results across studies. Filling such gaps will support future research to understand relationships among microbial exposures, buildings, and human health.
- A number of strategies can be used to translate knowledge into application, including expert reports and voluntary guidelines,

as well as standards and codes. Development of standards and adoption of codes can be important approaches for influencing building design and operation, although both have limitations.
- Communication and engagement with stakeholders, including professional bodies, occupants, facility owners and operators, and others, will be needed. Integrating expertise from the social and behavioral sciences can make important contributions to such efforts.

Advancing understanding of how microbial communities are influenced by building characteristics and applying that understanding requires undertaking studies that integrate information from the microbial and building science fields. This chapter reviews a range of tools needed to study the characteristics of buildings and building-associated microbiomes. Topics discussed include standardized methods for collecting data on built environments and microbial communities and tools for analyzing these data to improve understanding of microbiome–built environment interactions. The chapter considers a number of factors important to characterizing buildings and microbial communities. It also identifies areas in which progress will be necessary to realize the promise of future studies, including obtaining quantitative microbial information, understanding bioinformatics approaches and assumptions, improving study cross-comparison, and developing data-sharing infrastructures. Studies also will be needed that deepen knowledge of how indoor microbial exposures connect to effects on human health, and the chapter examines several approaches to obtaining health-focused information.

A number of different strategies will be needed to incorporate results of such research into integrated built environment–microbiome–human systems and to use the improved understanding of these systems gained from this research to manage indoor built environments to benefit the well-being of occupants. These strategies will necessarily include changing building design and operation, as well as communicating new guidance to occupants, building designers and owners, facilities managers, and others who design, build, occupy, and manage the built environment. The chapter describes those strategies and some of the challenges entailed in their implementation.

THE BUILT ENVIRONMENT AS A COMPLEX EXPERIMENTAL ENVIRONMENT

A number of studies characterizing microbial communities in different types of built environments have been conducted over the past decade to provide new information about indoor microbiomes. Of interest are not only active microorganisms but also those that are viable but dormant (which may represent a majority of intact microorganisms in a dry building), as well as microbial components and metabolites that may have health effects. Although results are difficult to generalize, these studies provide insight into the relative contributions of various sources to microbial communities in the built environment and similarities and differences among these communities associated with built environment features, along with information on spatial and temporal variation. Prior chapters present a number of these findings, which also have recently been reviewed by Adams and colleagues (2015, 2016) and Stephens (2016). The selection of tools and techniques with which to collect and analyze samples of built environment microbiomes needs to take account of the complexity and variation reflected in these dimensions.

Sources That Populate Indoor Microbial Communities

As discussed in Chapter 3, the microbial communities within built environments derive from a mix of outdoor microorganisms brought indoors through air, water, and occupants and microorganisms from indoor sources. Indoor microbial communities also include microorganisms shed from those who occupy the building, and they are influenced by microbial growth and death driven by indoor environmental conditions, such as the availability of moisture. How these sources contribute to shaping indoor microbial communities varies among buildings and occupants. Because outside air can be an important contributor to indoor fungi, buildings operated with natural ventilation—such as open windows, for example—may have indoor fungal microbial communities that are more similar to those of the environment outside relative to buildings that use mechanical ventilation and air filtration, which can remove some of these microorganisms. Humans reportedly account for 5 to 40 percent of identified microbial sequences across a sampling of built environment studies (Adams et al., 2016), reflecting their role as important but variable contributors to the indoor microbiome. Different materials used within a building also provide surface substrates of differing chemical and physical composition on which microbes settle, although the influence of various surface materials on indoor microbiomes remains unclear.

Spatial and Temporal Resolution

The microbial species and their abundances in outdoor air vary geographically, and they vary frequently by season and potentially over the course of a day. This variation in turn affects the composition of microorganisms entering a building through, for example, ventilation systems. Different areas in a building also show differences in microbial composition, which can reflect various locations within and uses of a room (e.g., floors versus door trim), differences among types of rooms (e.g., a bathroom versus an office), and the likely presence or absence of liquid water (e.g., sinks or showers versus walls of rooms without sources of water). The combination of properties of flooring material and human activity also strongly affects rates of microbial resuspension, which can have a significant impact on the composition of airborne microbes since "resuspended dust is estimated to constitute up to 60% of the total particulate matter in indoor air," as reported in several studies (Prussin and Marr, 2015, p. 6).

CHARACTERIZING BUILDINGS

A number of characteristics of buildings and occupants need to be addressed in studies of built environments and their microbiomes as a crucial complement to microbial information. The data collection effort undertaken for the Hospital Microbiome Project illustrates how such measurements can be implemented. This project explored the composition of microbial communities in 10 patient rooms and two nursing stations on two floors of a newly constructed hospital before and after occupancy by patients (Lax et al., 2017; Ramos et al., 2015; Shogan et al., 2013; Smith et al., 2013). As reported by Ramos and colleagues (2015), measurements were taken of more than 80 variables at 5-minute intervals over almost 12 months, including

- indoor environmental conditions, including air dry-bulb temperature, relative humidity, humidity ratio (a measure of absolute humidity or the moisture content of air), and illuminance (a measure of incident light) in the 10 patient rooms and two nursing stations;
- differential pressure between the 10 patient rooms and the hallways;
- surrogate measures of human occupancy and activity in the 10 patient rooms using both indoor carbon dioxide (CO_2) concentrations and infrared (IR) beam–break counters installed at the patient room doorways; and
- outdoor air intake fractions in the heating, ventilation, and air conditioning (HVAC) systems serving the two floors.

Accomplishing this data collection required placing a number of commercially available sensors and data storage systems in rooms, hallways, and nursing stations, as well as within HVAC and air-handling units. It also required personnel efforts over the year to check and maintain the sensors and to download and analyze the large amounts of data obtained. Efforts have recently been undertaken to develop an open-source platform that can aid researchers in creating their own systems of linked sensors (such as sensors for temperature, light intensity, and humidity) and data loggers to enable those conducting research on the built environment to design and undertake similar data collection efforts (Ali et al., 2016).

Decisions on which building characteristics to measure and how those decisions are driven by such factors as those highlighted in the prior section reflect the need to capture sufficient information to elucidate microbial sources and to understand factors that support microbial growth and activity, as well as the need to capture sufficient information to account for spatial and temporal variability in buildings and microbiomes. See Box 4-1 for a discussion of the use of longitudinal study designs to understand microbiome–built environment–human interactions.

Within the past 5 years, efforts have been made by those conducting research on the microbiome–built environment relationship to generate a "minimum information standard" for built environment samples. However, questions remain regarding how to balance the collection of sufficient data at sufficient frequency to understand and compare the results of built environment–microbiome studies with the challenges of time, cost, and feasibility. Decisions about which types of samples to collect and which building characteristics to measure also will need to consider applicability to understanding particular health effects. For example, collection of air, water, and dust samples and data on associated building parameters may be important for understanding inhalation exposures.

Building scientists employ a number of techniques to characterize indoor environments. Ramos and Stephens (2014) review many of these techniques, including those for collecting information on "(1) building characteristics and indoor environmental conditions, (2) HVAC system characterizations and ventilation rate measurements, (3) human occupancy measurements, (4) surface characterizations, and (5) air-sampling and aerosol dynamics" (p. 247). Collaborative consortia—such as SinBerBEST, which involves the University of California, Berkeley; Nanyang Technological University; and the National University of Singapore—are also working to develop and improve building sensors to support efficiency, sustainability, and indoor air quality.[1] The three areas highlighted below represent

---

[1] See http://sinberbest.berkeley.edu (accessed May 11, 2017).

> **BOX 4-1**
> **Design and Use of Longitudinal Building Studies to Understand Microbiome–Built Environment–Human Interactions**
>
> Longitudinal (spatiotemporal) sampling efforts for built environment–microbiome investigation have been undertaken in several recent studies. An example is the Home Microbiome Project (Lax et al., 2014), which collected samples with daily frequency, for 6 weeks, from 10 homes around the United States. In that study, sample sites were swabbed to collect microbial cells on different surfaces, including the nares, hand, and foot of the occupants and the counters, light switches, door handles, floors, etc., around the home. In three instances, individuals moved to another house during the study, and investigators were able to observe how the microbiome of the new home was affected by the incoming residents. The study involved a small cohort, and sampling occurred during one season, so it could not capture seasonal microbial variability that can occur (Moschandreas et al., 2008); further home studies will be useful. More recently, the Hospital Microbiome Project presented data from sampling of surfaces and occupants in 10 patient rooms and two nursing stations, daily for 365 consecutive days (Lax et al., 2017). Compared with homes, hospitals generally exercise greater control over and maintain more uniform environmental conditions over time. This project was similar in longitudinal design to the home project, but it allowed researchers to capture the dynamic exchange as patients changed occupancy in a room. It demonstrated the ongoing dynamic microbial exchange that patients have with their space, which helped researchers map the equilibrium point for each patient. Patients entering a room acquire microbes from the room environment, but after 24 hours, the room contains so many of the patient's own microbes that this signal of interaction is impossible to discern. A number of other factors can also affect room microbial distribution, includ-

several critical areas and measurement challenges for studies of the built environment.

Measuring Ventilation Type and Airflow Rate

As emphasized in Chapter 3, buildings vary greatly in design and operation, including in the HVAC systems they employ and how these systems are operated, the extent and timing of outdoor air intake versus indoor air recirculation, the ventilation rates, the use and efficiency of air filtration, and other measures. In the context of built environment–microbiome studies, there is no typical building, and one cannot assume ventilation and airflow parameters based only on building type. As a result, information on building systems and ventilation rates is among the data necessary to interpret sample results. In addition to basic characteristics of building

> ing entry and activities of visitors and staff, along with cleaning practices. These two studies have provided amplicon and metagenomic sequence data that can be used to help construct a platform for statistical analysis of the dynamic turnover of the microbial community in the built environment.
>
> Once researchers have a map of the dynamic flux of microbes between individuals and the built environments they occupy, it is possible to start asking questions about how such a flux could influence health outcomes for occupants. In addition to tracking pathogen exposure and transmission, it is also possible to explore the broader microbial exposure an individual patient receives, either at home, in a hospital, or in a work setting. Microbial exposure can stimulate immune responses and have impacts on the development or exacerbation of immune diseases such as atopy and asthma (Stein et al., 2016; see also Chapter 2). Through longitudinal investigation of microbial exposure and exchange across different building codes and buildouts, it might be possible to discern how these differences influence exposures and hence affect health outcomes. This research would provide a better mechanistic understanding of the influence of architectural and building management decisions on microbial exposure and human health, which could then be modeled to allow for optimization. Modeling could range from statistical inference models to agent-based flux modeling and could be integrated into physical dynamic models used to predict air and particle movement in spaces. Substantial advances in technology and knowledge will be required to make this happen at a scale that would be appropriate. For example, investigators need better sensor platforms that can enable higher-resolution investigation of microbial dynamics. They also require more extensive understanding of how surface material structure and composition can influence microbial survival and growth. The capacity to parameterize these models will enable improvements in understanding how to manipulate the built environment to manage microbial exposure and improve human health.

HVAC systems obtained from system specifications, such parameters as airflow rates can be measured in air-handling units, in ductwork, and at room inflow or outflow grills. Tracer gas methods can also be used to measure air change rates and to quantify air distribution, although there are some limitations. CO_2 concentrations are commonly measured, for example, but cannot be solely relied on for ventilation estimates (Mudarri, 1997; Persily, 1997).

Measuring Moisture in the Built Environment

As discussed in detail in Chapter 3, the availability of water to microorganisms is a critical factor supporting their growth and activity in the environment. Without available moisture, or at moisture levels too low to support such growth, the built environment can function more like a "microbial wasteland" (Chase et al., 2016; Gibbons, 2016). Where liquid water is present—for example, in plumbing and in areas that receive continual or periodic wetting, such as sinks—water samples can be collected for microbial analysis, and such characteristics as water temperature, pH, and chemistry can be analyzed. In many sampling locations, however, actual liquid water may not be present.

As noted in Chapter 3, relative humidity is commonly measured in studies of indoor air quality using off-the-shelf commercial sensors. However, relative humidity can vary spatially and temporally in a room and is also not the moisture parameter most relevant to building microorganisms on surfaces. Microorganisms entering through air systems, shed from occupants, or resuspended following human activities gradually settle and deposit onto surfaces.[2] A relative humidity measurement taken from the center of a room will not reflect the moisture available to support microbial growth on a surface across the room. The moisture actually present at these surfaces and available to the microorganisms is most relevant to understanding subsequent microbial activity. Moisture content within materials, such as drywall, can be measured using, for example, electrical conductivity sensors, but this internal moisture can be trapped in the bulk material or otherwise unavailable to a microbe. The measure of moisture associated most clearly with microbial growth is water activity (a_w) (Adan and Samson, 2011; Dedesko and Siegel, 2015; Flannigan et al., 2011; Harriman and Lstiburek, 2009; Macher et al., 2013). Because a_w is difficult to measure directly, however, the most commonly used approximation is equilibrium relative humidity (ERH). This quantity reflects the equilibration of moisture between the air and the material and can be measured using sensors placed in an enclosed space on top of the material, as described by Dedesko and Siegel (2015). ERH has been used in a number of microbiome–built environment studies (e.g., Chase et al., 2016, and many others). It is limited by the fact that it cannot detect the influence of adjacent areas of the material or their interaction with room air.

Measuring Occupancy and Human Activities

The role of humans and human activities in the built environment is complex and combines with building characteristics to affect microbial communities. The density of occupants in the environment affects not

---

[2] The dynamics of particles settling from air is dependent on such factors as the particles' aerodynamic diameter, which takes into account the effects of density. Different types of microorganisms (such as viruses, bacteria, and single and multicellular fungi) have different size ranges, although settling behavior is likely to be complicated by the fact that these organisms do not exist as single cells in culture, but may be associated in aerosols with dust, soil, water, and other microbial organisms or fragments (Qian et al., 2012).

only microbial shedding but also such variables as room temperature, humidity, and CO_2 content, which in turn may impact the operation of building systems designed to maintain environmental parameters and occupant comfort. Walking over different types of flooring resuspends settled microorganisms, as do such activities as vacuuming (Prussin and Marr, 2015; Qian et al., 2014). It is challenging to measure the multiple ways in which humans use indoor spaces. The Hospital Microbiome Project, for example, used such tools as IR beams in doorways to determine how many times people entered or left a room (Ramos et al., 2015). Other measures can be used to gain information on occupants, including activity questionnaires; manual and video observations; smartphone technologies to track and monitor occupants (Zou et al., 2017); and indirect measures of human density, such as indoor CO_2 concentrations (also employed in the Hospital Microbiome Project) (Ramos and Stephens, 2014). Approaches that connect chemical signatures with occupants and their activities and improve visualization to help make sense of the large amounts of interacting data and generate new hypotheses may also provide complementary information. Mass spectrometry data on chemicals from personal care products and microbial signatures collected from swab samples have been mapped to room surfaces and occupants to help visualize molecular distributions (Bouslimani et al., 2015; Dorrestein, 2016). Such approaches may represent another opportunity for integration of biological, chemical, computational, and other disciplines in understanding how humans, buildings, and microbial and chemical environments interact. Social and behavioral science research also provides theories and methods useful for studying individual and group behaviors (NASEM, 2017b); further engagement with these disciplines will be particularly valuable in understanding human interactions in built environments and the factors influencing human behaviors.

Building Simulation Tools

A number of design and analysis simulation tools can be useful in understanding the factors impacting environmental conditions that are potentially conducive or unfavorable to indoor microbial growth. Design tools are those used specifically to support building design, while analysis tools are also used to study building performance issues that are not necessarily part of the design process—for example, trying to understand indoor air quality problems in an existing building or to analyze experimental data. Table 4-1 identifies several existing simulation tools that are particularly relevant to addressing building energy use, airflow, contami-

TABLE 4-1 Selected Building Simulation Tools

Software Name	Summary
CHAMPS	Platform for combined building heat, air, moisture, and pollutant simulation and modeling
CONTAM	Multizone airflow and contaminant transport software
DesignBuilder	Whole-building energy-use simulation tool with graphical user interface to EnergyPlus; includes HVAC system selection and sizing
EnergyPlus	DOE's whole-building energy simulation engine; includes HVAC system selection and sizing as well as code compliance
eQUEST	Whole-building energy performance
ESP-r	Whole-building energy simulation program for integrated modeling of building energy performance; used primarily to support researchers undertaking detailed studies
IDA-ICE	Multizone simulation of indoor thermal climate and whole-building energy consumption
IES	Whole-building energy simulation, multizone and CFD airflow, HVAC system design
LoopDA3.0	Sizing of natural ventilation openings
OpenStudio	Open-source software development kit for building energy simulation
THERM	Two-dimensional heat transfer in building components such as windows, walls, foundations, and roofs, providing local temperature patterns that may relate to condensation and moisture damage
TRNSYS	Component-based simulation package capable of whole-building energy simulation and design optimization
WUFI	Heat and moisture transport through building assemblies such as walls and roofs; capable of one- and two-dimensional analyses

NOTE: CFD = computational fluid dynamics; CHAMPS = Combined Heat, Air, Moisture, and Pollutant Simulation; DOE = U.S. Department of Energy; eQUEST = Quick Energy Simulation Tool; HVAC = heating, ventilation, and air conditioning; IDA-ICE = IDA Indoor Climate and Energy; IES = Integrated Environmental Solutions; THERM = Two-Dimensional Building Heat-Transfer Modeling; TRNSYS = Transient System Simulation Tool; WUFI = Wärme und Feuchte Instationär.

nant transport, and moisture conditions.[3] An important task for future work will be to couple such simulation models for the built environment with explicit population dynamics models (alluded to in Box 1-2 in Chapter 1) for diverse, complex microbial communities. Nonlinear feedbacks within and among interacting populations can lead to surprising effects of

---

[3] More information on these and other tools is available from the International Building Performance Simulation Association (IBPSA) via its Building Energy Software Tools Directory (http://www.buildingenergysoftwaretools.com [accessed May 11, 2017]).

seemingly straightforward interventions in ecological systems, and being aware of such potential outcomes should be part of the conceptual toolkit for understanding microbiomes in the built environment (Abrams, 2009).

CHARACTERIZING INDOOR MICROBIAL COMMUNITIES

In addition to information about building design and operations, studies aimed at understanding the impacts of microbial communities in built environments rely on tools that characterize these microbiomes and their functional activities. The tools and techniques available for this purpose address three main types of questions.

First, which microorganisms are present in the community, and in what quantities? Answering this question requires tools that can detect types of organisms even when they are present at low abundance (sensitivity) and that can accurately detect a target organism in the presence of other types of organisms and confounding material (specificity). It also requires tools that can both identify diverse groups of target organisms to provide information on community composition and yield quantitative information on absolute abundance in the community.

Second, are these microorganisms active, and if so, what are they doing? Addressing this question requires an understanding of whether the microorganisms are capable of replicating (viability), as well as tools that can elucidate their functional activity (biological activity and functional coverage).

Third, what potential do these microorganisms have to cause health effects (negative or positive)? This question connects exposures to microorganisms in the built environment with occupant health effects. In addition to the information on viability, microbial tools that provide information on such molecules as toxins and allergens, epidemiologic investigations, and animal studies (such as dose-response studies) provide important information on relationships between exposures and outcomes (see Chapter 2).

Capabilities of the Current Molecular Toolkit for Characterizing Microbial Communities in the Built Environment

Studying microorganisms by culturing them has been undertaken for decades, but a large majority of microorganisms cannot be cultured, and culture-based approaches generally are too low-throughput to facilitate detailed community-level analysis. "Omics" approaches, a term referring broadly to approaches that yield collective measurements of sets of biologic molecules, have recently come to the forefront for microbial studies. These approaches include genomics (analysis of DNA), transcriptomics (information on mRNA, which reveals which genes are transcribed to be expressed by a cell), proteomics (data on the suites of proteins produced by cells), and metabolomics (focused on chemical metabolites). Such techniques frequently are used in combination with each other and with culturing to provide information on the presence of different taxonomic groups of microorganisms in a sample and to characterize microbial functions and activities. The parallel processing and computational/bioinformatics algorithms that underpin omics data analysis provide higher-throughput measurements relative to older generations of techniques.

The discussion below starts with a brief review of sample collection, handling, and analysis. It then focuses on information obtained through molecular characterization tools and highlights areas in which future development would support advances in the field. Appendix A contains a more detailed technical analysis of current and emerging molecular characterization tools and qualitative discussion of their performance in terms of features relevant to understanding microbial communities such as those in built environments. These features of sensitivity, specificity, organism coverage, taxonomic resolution, quantification, viability, functional coverage, and reproducibility are briefly introduced starting on page 160.

Sample Collection, Handling, and Analysis

Several aspects of sampling can affect the results obtained in indoor microbiome studies. For example, many studies aimed at characterizing indoor microbial communities employ analyses of microbial nucleic acids. However, recovering microbial DNA or RNA for quantitative analysis depends not only on the sampling method(s) used but also on how samples are handled, extracted, and processed.

Various methods can be used to collect samples from the indoor environment, including air sampling that pulls room air across dry filters or into liquid media, settling onto agar collection plates; vacuum sampling of settled dust; and surface swabbing. Obtaining representative samples of the air, water, and surfaces within the built environment is idiosyncratic, and different sample types have advantages and disadvantages for addressing different types of questions. To further characterize microbial communities and to explore fomite transmission, surface swabs may be useful. To gain insight into human respiratory exposures, however, it may be necessary to collect samples that more closely represent this exposure route, such as air samples or possibly settled dust. The use of long-term composite samples from indoor sources generally is the most advantageous collection paradigm for microbiome characterization. Such samples provide a time-averaged perspective on microbial composition compared with an instantaneous sample, and they may be necessary to obtain sufficient microbial biomass for analysis (Yooseph et al., 2013). Data from so-called grab samples should not be casually extrapolated into a perspective on exposure. However, which types of samples are most important to collect for purposes of characterizing relevant exposures of building occupants is not fully known, and further work in this area will be valuable in informing the design of future studies to test health connections.

As is the case with the analytical chain of custody for obtaining and handling gas, liquid, or surface samples for analysis of trace chemicals within the built environment, the "ultraclean" practice of using virgin plastics and glassware, all of which must be certified and rendered DNA-free, is essential for any indoor genetic characterizations. As with practices for chemical analysis, analytic blanks and controls (both positive and negative) must be included with the cohorts of indoor environment samples when genetic observations are the goal.

Sample handling and preservation can have a significant effect on the subsequent analysis of genetic material recovered from environmental samples, regardless of microbial origins (viral, fungal, or bacterial). Changes to microbial results have been reported as a consequence of storage conditions prior to analysis (temperature, use of buffers, or other factors). Lauber and colleagues (2010) found for human and soil microbiome samples that "because of the diversity of the samples, conditions tested and analytic methods used, we still lack a comprehensive understanding of how and whether storage of samples before DNA extraction impacts bacterial community analyses and the magnitude of these potential storage effects" (p. 80). Likewise, McKain and colleagues (2013) report changes to the measured proportions of *Bacteroidetes* versus *Firmicutes* as a result of differences in sample storage. Discussion continues in the built environment community on optimal sample handling and storage conditions. The most conservative sample preservation methods emphasize immediate storage in ultracold (<–20°C) and desiccating conditions until the samples can be processed in controlled ultraclean (DNA/RNA-free) circumstances. While it is widely accepted that dry, cryogenic storage (<–60°C) can preserve genetic materials for relatively long time periods, the shortest possible holding times are preferred prior to the extraction of genetic material from environmental samples. Similarly, extraction of microbial genomes from samples close to the point and time of sample collection is preferred if at all possible.

Diverse extraction protocols and commercial kits are available for recovering microbial genetic material from environmental samples, and recovery of genetic material is rapidly improving with the private sector's continued development of such protocols and kits. No standards exist in this arena as yet. However, extraction procedures and methods will affect how well microbial nucleic acid is recovered and thus will affect downstream results. As noted in a recent report,

> Spores are hardy, for example, but may require aggressive techniques to break them open and release sufficient amounts of an agent's DNA. Gram-negative bacteria are more easily lysed, but their genomic material also may be more easily sheared and degraded during extraction. As a result, the specific extraction methods used have the potential to bias the types of organisms that will be . . . most efficiently detected. (NRC and IOM, 2015, p. 102)

DNA/RNA extraction practices are expected to continue to develop for the foreseeable future, particularly with respect to recovery from indoor aerosols. Regardless of the diversity of extraction protocols being used, however, the use of parallel internal standards and controls represents best practice for the extraction of genetic material from any given environmental medium.

Identifying and Quantifying the Microorganisms Present in a Built Environment

Sensitivity, Specificity, Organism Coverage, and Taxonomic Resolution

To identify the microorganisms present in a built environment sample, detection technologies need to be able to identify the presence of a microbial group, even when at low abundance. In addition, they need to be able to identify the absence of a particular target when it is not present in the sample. For example, if DNA is extracted from an air filter that also contains pollen particles, a large portion of the recovered genomic material may yield plant genomes rather than the target genomes of the indoor microbial communities (Be et al., 2015).

Sensitivity is a critical parameter for detection, particularly for environmental samples in which many populations exist in low abundance (Rhee et al., 2004; Wu et al., 2001, 2006). Achievable sensitivity also varies for different types of omics technologies. For example, an array-based approach can have an advantage in detecting less abundant organisms compared with a sequencing-based approach (Zhou et al., 2015). Array approaches have not commonly been applied in the built environment setting, however, where studies generally draw on polymerase chain reaction (PCR) and sequencing (methods reviewed in Hoisington et al. [2014]). Improving sensitivity may be a particularly important challenge for the built environment because biomass collected from indoor samples, especially air samples, often is very low. Sample biomass generally is much lower, for example, than that obtained in other types of microbiome sampling for which tools and analysis platforms have been developed, such as samples taken from the human gut. A practical result is that microbial information from air samples often represents community integration over space and time, because common sampling methods rely on pumps to pull large volumes of air across a filter or analyze samples of dust that has settled over extended time periods.

Taxonomic resolution, on the other hand, is a measure of the information that can be obtained about each microbe in the sampled community (Hanson et al., 2012). Relevant questions include whether a tool enables the microorganisms to be identified at fine resolution, such as the level of an individual strain and species, or at higher taxonomic levels (coarser resolution), such as the level of genus and family. A number of genes are used as phylogenetic markers to provide information on taxonomic groupings. These genes include the 16S rRNA gene for prokaryotes (bacteria and archaea), the 18S rRNA gene for microbial eukaryotes such as protists, and the internally transcribed spacer (ITS) region for fungi. Functional marker genes (such as the genes *nifH*, *amoA*, and *nirS*) can also be used to provide information on microbes in a sample. Analysis of these phylogenetic and functional marker gene sequences often is based on short, several hundred base pair pieces, which limits the obtainable taxonomic resolutions (Jovel et al., 2016; Uyaguari-Diaz et al., 2016). The result is that many microbial ecology studies identify organisms at taxonomic levels coarser than individual species or strains. Some functional markers may be able to provide a finer level of taxonomic resolution than common phylogenetic markers, however.

All tools have strengths and limitations with regard to balancing sensitivity, specificity, organism coverage, and taxonomic resolution. For example, targeted (amplicon) sequencing relies on amplifying a section of DNA from a known gene, which can then be sequenced to obtain more information. The target amplicon can be chosen because it provides information on taxonomic groupings to help classify organisms present in a sample, or it can be chosen to identify a specific known gene. This form of targeted analysis can provide high sensitivity because it can pick up a gene that occurs in only a few organisms in a sample. Another approach to understanding taxonomic composition is to sequence genomic data broadly in a sample (shotgun metagenomic sequencing). Shotgun metagenomic sequence data can theoretically come from any part of each microbial community member's genome; thus, the information can facilitate tracking genetically differentiated types of organisms, such as strains of bacteria within a species (Greenblum et al., 2015). Metagenomic sequencing and microarray-based approaches can provide greater coverage of different types of organisms relative to amplicon sequencing of a specific gene,[4] but their effectiveness

---

[4]A caveat is that amplicon-based sequencing of common marker genes (such as 16S rRNA for bacteria) can provide even broader coverage for the set of microorganisms with that gene (e.g., in identifying bacterial taxa).

presents additional informatics challenges associated with assembly of the wealth of sequence information and new microarray design. The quality and information content of the reference databases needed to make taxonomic assignments represent another important issue. A number of microbial sequences have been deposited in reference databases that do not yet have clear taxonomic classifications.

Relative and Absolute Quantification[5]

It is useful to know not only which types of microorganisms are present in the microbial communities sampled from a built environment but also how abundant they are. The genomics information obtainable from most studies provides relative quantification, reflecting the abundance of a type of microorganism relative to the total microorganisms measured in the sample (as a fraction of the total). Information on relative abundance can be useful in some characterization studies. However, having information on absolute abundance is also important to enable knowledge to move toward practical application. This information is needed, for example, as part of exposure assessments to better understand dose-response relationships and connections between exposures and human outcomes. Information on both relative and absolute abundance will also be useful in evaluating interventions in the built environment and how they affect microbial communities (a topic discussed further in Chapter 5).

Obtaining absolute quantification information is challenging, however. All measurements are made in comparison with standards, and defining and developing appropriate, validated standards for measurement of microbial communities remains an issue for microbiome studies. Even were such standards available, it is difficult to provide quantitative information with amplicon sequencing and shotgun sequencing approaches (Nayfach et al., 2016; Zhou et al., 2011). Such methods as quantitative PCR (qPCR) would need to be combined with sequencing data to incorporate quantitative or semiquantitative detail. Array-based analysis may also be of use in obtaining quantitative information (Nayfach and Pollard, 2016; Zhou et al., 2015). Moreover, given inherent variations in experimental protocols and bioinformatics analyses, abundance measurements obtained through omics technologies can differ among samples even under identical conditions. There are also technical challenges associated with environmental sampling

---

[5]Relative abundance of a microorganism in a sample refers to the percentage of that type of microorganism that was identified relative to the total microorganisms identified in that sample. Absolute abundance, on the other hand, would reflect the actual number of that type of microorganism that was in the substrate (surface, air, water, or bulk material) in the built environment from which the sample was collected.

and nucleic acid extraction that make it difficult to ensure that the genomic information obtained matches the communities that exist in the original built environment. A variety of interfering substances, such as chemicals in dust, can be present in built environment samples that complicate the ability to successfully extract and amplify DNA or that have genomic material of their own (such as pollen) that may make it challenging to pick up information from rare microbial species. In addition, it is easier to obtain DNA from some types of microorganisms than others using standard extraction protocols, a factor that affects the abundances of microorganisms detected in the sample (Peccia and Hernandez, 2006). A further biological issue, particularly for fungi, is that microorganisms may vary in the number of rRNA copy numbers present per cell, making it challenging to obtain accurate information if such genes are used to assess population numbers (Taylor et al., 2016).

These issues all represent important impediments to the ability to relate microbiome data to fundamental models of population, community, and ecosystem ecology. They also hinder health risk assessment, where absolute numbers matter.

Understanding the Viability and Functions of Microorganisms

Information about the viability and functional activities of indoor microorganisms can be obtained in several ways. Viable microorganisms are those that maintain the ability to replicate in the built environment under suitable environmental conditions. Because the built environment generally has limited moisture, particularly in the air and on surfaces, some microorganisms may be viable but inactive (e.g., as fungal spores) until conditions change. Alternatively, microorganisms detected by DNA sequencing may be "dead" or may exist only as partial microbial fragments and components. The traditional approach for assessing viability has been to culture microorganisms—for example, on agar culture plates—and this remains the standard for pure culture experiments or research involving known pathogens or specific microorganisms that can be cultured in these ways. Yet, while culture-based approaches remain an important complementary technology to such tools as genomics, suitable culture conditions for many microorganisms are not known, or do not account for the range of environments that prompt microbial metabolic activity and reproduction, or may miss the activity of less abundant taxa. Culture-based measures thus are limited as to the information about microbial communities they can provide.

Crucially, researchers also need to know what the microorganisms in a built environment are doing. The biological activity of a microorganism refers to its metabolic functions as a living entity, regardless of its ability to propagate. As discussed above, metagenomics can provide a "snapshot" of the diversity of a microbial population and thus some information on functional potential in a community, but this DNA-based information does not reveal whether the microorganisms are actively engaged in metabolic activity. Other omics approaches, including metatranscriptomics, metaproteomics, and metametabolomics, provide additional information for characterizing microbial communities functionally (Gutarowska et al., 2015; Zhou et al., 2015).

Additionally, microorganisms produce many molecules that can be measured in the built environment. These include bacterial endotoxin, fungal mycotoxins, and a variety of bacterial and fungal cell wall components that may have toxic effects on cells as a result of exposure. A variety of microbial volatile organic compounds can also be measured in indoor air (Araki et al., 2009). Such techniques as mass spectrometry and associated variations, used to analyze chemicals based on their mass and charge, are valuable for identifying microbial molecules in buildings (Saraf et al., 1997). Measuring these molecules in the built environment can provide markers for the presence of the respective microorganisms (which may not be culturable), as well as yield information relevant to understanding the potential health effects of bioaerosols and microbial samples collected from the built environment. For example, *Bordetella pertussis*, the airborne bacterium responsible for whooping cough, has not been recovered from ambient aerosol by conventional culture techniques, but it produces an exotoxin that can be measured (Yao et al., 2009). Where particular molecules can be linked to health or other effects, it may be possible to incorporate future monitoring. The use of microbial molecules as markers also has some limitations, since a given molecule may be produced by multiple microbial species, limiting taxonomic resolution. Ergosterol, for example, has been used as a surrogate measure of fungal biomass. Some molecules also may be carried in or persist in the environment even if the producing microorganism is no longer present.

Ideally, a combination of DNA- and mRNA-based measurements, as well as protein- and metabolite-based measurements, would be used to assess the presence and activities of microbial populations in a community in complementary and mutually reinforcing ways. Understanding the functional activities of microorganisms in microbial communities in a built environment remains a particular challenge. Because the functions associated with particular genes and molecules may remain unknown, even recovery of the complete complement of proteins, genes, or metabolites does not automatically yield an accurate functional assessment. However, integrating molecular data from multiple sources—genome, transcript, protein, and metabolite—presents an important opportunity to identify more accurately the biological processes that explain how the diverse elements of the microbiome persist in a specific environment (Jansson and Baker, 2016; Quinn et al., 2016).

Reproducibility and Development of Reference Materials

Understanding variability and reproducibility among microbial samples in a built environment is a challenge for reasons noted, including low biomass, environmental conditions, and an imperfect ability to extract and measure information and link it to microbial species. However, understanding the strengths and limitations of existing studies will enable comparisons across study results. Given the high community complexity and potentially dynamic nature of microorganisms, natural biologic variation can prevent two laboratories from producing the same results even when they control for technical variation. For example, microscale variation in the composition of microbial communities may result in differences (Kauserud et al., 2012; Pinto and Raskin, 2012; Zhou et al., 2011) even among samples collected by the same laboratory from proximate locations. In genomics studies, part of the reproducibility challenge is also due to technical variation and to the technologies themselves because of inherent measurement errors and biases. Part of the variation among samples can be laboratory-based as well, resulting from differences in sequencing depths (Bartram et al., 2011; Lemos et al., 2012; Zhou et al., 2011) or from variations in sequencing and sequence preprocessing approaches (Pinto and Raskin, 2012; Schloss et al., 2011).

To help address reproducibility and cross-experiment comparison, several benchmarking efforts and efforts to develop reference microbial communities have been and continue to be undertaken. The National Institute of Standards and Technology (NIST), for example, has been active in efforts to standardize microbiome measures, although attempts to apply this work to the built environment remain nascent (NIST, 2012).[6] Nascent efforts are also focused on designing mock microbial communities and benchmarking standards. Opportunities to develop reference materials that better capture living biologic material in a controlled environment will further enhance existing reference material resources. For example, a recent effort—Mock Bacteria ARchaea Community (MBARC-26) (Singer et al., 2016)—involves attempting to construct a mock community with representation from environmental habitats, although this work does not encompass the built environment. The EcoFAB initiative being carried out through Lawrence Berkeley National Laboratory is also focused on

---

[6]See https://www.nist.gov/news-events/events/2016/08/standards-microbiome-measurements-workshop (accessed July 16, 2017); https://www.nist.gov/programs-projects/microbiome-community-measurements (accessed July 16, 2017).

developing model microbial ecosystems to improve understanding of microbial communities in humans and animals and in environments such as soil (Berkeley Lab, 2015). Although this effort has not yet incorporated the built environment, this may be an area for further development. In addition, NIST recently cofounded the International Metagenomics and Microbiome Standards Alliance, which may serve as a consortium to help organize future efforts in designing standards and reference materials. It is important to note that model microbial communities may not capture all in situ interactions and activities, and validation in human populations and built environments will be necessary to confirm their utility. However, efforts to define and develop mock communities for the built environment would be helpful in improving standard metrics and benchmarks.

Open questions remain with respect to how best to build on existing benchmarking efforts to guide validation, modeling, and cross-study comparisons in support of future work on the built environment. Developing a better understanding of the conditions under which accurate and reproducible microbiome measurements in built environments can be made will be a foundational requirement for moving investigative research in this area toward practical applications.

LINKING ANALYSIS OF MICROBIAL COMMUNITIES TO BUILDING CHARACTERISTICS AND HUMAN HEALTH IMPACTS

Gaining a holistic picture of how the design and operations of built environments, the identities and functions of microbial communities, and potential impacts on humans and the environment are interconnected will require new studies and study designs, particularly as research moves beyond ecological characterization and toward translation and application.

Elucidating Causal Connections Between Microbial Exposures and Human Health Outcomes

Supporting or promoting health is a key motivation for understanding indoor microbiomes and using that knowledge to inform how buildings are designed, built, maintained, and operated. To make progress toward such practical applications, researchers will need to build on the existing base of studies to develop and test hypotheses. A number of steps are needed to determine the public health relevance of interrelationships among built environments, indoor microbiomes, and humans. The general steps summarized below involve the collection and analysis of data in a manner aimed at demonstrating relationships in a clinically relevant framework. The committee is not suggesting that all of these types of information need to be collected for all studies, but rather that these are considerations:

- Define the objective to be tested in the study, such as the hypothesis that an indoor microbial exposure relates to a certain health outcome.
- Identify the microbial exposure or exposures of interest by taxa (where they are on the biologic tree of microbes) and by function (what are they doing individually and together). For example, are microorganisms producing molecules that could interact with human cells in the airways? It will be important to identify techniques for measuring these microbial exposures and their functions.
- Identify clinically relevant measures of the health outcomes being assessed.
- Identify features of the built environment hypothesized to be relevant through their effects on indoor microbial communities or their direct impacts on human health. Identify strategies and techniques for measuring these features of the built environment.
- Identify the relationships among microbial exposures, features of the built environment, and relevant health outcomes. These relationships may vary for different types of microorganisms or microbial communities, types of exposure, doses and stages of life, individual susceptibility, such cofactors as stress, or other factors.
- Collect information on appropriate building data, such as temperature, light, airflow, and other building system characteristics. Also collect information on such occupant factors as activities, cleaning practices, health status, and social factors.
- Analyze the data in a manner that sheds light on the plausibility, consistency, and reproducibility of results. For example, longitudinal data may highlight an increased risk associated with higher baseline exposure. If feasible to study, does the outcome change if the exposure being tested is removed? For certain types of exposures, it may be possible to establish dose-response relationships through appropriate study design.
- Examine potential confounders and effect modifiers to understand their role in observed associations between built environment microbial exposures and human health outcomes. Continue the process until the totality of evidence is strong enough to support informed decision making.

Uses and Limitations of Epidemiologic and Animal Studies in Understanding Health Effects

A variety of study designs can be useful in increasing evidence for correlations already suggested between built environment microbial exposures and health outcomes, in generating evidence to support or refute additional links, and in extending observed associations and animal model results to causal connections in humans. Generating evidence may require iterations between longitudinal epidemiologic models in humans, validation in animal models, and testing through intervention studies, described briefly below:

- **Longitudinal cohort (observational) studies:** These studies follow groups of subjects, ideally comparing those exposed and not exposed to a hypothesized risk factor with respect to the occurrence of a health outcome. A subtype of studies follows individuals from around birth (birth cohort studies), which can be particularly useful in understanding effects of early-life exposures on health later in life. Longitudinal cohort studies are preferred over retrospective case-control or cross-sectional studies because they can be used to identify correlations between exposure and disease over time. Investigators may take advantage of longitudinal birth cohort studies to assess, for example, whether studies showing protective properties arising from exposure to microbes in farm environments are reproducible or generalizable to other settings. The new U.S. national consortium on birth cohort studies (Environmental Influences on Child Health Outcomes [ECHO]) may offer one type of study infrastructure that can provide opportunities to facilitate such research.
- **Animal validation studies:** Controlled studies in animals can be useful in testing and refining observational correlations. One study, for example, tested the observation that exposure to a diverse microbial community was associated with reduced allergic responses by feeding dust from a house with a dog to mice and then studying changes in their gut microbiomes and how those changes affected immune response (Fujimura et al., 2014). Animal studies can both provide greater control over exposures and environmental conditions relative to human observational studies and help establish important dose-response relationships.
- **Intervention studies:** Such studies categorize participants into groups that do or do not experience a particular intervention to examine its effects. For example, one can envision built environment interventions that change cleaning practices, water temperatures, indoor humidity levels, or other building systems and conditions and occupant behaviors in order to test whether these changes affect a heath outcome of interest. Given the multitude of factors influencing microbial communities in buildings and human exposures to those communities and the difficulties of trying to disentangle effects on health, intervention studies can be useful in further testing relationships hypothesized or observed through observational studies and in animal models.

Testing a well-defined exposure and well-defined health outcome for which clear assessment measures are available improves the statistical power of a study. However, epidemiologic and animal studies have important limitations when applied to the built environment field. In this field, data collection will incorporate not only the building and microbial features discussed in earlier sections (such as ventilation systems, chemical emissions from building materials, physicochemical variation in the built environment, and microbial proliferation), it also will take into account occupant factors that can complicate results. Applying epidemiologic and animal studies to understand exposure to a specific microorganism, such as a pathogen, is more straightforward than applying such studies to tease out multiple microbial exposures, such as those that occur in most built environments. For example, to understand how characteristics of the built environment and its microbiome influence childhood neurodevelopment and neurocognitive outcomes, it will be essential to consider the sample types that should be collected and the microbial parameters that should be measured to characterize the relevant exposures. It will also be essential to consider the roles of factors beyond microbial exposure, including socioeconomic status, diet, environmental pollutants, and such social parameters as educational background. For noninfectious outcomes, there may be substantial time between an exposure and its observed effect (e.g., if there is an important early-life window that influences later development). And people generally do not experience a single built environment but rather multiple environments as they move from home to car to office to gym or movie theater and back to home. Deriving associations among this array of factors will be highly challenging, as will identifying their relationships to specific outcomes.

Both strategies for improving the integration of such data into estimates of disease burden and the use of appropriate sensors, which can serve as automated endpoint technologies for monitoring of building microenvironments and other factors, can contribute to meeting these challenges. A number of sensor technologies already exist, although greater use of personal sensor systems could be explored to enable improved monitoring of the personal activities and exposures of individuals under study. Observational studies with longitudinal and cross-sectional analyses can help define physical, chemical, and biological markers with associations with health outcomes. Identifying the microbial and metabolic biomarkers associated with disease burden, disease onset, and disease or treatment outcomes will be particularly important in connecting environmental microbiome exposures to health effects. These biomarkers and their mechanisms of action can be one focus of future studies under controlled conditions (such as animal studies) to gain clearer understanding of specific microbial exposures, dose-response relationships, and physiologic outcomes. Further improvements in such areas as transport modeling can also contribute. Together, these elements can allow for studying components of systems biology when multiple interacting effectors and outcomes are involved. Teasing out these relationships may require network and machine learning approaches (and additional statistical tools), which would help support further development of the field, although important work remains to be done in validating such approaches for built environment analyses.

Assessing the Utility of Prior Study Data and Stored Samples

Some samples currently in storage from prior studies conducted for other purposes might be useful resources for the microbiome–built environment research community. For example, the 2006 National Health and Nutrition Examination Survey (NHANES) of the National Children's Study (NCS) archive included dust and blood samples along with questionnaires covering a variety of health issues.[7] To understand whether and how prior study samples could be reanalyzed, the community will need to define the characteristics that determine sample quality and utility with regard to the particular study questions being asked. For microbiome analysis, a sample needs to have been handled and stored appropriately, and its utility can be assessed qualitatively based on the expectation for microbial profiles associated with similar built environments. Having a high-quality sample may not be absolutely necessary if the degradation of the community signature was not so complete that it impaired the ability to detect trend differences among different conditions. Care must be taken, however, to ensure that the sample is sufficient to address the hypotheses being tested and the relevance of the findings can be confirmed. In addition, the anticipated associations and evaluation of sample-derived data are based on current assumptions, which may not turn out to be appropriate. Sample utility is also affected by knowing whether the associative variables, such as health outcomes and building measurements, were collected appropriately to enable testing of associations with the microbial and metabolic profiles that can be derived from these samples. For example, asthma research studies often are well suited to helping to derive health outcomes associated with microbial profiles, but studies on dust chemistry and childhood development are likely to have limited applicability because of a lack of control associated with variable measurement and a sparse matrix of sample acquisition that reduces the relevance for interpretation of microbial exposure.

---

[7]Information on the NCS repository is available at https://www.nichd.nih.gov/research/NCS/Pages/researchers.aspx#data (accessed July 16, 2017). Information on dust samples collected as part of NHANES is available at https://wwwn.cdc.gov/nchs/nhanes/2005-2006/ALDUST_D.htm (accessed May 11, 2017).

One important aspect of assessing prior studies for potential analysis of microbiome–built environment samples is determining the list of collected variables that could be potential confounding influences for the particular outcome of interest (e.g., pet ownership, household occupancy, diet and lifestyle, educational profiles, pollution exposure), and these relevant factors may not have been recorded in prior studies. As a result, applying new microbiome analysis to previously collected samples may be possible but will likely be restricted to testing hypotheses contextualized by the original focus of the study, particularly for case-control studies. Thus, population-based epidemiologic studies that have collected a large amount of exposure and outcome information, which can be analyzed in combination with analysis of microbial samples, are likely to be most useful.

Taking Advantage of Near-Term Opportunities

While time, effort, and significant fiscal commitment from public and private entities will be required for many of the areas of investigation identified in this chapter to come to fruition, the community of microbiome–built environment researchers could leverage near-term opportunities to study linkages among building conditions, building microbiomes, and humans. For example, opportunities for such studies may arise during disaster response and in efforts to support resilience planning. Acute events such as Hurricanes Katrina and Sandy resulted in widespread flooding of homes and underground subway stations, and they provided an opportunity to examine how microbial communities changed when exposed to these extreme conditions. The U.S. National Response Team,[8] which "provides technical assistance, resources and coordination on preparedness, planning, response and recovery activities for emergencies," and NIST investigative teams that enforce the National Construction Safety Team Act, which "establishes investigative teams to assess building performance and emergency response and evacuation procedures in the wake of any building failure,"[9] could provide potential opportunities for the involvement of researchers exploring microbiome–built environment–health interactions by fostering appropriate agency and organizational connections.

---

[8] See https://www.nrt.org (accessed July 16, 2017).
[9] Public Law 107-231, October 1, 2002 (https://www.gpo.gov/fdsys/pkg/PLAW-107publ231/pdf/PLAW-107publ231.pdf [accessed May 11, 2017]).

NEEDS FOR FUTURE PROGRESS

Obtaining Quantitative Microbial Information

As discussed above, many biologic and technical challenges are entailed in obtaining data on relative and absolute abundance for built environment microbiome samples. Achieving advances in methods for improving the quantitative information that can be obtained from samples is important for future progress in the field. Not only is quantitative information important for establishing connections with health outcomes, it also helps underpin the development and interpretation of models for potential interventions in the built environment and analysis of their impacts.

Improving Comparison Across Studies

Multiple challenges are also entailed in drawing more effective comparisons across the results of existing microbiome–built environment studies. Different groups may collect samples from the built environment in different ways, may collect different sets of building and occupant data to accompany samples, may use different characterization tools, or may undertake sample analysis and data interpretation differently. Genomics tools, for example, identify a diverse set of bacterial organisms in the built environment, and the number of catalogued organisms continues to grow as databases are updated to reflect ongoing microbial sequencing. Several different databases are used in this context, each of which has different levels of deposit criteria, quality control, and curation. However, the reproducibility of measurements of microbiome composition can also vary depending on experimental conditions. The ability to compare results across studies enables researchers to better assess converging lines of independent evidence in parsing the factors that affect the formation and functions of microbial communities in built environments. Efforts to generate standards for the collection of data and metadata, common storage formats and resources, and microbial reference materials that can be used to calibrate results across laboratories are important in addressing this need. Galaxy (Afgan et al., 2016) and Cyverse (Merchant et al., 2016) are examples of ongoing efforts to build software and infrastructure for complex and computationally expensive omic analytic pipelines that are intended to be scalable, shareable, and reproducible through the use of version control, detailed scientific workflow[10] logs, common data standards, and access to large-scale computing resources. These

---

[10] Additional background on scientific workflows can be found at http://cnx.org/contents/j-3C75Ok@3/Scientific-Workflows (accessed July 16, 2017).

examples demonstrate the work needed to generate reproducible analysis with microbiome–built environment datasets.

Understanding Bioinformatics Assumptions and Limitations

The omics technologies now used to help characterize microbiomes rely on bioinformatics algorithms to make sense of the information obtained and to determine the reference databases needed to link, for example, the obtained sequence information to the taxonomic identification of the microorganisms detected in built environment samples. Sequence data generally are interpreted using programs that compare sequences for similarity and assign them to operational taxonomic units (OTUs) based on a predesignated similarity for a given level of ecological resolution (such as 97 percent similar). Different underlying assumptions encoded in the informatics software, however, can affect how these sequences are clustered into OTUs and how they result in particular microbial taxonomic assignments. Efforts to develop simulated datasets with which to benchmark computational tools have demonstrated that different descriptions of community structure may be found for the same input data depending on the analysis tool and parameters selected (Lindgreen et al., 2016; Randle-Boggis et al., 2016; Weiss et al., 2016).[11] Understanding the assumptions underlying bioinformatics software and how results compare across different informatics packages is also foundational in understanding built environment–microbiome results and moving toward their application. The further development and adoption of benchmarking and reference standards will be valuable in this regard. This point applies to the methods used to recover DNA from environmental samples, as well as the primers used to amplify specific DNA sequences, which typically are used as ecological identifiers for phylogenetic comparisons. Also important is requiring that the software tools and detailed scientific workflows used to generate an analysis be made available for peer review to help ensure that results are independently verifiable.

Supporting Sharing of Data on Microbial Communities and Metadata on Buildings and Building Systems Through the Use of Data Commons

A data commons is a collection of computational resources that provide a common platform for access to data sources for analysis, supporting a community of researchers. Key components include storage for generated data, metadata integration, and retrieval of data in forms that enable downstream analysis. Developing this resource for microbiome–built environ-

---

[11] See http://www.cami-challenge.org (accessed May 11, 2017).

ment research will require publicly accessible data repositories, such as the Sequence Read Archive, where raw genomic data or raw metabolomics data (Wang et al., 2016) can be housed. It also will require data-sharing standards to ensure sufficiently complete and accurate descriptions of the experimental conditions used to generate new microbiome data (Leinonen et al., 2011). Ensuring that experimental data and software are accessible to the research community is key to supporting the microbiome–built environment field.

A data commons can be used to ensure that differences in results obtained with computational analysis tools are understood more clearly by enabling comparisons across common datasets. Data commons also encourage the development of new modeling or analysis tools by providing data access to tool developers who are not data generators using common data formats. Moreover, microbiome–built environment data are collected from a diverse array of research efforts across multiple institutions and research disciplines, all with the need for analysis tools that can operate on these data collections using a potentially distributed set of computing resources to support scalable and independent analysis, and in ways that enable individual investigators to contribute to the field as a whole. A data commons can also provide access to common analytic pipelines, the contents, logic, and algorithms of which are public and that provide standards for analysis, such as in community profiling.[12]

A previous effort to develop minimum requirements for building metadata led to a defined set of data to be collected in conjunction with experimental studies of indoor microbial communities—the MIxS-BE package. This standard currently includes parameters such as air and surface temperatures; measures of air and surface humidity; surface type, material, and pH; type of HVAC and filter system; and a number of other details (Glass et al., 2014).[13] While commendable in addressing the need for such information, these data are also somewhat limited in detail and extent. For example, the metadata template includes type of heating and cooling system, but the listed options encompass no information on how the system is configured or controlled or other key information needed to characterize these systems more fully. Similarly, indoor surfaces are described in terms of location or type but without greater detail on the material or its likelihood of retaining moisture.

---

[12]Examples of data commons that help support various research communities include National Institutes of Health (NIH) metabolomics and microbiome data-sharing requirements (http://www.metabolomicsworkbench.org/nihmetabolomics/datasharing.html [accessed July 25, 2017]); the Nephele cloud-based microbiome analysis pipeline (https://nephele.niaid.nih.gov [accessed July 25, 2017]); the Stanford Data Science Initiative (https://sdsi.stanford.edu/data-commons [accessed July 25, 2017]); MassIVE (http://gnps.ucsd.edu [accessed July 25, 2017]); and QIITA (http://qiita.microbio.me [accessed July 25, 2017]).

[13]See http://gensc.org/mixs (accessed July 16, 2017).

A similar building and system data definition effort was undertaken in conjunction with the U.S. Environmental Protection Agency (EPA) Building Assessment Survey and Evaluation (BASE) study of indoor air quality conditions and outcomes in 100 U.S. office buildings (EPA, 2006). That study included protocols for collecting more detailed information than called for in MIxS-BE on the building, the spaces being studied, and the HVAC systems serving those spaces, as well as for conducting measurements of environmental conditions and ventilation system performance. It addressed the condition of many system components, including their functions, state of repair, and dirt and moisture levels, and it contained a more detailed description of HVAC system type and control. The BASE protocol is more informative than the MIxS-BE package, although it may be more detailed than is needed for many indoor microbial studies. Given that the BASE protocol was developed more than 20 years ago, it may be useful to update it for its potential application to indoor microbial studies and to continue efforts to develop common data templates. The trade-offs between obtaining as much useful information as possible to characterize buildings, occupants, and their environments and the volume of information to be collected and analyzed also will require further discussion and agreement. An updated and consensus-oriented protocol for building and system characterization would be useful for future studies of the design, condition, and performance of buildings to advance understanding of a range of indoor microbial issues.

Thus, opportunities to refine data specification frameworks are needed, as are efforts to ensure that data can be accessible across publicly searchable databases, which have open curation standards. For sequence information, databases include those maintained by the European Bioinformatics Institute (EBI) and the National Institutes of Health's (NIH's) U.S. National Center for Biotechnology Information (NCBI), although efforts to make nongenomic metadata widely available through central repositories appear to be less well developed (Dorrestein, 2017; Vizcaíno et al., 2016; Wang et al., 2016). Long-term support is needed for large-scale data repositories that store both raw and processed forms of omics data. Storing raw data output will be critical to ensure that more complete and accurate recovery of genomic, proteomic, and metabolomic data can be obtained through algorithmic improvements well after the data have been collected. All of these efforts will require the engagement of researchers with built environment, building science, and engineering expertise, along with microbial ecologists and other researchers that are experts in microbiome measurement data (Abarenkov et al., 2016). Potential partners for further development of a data commons may exist in multiple federal agencies, including NIST, EPA, the National Institute for Occupational Safety and Health (NIOSH), and NCBI; professional societies, including the American Conference of Governmental Industrial Hygienists (ACGIH), the American Society of Heating, Refrigerating and Air-Conditioning Engineers (ASHRAE), the American Society for Microbiology (ASM), the Indoor Air Quality Association (IAQA), and the International Society of Indoor Air Quality and Climate (ISIAQ); and others.

MOVING FROM RESEARCH TOWARD PRACTICE

The results of microbiome–built environment–health research can be translated into practical action through multiple avenues, reviewed below.

Effecting Change in Building Practices

As new levels of understanding are achieved, a variety of strategies can be used to translate that knowledge into practice in building design and operation. These strategies are listed from fastest and easiest to those that will be slower to implement but may have the broadest impact.

- Best practices for building design and operation can be described in reports and other documents written by researchers and other experts and intended for practitioners.
- Voluntary guidance on how to design and operate buildings to support improved indoor microbial environments can be produced by engineering and professional societies, such as the American Industrial Hygiene Association (AIHA); ASHRAE; the Building Owners and Managers Association (BOMA); and government agencies such as EPA, the General Services Administration (GSA), NIOSH, and state public health departments.
- Voluntary building rating, labeling, and certification programs and green building programs such as Leadership in Energy and Environmental Design (LEED), Green Globes, WELL, and Fitwel may be particularly appropriate in the context of attempting to advance practice toward a higher level of building performance.
- Industry consensus standards for the design and operation of buildings to support improved indoor microbial environments, written by standards development organizations such as ASHRAE, include minimum standards intended for wide application. They include ASHRAE Standard 62.1 (ventilation for indoor air quality) and standards specifically directed at high-performance buildings such as ASHRAE/Illuminating Engineering Society (IES)/U.S. Green Building Council (USGBC) Standard 189.1 (on design of high-performance green buildings).
- Model building codes, such as those promulgated by the International Association of Plumbing and Mechanical Officials (IAPMO),

the International Code Council (ICC), and others, generally are based on consensus standards and subsequently adopted by local jurisdictions in their building codes and regulations (often with modifications based on local needs and priorities). Similarly, requirements for building design and operation can be developed for federal agencies that design and operate their own buildings, such as the U.S. Department of Defense (DOD) and GSA.

Standards and Their Limitations: The Example of Ventilation

One of the approaches listed above that can help effect change is embedding knowledge in standards. The section reviews the uses and limitations of ventilation standards as an example.

Ventilation requirements in standards and building regulations are essential to the design of buildings, yet they have several limitations. For example, using the same outdoor air requirement for all spaces of the same type ignores important differences among occupants and their activities, the materials and furnishings in the spaces, and the quality of the outdoor air. These requirements are also intentionally minimum values, meaning that anything lower would violate the standard or regulation. In practice, these minimum values are used without considering the potential benefits of higher levels of outdoor air intake.

It is important to bear in mind that while ventilation requirements are important for establishing building design goals, they are only a first step in the process of achieving effective ventilation in buildings. Once the outdoor air (and exhaust) ventilation requirements for a building and its spaces have been determined, these requirements need to be incorporated into the design of the building and its ventilation systems and properly documented so contractors and building operators will understand the assumptions on which the design is based. Following design, the system needs to be properly installed and commissioned to ensure that it is complying with the design requirements. This latter step involves testing the installed system under a range of operating conditions, including different internal loads, control sequences, and outdoor weather conditions. Finally, the system needs to be operated and maintained effectively over the life of the building to ensure that the system continues to perform as intended. System operation and maintenance involves periodic inspection of system components, calibration of sensors used to control the system, and many other steps, which are described in standards and other documents (ASHRAE, 2012, 2016). Given that building and space uses (including occupancy) often change during the life of a building, it is critical that the ventilation design requirements be reevaluated when such changes occur to ensure that the system can continue to meet the building's ventilation needs. These points highlight the fact that even after knowledge needed to support changes in practice has been identified and translated to formal requirements, many other factors involving building designers, owners, operations and maintenance personnel, and occupants come into play.

Communication and Engagement

Designers, professional societies, owners, operators, and occupants all need to be engaged to support an effective translation of built environment–microbiome information from research into practice. The results of studies on the microbial communities that surround people every day in their homes, schools, and offices and what impacts these communities may be having on people's health and well-being can be of wide public interest. It will be important to communicate effectively about the results of ongoing studies, as well as the caveats on and limitations of that knowledge. Communicating to people that they are surrounded by microbial communities whose effects may include beneficial, neutral, or harmful interactions and providing people with information they need to make choices about their built environments are important goals. At the same time, investigators will want to avoid promoting unjustified fears about the microbial ecosystems that coexist with humans or overselling the strength of available evidence.

The importance of public engagement with and communication about science continues to gain recognition. A recent report from the National Academies of Sciences, Engineering, and Medicine notes that those communicating about science need to consider the goals of the communication—for example, whether it is intended primarily to provide information or to influence behavior—and to align the communication approaches used accordingly. The report also emphasizes the limitations of the "deficit model" of science communication and makes suggestions for a research agenda to improve effective science communication practices (NASEM, 2017a). The "deficit model" assumes that if people only had more factual information about a topic, they would behave in a manner consistent with the scientific evidence. This model has repeatedly been shown to be wrong; however, decision making and behavior are influenced by many factors other than scientific evidence. Translating the findings of microbiome–built environment research into policy and practice will require not only the integration of scientific and clinical information but also consideration of such factors as economic costs and benefits, personal values, and social and political realities. Different potential audiences may be interested in what they can do based on the knowledge communicated, but they are likely to have varying resources, values, and competing priorities (Kahlor, 2016). The involvement of experts from the social and behavioral sciences in microbiome–built environment–health studies can be a useful strategy for elucidating the many factors relevant to stakeholder communities and effectively designing and undertaking engagement.

Citizen science efforts are another approach to building interest in and awareness about scientific topics, and the microbiome–built environment field is well suited to such efforts. For example, the Wild Life in Our Homes project describes the hypotheses being tested in understandable language, invites people to collect and send samples from their homes, and has analyzed the microbial diversity those samples contain.[14] Similarly, Project MERCCURI (Microbial Ecology Research Combining Citizen and University Researchers on ISS) sent publicly collected microbial samples into space and examined their growth (Coil et al., 2016).[15]

SUMMARY OBSERVATIONS AND KNOWLEDGE GAPS

Summary Observations

Useful suites of tools exist with which to characterize building and occupant factors and microbial communities. Researchers can draw on a variety of omics tools and bioinformatics approaches to characterize indoor microorganisms and study their activities. In addition, various sensors and simulation tools are available for gathering data on building systems and occupant activity.

Both experiments and modeling will help the research community better understand the interrelationships among buildings, microbial communities, and human occupants, and this understanding will support eventual application of the knowledge gained through research. Information from complementary approaches identifying and characterizing indoor microorganisms and microbial products, describing building parameters, capturing occupant behaviors, and collecting human exposure and health data will need to be integrated by the field.

Further efforts in foundational areas that support the research infrastructure for built environment–microbiome studies are needed. The research infrastructure that supports the field includes components that affect the ability of investigators to collect, analyze, store, share, and compare information. Important aspects of this system include continued improvements in microbial and building characterization tools, data collection standards, reference materials, and benchmarking efforts, such as the development of mock microbial communities, validation of experimental approaches, and resources for accessible data storage and sharing to facilitate cross-study comparison and the generation of new hypotheses.

---

[14] See http://robdunnlab.com/projects/wild-life-of-our-homes (accessed May 11, 2017).
[15] See http://spacemicrobes.org (accessed May 11, 2017).

Interest in connecting microbial characterization of the built environment to an improved understanding of human health impacts will benefit from studies designed to address health-relevant hypotheses. A number of considerations need to be incorporated into the design and conduct of studies aimed at clarifying potential health effects. These considerations include identifying and collecting the types of built environment, microbial, and occupant samples and data most relevant for understanding exposures; identifying appropriate measures for assessing the health outcome(s) of interest; and developing improved and validated approaches to exposure assessment. A variety of study types, including observational epidemiologic (longitudinal) studies, animal model studies, and intervention studies, will be useful.

Many groups are involved in conducting microbiome–built environment research and moving the knowledge thereby gained toward practical changes in such areas as building and indoor air quality codes and standards. The communities that will need to be engaged in this process include building, microbial, and clinical and public health scientists conducting investigations; chemical and materials scientists; building designers; and communities of professional practitioners. Once sufficient knowledge has been gained, a number of strategies can aid in translating that knowledge into practice, from voluntary guidance and descriptions of best practices to formal codes. However, developing the actual and virtual infrastructures needed to promote effective interdisciplinary research and communication in the field will require sustained engagement and funding.

Knowledge Gaps

On the basis of the summary observations above and the information developed in this chapter on the sets of available tools, the committee identified the following goals to address capability gaps and advance the field:

1. **Develop the research infrastructure in the microbiome–built environment–human field needed to promote reproducibility and enhance cross-study comparison.** A framework for establishing further infrastructure to support this field will usefully include the development of shared understandings among investigators on sample and metadata collection and on sample handling, storage, and processing conditions to support effectively addressing different types of research questions, along with the promulgation of best practices and metrics for analysis. The research infrastructure will need to encompass the use of a variety of complementary experimental, modeling, and analysis tools to understand the composition and function of microbial communities and to connect such research to

impacts of interest, such as human health effects, materials degradation, changes in energy usage, and others. The development of community standards, reference materials, and benchmarking materials will be valuable, as will the development of a broader data commons that includes the ability to share data in accessible ways to facilitate integrated data analyses and cross-study comparisons. These efforts are not trivial, and the committee does not mean to imply that the community is unaware of these needs, but to highlight that further progress in establishing this fundamental infrastructure will contribute to the advancement of the field.

2. **Develop infrastructures and practices to support effective communication and engagement with those who own, operate, occupy, and manage built environments.** This will be an important area for attention, especially as the field continues to advance toward application. Social and behavioral scientists expert in such areas as communication can provide insights to inform these efforts.

REFERENCES

Abarenkov, K., R. I. Adams, I. Laszlo, A. Agan, E. Ambrosio, A. Antonelli, M. Bahram, J. Bengtsson-Palme, G. Bok, P. Cangren, V. Coimbra, C. Coleine, C. Gustafsson, J. He, T. Hofmann, E. Kristiansson, E. Larsson, T. Larsson, Y. Liu, S. Martinsson, W. Meyer, M. Panova, N. Pombubpa, C. Ritter, M. Ryberg, S. Svantesson, R. Scharn, O. Svensson, M. Töpel, M. Unterseher, C. Visagie, C. Wurzbacher, A. F. S. Taylor, U. Kõljalg, L. Schriml, and R. H. Nilsson. 2016. Annotating public fungal ITS sequences from the built environment according to the MIxS-Built Environment standard: A report from a May 23–24, 2016 workshop (Gothenburg, Sweden). *MycoKeys* 16:1-15.

Abrams, P. A. 2009. When does greater mortality increase population size? The long history and diverse mechanisms underlying the hydra effect. *Ecology Letters* 12:462-474.

Adams, R. I., A. C. Bateman, H. M. Bik, and J. F. Meadow. 2015. Microbiota of the indoor environment: A meta-analysis. *Microbiome* 3:49. https://microbiomejournal.biomedcentral.com/articles/10.1186/s40168-015-0108-3 (accessed July 14, 2017).

Adams, R. I., S. Bhangar, K. C. Dannemiller, J. A. Eisen, N. Fierer, J. A. Gilbert, J. L. Green, L. C. Marr, S. L. Miller, J. A. Siegel, B. Stephens, M. S. Waring, and K. Bibby. 2016. Ten questions concerning the microbiomes of buildings. *Building and Environment* 109:224-234.

Adan, O. C. G., and R. A. Samson. 2011. *Fundamentals of mold growth in indoor environments and strategies for healthy living*. Wageningen, The Netherlands: Wageningen Academic Publishers.

Afgan, E., D. Baker, M. van den Beek, D. Blankenberg, D. Bouvier, M. Čech, J. Chilton, D. Clements, N. Coraor, C. Eberhard, B. Grüning, A. Guerler, J. Hillman-Jackson, G. Von Kuster, E. Rasche, N. Soranzo, N. Turaga, J. Taylor, A. Nekrutenko, and J. Goecks. 2016. The Galaxy platform for accessible, reproducible and collaborative biomedical analyses: 2016 update. *Nucleic Acids Research* 44(W1):W3-W10.

Ali, A. S., Z. Zanzinger, D. Debose, and B. Stephens. 2016. Open Source Building Science Sensors (OSBSS): A low-cost Arduino-based platform for long-term indoor environmental data collection. *Building and Environment* 100:114-126.

Araki, A., Y. Eitaki, T. Kawai, A. Kanazawa, M. Takeda, and R. Kishi. 2009. Diffusive sampling and measurement of microbial volatile organic compounds in indoor air. *Indoor Air* 19(5):421-432.

ASHRAE (American Society of Heating, Refrigerating and Air-Conditioning Engineers). 2012. *ASHRAE, ANSI/ASHRAE/ACCA Standard 180-2012. Standard practice for inspection and maintenance of commercial building HVAC systems*. Atlanta, GA: American Society of Heating, Refrigerating and Air-Conditioning Engineers, Inc.

ASHRAE. 2016. *ANSI/ASHRAE Standard 62.1: Ventilation for acceptable indoor air quality*. Atlanta, GA: ASHRAE.

Bartram, A. K., M. D. Lynch, J. C. Stearns, G. Moreno-Hagelsieb, and J. D. Neufeld. 2011. Generation of multimillion-sequence 16S rRNA gene libraries from complex microbial communities by assembling paired-end illumina reads. *Applied and Environmental Microbiology* 77(11):3846-3852.

Be, N. A., J. B. Thissen, V. Y. Fofanov, J. E. Allen, M. Rojas, G. Golovko, Y. Fofanov, H. Koshinsky, and C. J. Jaing. 2015. Metagenomic analysis of the airborne environment in urban spaces. *Microbial Ecology* 69(2):346-355.

Berkeley Lab. 2015. EcoFAB: Initiative researchers will model ecosystem biology on the bench. *BioSciences*. September 25. http://biosciences.lbl.gov/2015/09/25/ecofab-initiative-researchers-will-model-ecosystem-biology-on-the-bench (accessed July 16, 2017).

Bouslimani, A., C. Porto, C. M. Rath, M. Wang, Y. Guo, A. Gonzalez, D. Berg-Lyon, G. Ackermann, G. J. Moeller Christensen, T. Nakatsuji, L. Zhang, A. W. Borkowski, M. J. Meehan, K. Dorrestein, R. L. Gallo, N. Bandeira, R. Knight, T. Alexandrov, and P. C. Dorrestein. 2015. Molecular cartography of the human skin surface in 3D. *Proceedings of the National Academy of Sciences of the United States of America* 112(17):E2120-E2129.

Chase, J., J. Fouquier, M. Zare, D. L. Sonderegger, R. Knight, S. T. Kelley, J. Siegel, and J. G. Caporaso. 2016. Geography and location are the primary drivers of office microbiome composition. *mSystems* 1(2):e00022-16. doi:10.1128/mSystems.00022-16.

Coil, D. A., R. Y. Neches, J. M. Lang, W. E. Brown, M. Severance, D. Cavalier, and J. A. Eisen. 2016. Growth of 48 built environment bacterial isolates on board the International Space Station (ISS). *PeerJ* 4:e1842.

Dedesko, S., and J. A. Siegel. 2015. Moisture parameters and fungal communities associated with gypsum drywall in buildings. *Microbiome* 3(1):71.

Dorrestein, P. 2016. *Mass spectrometry-based visualization of molecules associated with human habits*. Presentation to the Committee on Microbiomes of the Built Environment: From Research to Application, October 17.

Dorrestein, P. 2017. Digitizing the chemistry associated with microbes: Importance, current status, and opportunities. In *The Chemistry of Microbiomes: Proceedings of a Seminar Series*. Washington, DC: The National Academies Press.

EPA (U.S. Environmental Protection Agency). 2006. *Building Assessment Survey and Evaluation (BASE) study data on indoor air quality in public and commercial buildings*. 402-C-06-002. Washington, DC: EPA.

Flannigan, B., R. A. Samson, and J. D. Miller. 2011. *Microorganisms in home and indoor work environments: Diversity, health impacts, investigation and control* (2nd ed.). Boca Raton, FL: CRC Press.

Fujimura, K. E., T. Demoor, M. Rauch, A. A. Faruqi, S. Jang, and C. C. Johnson. 2014. House dust exposure mediates gut microbiome Lactobacillus enrichment and airway immune defense against allergens and virus infection. *Proceedings of the National Academy of Sciences of the United States of America* 111(2):805-810.

Gibbons, S. 2016. The built environment is a microbial wasteland. *mSystems* 1(2):e00033-16. doi:10.1128/mSystems.00033-16.

Glass, E. M., Y. Dribinsky, P. Yilmaz, H. Levin, R. Van Pelt, D. Wendel, A. Wilke, J. A. Eisen, S. Huse, A. Shipanova, M. Sogin, J. Stajich, R. Knight, F. Meyer, and L. M Schriml. 2014. MIxS-BE: A MIxS extension defining a minimum information standard for sequence data from the built environment. *The ISME Journal* 8:1-3. http://www.nature.com/ismej/journal/v8/n1/full/ismej2013176a.html (accessed July 16, 2017).

Greenblum, S., R. Carr, and E. Borenstein. 2015. Extensive strain-level copy-number variation across human gut microbiome species. *Cell* 160(4):583-594.

Gutarowska, B., S. Celikkol-Aydin, V. Bonifay, A. Otlewska, E. Aydin, A. L. Oldham, J. I. Brauer, K. E. Duncan, J. Adamiak, J. A. Sunner, I. B. Beech. 2015. Metabolomic and high-throughput sequencing analysis—modern approach for the assessment of biodeterioration of materials from historic buildings. *Frontiers in Microbiology* 6:979.

Hanson, C. A., J. A. Fuhrman, M. C. Horner-Devine, and J. B. H. Martiny. 2012. Beyond biogeographic patterns: Processes shaping the microbial landscape. *Nature Reviews Microbiology* 10(7):497-506.

Harriman, L. G., and J. W. Lstiburek. 2009. *The ASHRAE guide for buildings in hot and humid climates* (2nd ed.). Atlanta, GA: ASHRAE.

Hoisington, A., J. P. Maestre, J. A. Siegel, and K. A. Kinney. 2014. Exploring the microbiome of the built environment: A primer on four biological methods available to building professionals. *HVAC&R Research* 20(1):167-175.

Jansson, J. K., and E. S. Baker. 2016. A multi-omic future for microbiome studies. *Nature Microbiology* 1(5):16049.

Jovel, J., J. Patterson, W. Wang, N. Hotte, S. O'Keefe, T. Mitchel, T. Perry, D. Kao, A. L. Mason, K. L. Madsen, and G. K. S. Wong. 2016. Characterization of the gut microbiome using 16S or shotgun metagenomics. *Frontiers in Microbiology* 7:459.

Kahlor, L. 2016. *Risk communication and information seeking*. Presentation to the Committee on Microbiomes of the Built Environment: From Research to Application, June 20.

Kauserud, H., S. Kumar, A. K. Brysting, J. Norden, and T. Carlsen. 2012. High consistency between replicate 454 pyrosequencing analyses of ectomycorrhizal plant root samples. *Mycorrhiza* 22(4):309-315.

Lauber, C. L., N. Zhou, J. I. Gordon, R. Knight, and N. Fierer. 2010. Effect of storage conditions on the assessment of bacterial community structure in soil and human-associated samples. *FEMS Microbiology Letters* 307(1):80-86.

Lax, S., D. P. Smith, J. Hampton-Marcell, S. M. Owens, K. M. Handley, N. M. Scott, S. M. Gibbons, P. Larsen, B. D. Shogan, S. Weiss, J. L. Metcalf, L. K. Ursell, Y. Vazquez-Baeza, W. Van Treuren, N. A. Hasan, M. K. Gibson, R. Colwell, G. Dantas, R. Knight, and J. A. Gilbert. 2014. Longitudinal analysis of microbial interaction between humans and the indoor environment. *Science* 345(6200):1048-1052.

Lax, S., N. Sangwan, D. Smith, P. Larsen, K. M. Handley, M. Richardson, K. Guyton, M. Krezalek, B. D. Shogan, J. Defazio, I. Flemming, B. Shakhsheer, S. Weber, E. Landon, S. Garcia-Houchins, J. Siegel, J. Alverdy, R. Knight, B. Stephens, and J. A. Gilbert. 2017. Bacterial colonization and succession in a newly opened hospital. *Science Translational Medicine* 9(391):eaah6500. http://stm.sciencemag.org/content/9/391/eaah6500 (accessed July 16, 2017).

Leinonen, R., H. Sugawara, and M. Shumway. 2011. The sequence read archive. *Nucleic Acids Research* 39:D19-D21.

Lemos, L. N., R. R. Fulthorpe, and L. F. Roesch. 2012. Low sequencing efforts bias analyses of shared taxa in microbial communities. *Folia Microbiologica* 57(5):409-413.

Lindgreen, S., K. L. Adair, and P. P. Gardner. 2016. An evaluation of the accuracy and speed of metagenome analysis tools. *Scientific Reports* 6:19233.

Macher, J. M., J. Douwes, B. Prezant, and T. Reponen. 2013. Bioaerosols. In *Aerosols handbook: Measurement, dosimetry, and health effects*, 2nd ed., edited by L. S. Ruzer and N. H. Harley. Boca Raton, FL: CRC Press. Pp. 285-344.

McKain, N., B. Genc, T. J. Snelling, and R. J. Wallace. 2013. Differential recovery of bacterial and archaeal 16S rRNA genes from ruminal digesta in response to glycerol as cryoprotectant. *Journal of Microbiological Methods* 95(3):381-383.

Merchant, N., E. Lyons, S. Goff, M. Vaughn, D. Ware, D. Micklos, and P. Antin. 2016. The iPlant collaborative: Cyberinfrastructure for enabling data to discovery for the life sciences. *PLOS Biology* 14(1):e1002342.

Moschandreas, D. J., K. R. Pagilla, and L. V. Storino. 2008. Time and space uniformity of indoor bacteria concentrations in Chicago area residences. *Aerosol Science and Technology* 37:899-906.

Mudarri, D. H. 1997. Potential correction factors for interpreting CO_2 measurements in buildings. *ASHRAE Transactions* 103(2):244-255.

NASEM (National Academies of Sciences, Engineering, and Medicine). 2017a. *Communicating science effectively: A research agenda*. Washington DC: The National Academies Press.

NASEM. 2017b. *The value of social, behavioral, and economic sciences to national priorities: A report for the National Science Foundation*. Washington DC: The National Academies Press.

Nayfach, S., and K. S. Pollard. 2016. Toward accurate and quantitative comparative metagenomics. *Cell* 166(5):1103-1116.

Nayfach, S., B. Rodriguez-Mueller, and K. S. Pollard. 2016. An integrated metagenomics pipeline for strain profiling reveals novel patterns of bacterial transmission and biogeography. *Genome Research* 26(11):1612-1625.

NIST (National Institute of Standards and Technology). 2012. *Challenges in microbial sampling in indoor environments: Workshop report summary*. NIST Technical Note 1737. http://www.microbe.net/wp-content/uploads/2012/03/Sloan-Roport-Final-TN-1737.pdf (accessed July 16, 2017).

NRC (National Research Council) and IOM (Institute of Medicine). 2015. *BioWatch PCR assays: Building confidence, ensuring reliability*. Washington, DC: The National Academies Press.

Peccia, J., and M. Hernandez. 2006. Incorporating polymerase chain reaction-based identification, population characterization, and quantification of microorganisms into aerosol science: A review. *Atmospheric Environment* 40:3941-3961.

Persily, A. K. 1997. Evaluating building IAQ and ventilation with carbon dioxide. *ASHRAE Transactions* 103(2):193-204.

Pinto, A. J., and L. Raskin. 2012. PCR biases distort bacterial and archaeal community structure in pyrosequencing datasets. *PLOS ONE* 7(8):e43093.

Prussin, A. J., and L. C. Marr. 2015. Sources of airborne microorganisms in the built environment. *Microbiome* 3(1):78.

Qian, J., D. Hospodsky, N. Yamamoto, W. W. Nazaroff, and J. Peccia. 2012. Size-resolved emission rates of airborne bacteria and fungi in an occupied classroom. *Indoor Air* 22(4):339-351.

Qian, J., J. Peccia, and A. R. Ferro. 2014. Walking-induced particle resuspension in indoor environments. *Atmospheric Environment* 89:464-481.

Quinn, R. A., J. A. Navas-Molina, E. R. Hyde, S. J. Song, Y. Vázquez-Baeza, G. Humphrey, J. Gaffney, J. J. Minich, A. V. Melnik, J. Herschend, J. DeReus, A. Durant, R. J. Dutton, M. Khosroheidari, C. Green, R. da Silva, P. C. Dorrestein, and R. Knight. 2016. From sample to multi-omics conclusions in under 48 hours. *mSystems* 1(2):e00038-16. doi:10.1128/mSystems.00038-16.

Ramos, T., and B. Stephens. 2014. Tools to improve built environment data collection for indoor microbial ecology investigations. *Building and Environment* 81:243-257.

Ramos, T., S. Dedesko, J. A. Siegel, J. A. Gilbert, and B. Stephens. 2015. Spatial and temporal variations in indoor environmental conditions, human occupancy, and operational characteristics in a new hospital building. *PLOS ONE* 10(3):e0118207.

Randle-Boggis, R. J., T. Helgason, M. Sapp, and P. D. Ashton. 2016. Evaluating techniques for metagenome annotation using simulated sequence data. *FEMS Microbiology Ecology* 92(7). doi:10.1093/femsec/fiw095.

Rhee, S.-K., X. Liu, L. Wu, S. C. Chong, X. Wan, and J. Zhou. 2004. Detection of genes involved in biodegradation and biotransformation in microbial communities by using 50-mer oligonucleotide microarrays. *Applied and Environmental Microbiology* 70(7):4303-4317.

Saraf, A., L. Larsson, H. Burge, and D. Milton. 1997. Quantification of ergosterol and 3-hydroxy fatty acids in settled house dust by gas chromatography-mass spectrometry: Comparison with fungal culture and determination of endotoxin by a Limulus amebocyte lysate assay. *Applied Environmental Microbiology* 63(7):2554-2559.

Schloss, P. D., D. Gevers, and S. L. Westcott. 2011. Reducing the effects of PCR amplification and sequencing artifacts on 16S rRNA-based studies. *PLOS ONE* 6(12):e27310.

Shogan, B. D., D. P. Smith, A. I. Packman, S. T. Kelley, E. M. Landon, S. Bhangar, G. J. Vora, R. M. Jones, K. Keegan, B. Stephens, T. Ramos, B. C. Kirkup, Jr., H. Levin, M. Rosenthal, B. Foxman, E. B. Chang, J. Siegel, S. Cobey, G. An, J. C. Alverdy, P. J. Olsiewski, M. O. Martin, R. Marrs, M. Hernandez, S. Christley, M. Morowitz, S. Weber, and J. Gilbert. 2013. The Hospital Microbiome Project: Meeting report for the 2nd Hospital Microbiome Project, Chicago, USA, January 15, 2013. *Standards in Genomic Sciences* 8(3):571-579.

Singer, E., B. Andreopoulos, R. M. Bowers, J. Lee, S. Deshpande, J. Chiniquy, D. Ciobanu, H.-P. Klenk, M. Zane, C. Daum, C., A. Clum, J.-F. Cheng, A. Copeland, and T. Woyke. 2016. Next generation sequencing data of a defined microbial mock community. *Scientific Data* 3:160081.

Smith, D., J. Alverdy, G. An, M. Coleman, S. Garcia-Houchins, J. Green, K. Keegan, S. T. Kelley, B. C. Kirkup, L. Kociolek, H. Levin, E. Landon, P. Olsiewski, R. Knight, J. Siegel, S. Weber, and J. Gilbert. 2013. The Hospital Microbiome Project: Meeting report for the 1st Hospital Microbiome Project Workshop on sampling design and building science measurements, Chicago, USA, June 7–8, 2012. *Standards in Genomic Sciences* 8(1):112-117.

Stein, M. M., C. L. Hrusch, J. Gozdz, C. Igartua, V. Pivniouk, S. E. Murray, N., J. G. Ledford, M. Marques dos Santos, R. L. Anderson, N. Metwali, J. W. Neilson, R. M. Maier, J. A. Gilbert, M. Holbreich, P. S. Thorne, F. D. Martinez, E. von Mutius, D. Vercelli, C. Ober, and A. I. Sperling. 2016. Innate immunity and asthma risk in Amish and Hutterite farm children. *New England Journal of Medicine* 375(5):411-421.

Stephens, B. 2016. What have we learned about the microbiomes of indoor environments? *mSystems* 1(4):e00083-e00116. doi:10.1128/mSystems.00083-16.

Taylor, D. L., W. A. Walters, N. J. Lennon, J. Bochicchio, A. Krohn, J. G. Caporaso, and T. Pennanen. 2016. Accurate estimation of fungal diversity and abundance through improved lineage-specific primers optimized for illumina amplicon sequencing. *Applied and Environmental Microbiology* 82(24):7217-7226.

Uyaguari-Diaz, M. I., M. Chan, B. L. Chaban, M. A. Croxen, J. F. Finke, J. E. Hill, M. A. Peabody, T. Van Rossum, C. A. Suttle, F. S. L. Brinkman, J. Isaac-Renton, N. A. Prystajecky, and P. Tang. 2016. A comprehensive method for amplicon-based and metagenomic characterization of viruses, bacteria, and eukaryotes in freshwater samples. *Microbiome* 4(1):20.

Vizcaíno, J. A., A. Csordas, N. del-Toro, J. A. Dianes, J. Griss, I. Lavidas, G. Mayer, Y. Perez-Riverol, Y., F. Reisinger, T. Ternent, Q.-W. Xu, R. Wang, and H. Hermjakob. 2016. 2016 update of the PRIDE database and its related tools. *Nucleic Acids Research* 44(22):D447-D456.

Wang, M., J. J. Carver, V. V. Phelan, L. M. Sanchez, N. Garg, Y. Peng, D. D. Nguyen, J. Watrous, C. A. Kapono, T. Luzzatto-Knaan, C. Porto, A. Bouslimani, A. V. Melnik, M. J. Meehan, W. T. Liu, M. Crüsemann, P. D. Boudreau, E. Esquenazi, M. Sandoval-Calderón, R. D. Kersten, L. A. Pace, R. A. Quinn, K. R. Duncan, C. C. Hsu, D. J. Floros, R. G. Gavilan, K. Kleigrewe, T. Northen, R. J. Dutton, D. Parrot, E. E. Carlson, B. Aigle, C. F. Michelsen, L. Jelsbak, C. Sohlenkamp, P. Pevzner, A. Edlund, J. McLean, J. Piel, B. T. Murphy, L. Gerwick, C. C. Liaw, Y. L. Yang, H. U. Humpf, M. Maansson, R. A. Keyzers, A. C. Sims, A. R. Johnson, A. M. Sidebottom, B. E. Sedio, A. Klitgaard, C. B. Larson, P. C. A. Boya, D. Torres-Mendoza, D. J. Gonzalez, D. B. Silva, L. M. Marques, D. P. Demarque, E. Pociute, E. C. O'Neill, E. Briand, E. J. Helfrich, E. A. Granatosky, E. Glukhov, F. Ryffel, H. Houson, H. Mohimani, J. J. Kharbush, Y. Zeng, J. A. Vorholt, K. L. Kurita, P. Charusanti, K. L. McPhail, K. F. Nielsen, L. Vuong, M. Elfeki, M. F. Traxler, N. Engene, N. Koyama, O. B. Vining, R. Baric, R. R. Silva, S. J. Mascuch, S. Tomasi, S. Jenkins, V. Macherla, T. Hoffman, V. Agarwal, P. G. Williams, J. Dai, R. Neupane, J. Gurr, A. M. Rodríguez, A. Lamsa, C. Zhang, K. Dorrestein, B. M. Duggan, J. Almaliti, P. M. Allard, P. Phapale, L. F. Nothias, T. Alexandrov, M. Litaudon, J. L. Wolfender, J. E. Kyle, T. O. Metz, T. Peryea, D. T. Nguyen, D. VanLeer, P. Shinn, A. Jadhav, R. Müller, K. M. Waters, W. Shi, X. Liu, L. Zhang, R. Knight, P. R. Jensen, B. Ø. Palsson, K. Pogliano, R. G. Linington, M. Gutiérrez, N. P. Lopes, W. H. Gerwick, B. S. Moore, P. C. Dorrestein, and N. Bandeira. 2016. Sharing and community curation of mass spectrometry data with Global Natural Products Social Molecular Networking. *Nature Biotechnology* 34(8):828-837.

Weiss, S., W. Van Treuren, C. Lozupone, K. Faust, J. Friedman, Y. Deng, L. C. Xia, Z. Z. Xu, L. Ursell, E. J. Alm, A. Birmingham, J. A. Cram, J. A. Fuhrman, J. Raes, F. Sun, J. Zhou, and R. Knight. 2016. Correlation detection strategies in microbial data sets vary widely in sensitivity and precision. *The ISME Journal* 10(7):1669-1681.

Wu, L., D. K. Thompson, G. Li, R. A. Hurt, J. M. Tiedje, and J. Zhou. 2001. Development and evaluation of functional gene arrays for detection of selected genes in the environment. *Applied and Environmental Microbiology* 67(12):5780-5790.

Wu, L., X. Liu, C. W. Schadt, and J. Zhou. 2006. Microarray-based analysis of subnanogram quantities of microbial community DNAs by using whole-community genome amplification. *Applied and Environmental Microbiology* 72(7):4931-4941.

Yao, M., T. Zhu, K. Li, S. Dong, Y. Wu, X. Qiu, B. Jiang, L. Chen, and S. Zhen. 2009. Onsite infectious agents and toxins monitoring in 12 May Sichuan earthquake affected areas. *Journal of Environmental Monitoring* 11(11):1993-2001.

Yooseph, S., C. Andrews-Pfannkoch, A. Tenney, J. McQuaid. S. Williamson, M. Thiagarajan, D. Brami, L. Zeigler-Allen, J. Hoffman, J. B. Goll, D. Fadrosh, J. Glass, M. D. Adams, R. Friedman, and J. C. Venter. 2013. A metagenomics framework for the study of airborne microbial communities. *PLOS ONE* 8(12):e81862.

Zhou, J. Z., L. Wu, Y. Deng, X. Zhi, Y.-H. Jiang, Q. Tu, J. Xie, J. D. Van Nostrand, Z. He, and Y. Yang. 2011. Reproducibility and quantitation of amplicon sequencing-based detection. *The ISME Journal* 5(8):1303-1313.

Zhou, J. Z., Z. He, Y. Yang, Y. Deng, S. G. Tringe, and L. Alvarez-Cohen. 2015. High-throughput metagenomic technologies for complex microbial community analysis: Open and closed formats. *mBio* 6(1). doi:10.1128/mBio.02288-14.

Zou, H., Z. Chen, H. Jiang, L. Xie, and C. Spanos. 2017. Accurate indoor localization and tracking using mobile phone inertial sensors, WiFi and iBeacon. In *2017 IEEE International Symposium on Inertial Sensors and Systems (INERTIAL)*, Kauai, HI. Pp. 1-4. doi:10.1109/ISISS.2017.7935650.

5

Interventions in the Built Environment

> **Chapter Highlights**
> - Several possible interventions in built environments are available to reduce exposures to harmful microorganisms. These interventions include increasing outdoor ventilation rates; increasing air filtration efficiency; and employing air, water, and surface disinfection strategies.
> - A newer intervention approach promotes indoor exposures to beneficial microorganisms.
> - Quantitative frameworks can be used to understand the factors associated with microbial transport and how interventions may affect it. In enabling a better understanding of the effects of different interventions, such frameworks can aid in the design of future intervention approaches.
> - Models used in studying, designing, and making decisions about built environment interventions will need to include not only aspects of the built environment and the microbiome but also occupant behavior, health impacts, and potential trade-offs with energy consumption and economic factors.

The relationships between buildings and microbes discussed in this report suggest that human microbial exposures may be modulated and controlled through interventions related to building design, construction, and operation. Such interventions are intended to improve human health and have two broad goals: (1) reducing human exposure to harmful microbes and (2) encouraging human exposure to beneficial microbes. This chapter describes existing and potential interventions for modifying human microbial exposures to improve health. The discussion identifies potential trade-offs associated with such interventions, such as increased energy consumption or building costs. The focus is on control of air- and surface-borne microbes in buildings.[1] Where relevant, selected results on the effectiveness of common interventions for the built environment with respect to the microbiome and human health are presented.

There are a wide variety of biological particles and chemicals of both biological and nonbiological origin inside buildings, multiple sources of these agents, and varied exposure routes, leading to myriad intervention approaches that merit consideration. Both physical- and chemical-based interventions exist or have been proposed for controlling or reducing microbial exposures in buildings. These types of interventions include changes to building design and operation, such as control of ventilation rates or the use of air filtration systems, and the use of disinfectants to inactivate viable microorganisms. Other types of interventions focus on the promotion of exposures to potentially beneficial microorganisms, such as by enhancing building and human connections with outdoor microbial diversity. Assessment of the appropriateness of these interventions will need to go beyond their effectiveness at changing building microbial exposures to include broader cost-benefit analyses that consider positive outcomes; weigh potential negative outcomes; and include economic factors, environmental factors, and effects on human health. The chapter includes a discussion of approaches to better understanding and optimizing interventions and their potential trade-offs in the framework of building engineering controls.

PHYSICAL INTERVENTIONS TO REDUCE EXPOSURE TO HAZARDOUS MICROBES

Physical interventions are most commonly practiced through engineering controls in the built environment. These typically include local or systemic changes to temperature, ventilation, moisture, and light (including ultraviolet [UV]).

---

[1]Most traditional control measures for microorganisms in water supplies, such as filtration and inclusion of disinfectants at water treatment facilities, are conducted outside of the building.

Changes to Ventilation Practice

Increasing outdoor air ventilation rates is a common strategic intervention intended to reduce occupants' microbial exposure in buildings, either through dilution of indoor air with outdoor air of different or lower microbial loads, or by supporting indoor environmental conditions (i.e., changes to relative humidity) that are less conducive to microbial growth. All buildings experience some degree of indoor–outdoor air exchange, some of which is purposeful (ventilation) and some unintended (infiltration). Mechanical ventilation approaches, whether natural or hybrid, are preferred to infiltration because they provide better control of the ventilation, the ability to treat outdoor contaminants and dehumidify outdoor air, and potentially the ability to reduce energy impacts by recovering or discharging heat from outgoing air.

A simplified material balance equation for a single-zone space and steady-state contaminant concentration with no internal loss terms due to deposition or engineered control systems is shown below:

$$C_{in} = PC_{out} + \frac{G}{Q_o} \qquad \text{(Equation 5.1)}$$

where C_{in} (mass or number per volume) is the steady-state indoor contaminant concentration, C_{out} (mass or number per volume) is the outdoor concentration, P (unitless) is the penetration factor for outdoor contaminants, G is the indoor contaminant generation rate (mass or number per time), and Q_o (volume per time) is the outdoor airflow rate into the building being considered.

The impacts of ventilation can be understood by considering the terms in Equation 5.1. In this case, airborne microorganisms are considered the "contaminant" in the equation. Assuming for purposes of discussion that all terms in the equation are constant, increasing the outdoor ventilation rate (Q_o) will decrease the term G/Q_o on the right side of the equation, decreasing the level of the indoor contaminant. However, if the outdoor concentration of this contaminant is greater than zero, the indoor concentration can never be lower than PC_{out} (the amount of the contaminant that penetrates indoors from outside). Increasing the rate of outdoor air ventilation is effective as an intervention only when the air contaminant concentration outdoors is lower than that indoors. A possible practical effect, for example, could be to decrease the concentration of bacteria with an indoor source (such as shedding from humans) while increasing the concentration of fungi introduced from outdoor air.

This same conclusion is valid when considering water vapor as a contaminant. In that case, the material balance is more complicated than reflected in Equation 5.1, as it needs to consider temperature effects; moisture removal by the heating, ventilation, and air conditioning (HVAC) system; and moisture storage in building materials. Because microorganisms require moisture to grow, increased ventilation with humid outdoor air is not likely to be an effective intervention without careful consideration of the impact on indoor moisture levels or the addition of dehumidification (possibly in association with air conditioning).

The paradigm of using ventilation for treatment of indoor air has been applied mainly to removal of chemical contaminants with exclusive or primarily indoor sources. This model can be extended to many infectious viruses and other microorganisms, where humans are the sources and the microbial concentrations outdoors are considered to be very low. However, a number of hazardous microbes, such as plant-derived and fungal allergens, commonly have an outdoor source that is more important than any indoor sources. Thus, increasing ventilation would be an effective strategy for such species only if the outdoor air were subjected to filtration or other treatment.

In considering the impact of increased outdoor air ventilation on levels of indoor airborne contaminants, it is also important to consider the loss mechanisms that could be added to Equation 5.1 and the generation mechanisms embodied in the term G. Loss mechanisms include removal through particle filtration in the ventilation system or in-room filtration devices, as well as the decay of microbial infectivity and particle and gas deposition onto surfaces. It is important to recognize here that hypersensitivity potential (e.g., allergenicity) is often not related to the decay of infectious potential. Generation mechanisms include human microbial shedding, resuspension of microorganisms from surfaces, and desorption of volatile organic compounds (VOCs) from surfaces to air. Semivolatile organic compounds (SVOCs) adsorbed by building materials, including SVOCs produced by microbes, are very slow to release or partition into the air, making ventilation less effective in their removal than removal of more volatile compounds. The relative magnitude of these mechanisms compared with dilution through ventilation will determine the overall effectiveness of increased ventilation as an intervention.

Analyses of material and mass balance need to take particle size into account because deposition rates, resuspension rates, filtration efficiencies, and building penetration rates depend strongly on this parameter (Nazaroff, 2004). The average geometric mean diameters for bioaerosol particles that contain bacteria and fungi (which may be in the form of aggregated cells and/or spores attached to particulate matter) have been reported to be 5.5 and 5.9 µm, respectively (Hospodsky et al., 2015). Particles in this size range penetrate building envelopes inefficiently and can be removed effectively by most ventilation system filters. For viruses, a study of day care centers and airplanes found that 64 percent of influenza genomes in air were associated with particles smaller than 2.5 µm in diameter (Yang et al., 2011), corresponding to lower deposition rates and effective removal by ventilation. However, resuspension of microbes in densely occupied settings, such as classrooms, can dramatically increase indoor air concentrations of bacteria and fungi, even at high outdoor air ventilation rates (Hospodsky et al., 2015). To understand the more complex systems that represent real buildings and built environment microbiomes, it would be necessary to develop separate mass balance equations that appropriately treat generation and loss rates for each microbe or contaminant of interest. Given the diversity of microorganisms that form built environment microbiomes, developing and integrating individual mass balance equations for each can be challenging, although the creation of models based on this concept that are more precise than the simplified Equation 5.1 will be useful.

Natural Ventilation and Envelope Tightness

Employing natural ventilation (i.e., ventilation without use of mechanical systems) instead of or in addition to mechanical ventilation has been advocated as an intervention for its potential benefits for occupants' health and comfort. Studies have shown that the prevalence of symptoms of sick building syndrome is lower in naturally ventilated than in mechanically ventilated buildings, but the mechanism for this finding has not been established (Seppänen and Fisk, 2002). Natural ventilation often is proposed to increase outdoor ventilation rates, which will decrease the levels of internally generated airborne microbes. However, outdoor air entering buildings through natural ventilation may not be (cost) effectively captured for filtration and conditioning. As a result, such practices can introduce outdoor contaminants, including allergens and moisture, at undesirable levels, and therefore may not necessarily provide net improvements with regard to indoor environmental conditions. Moreover, the rates of outdoor air entry and the distribution of this air within a building must be carefully considered for appropriate building analyses in this context. More advanced systems that provide better control of ventilation rates and air distribution than those often found in the United States (CIBSE, 2005; Schulze and Eicker, 2013) may warrant further study for domestic implementation.

Increasing the tightness of the building envelope is another way to affect ventilation and exposures in buildings. In both mechanically and naturally ventilated buildings, the entry of unconditioned, unfiltered, and infiltrated air will interfere with the performance of the ventilation system. Reducing infiltration results in better indoor temperature and humidity control, lower likelihood of indoor moisture accumulation, and lower entry rates of outdoor microbes, all of which can reduce the likelihood of elevated indoor microbe levels. In buildings ventilated predominantly by infiltration through spurious air leakage, which includes most U.S. residences, envelope tightening provides for better airflow control. However, tighter building envelopes need to be accompanied by reliable ventilation, mechanical or natural, that meets the outdoor air exchange and humidity control requirements of a building.

Studies on Building Ventilation and Microbial Exposures

Associations between increased outdoor ventilation rates (typically >10–15 liters/second per person) and improved human health are well established (Bornehag et al., 2005; Menzies and Bourbeau, 1997). Common symptoms noted in ventilation and health studies include allergies and other hypersensitivity responses, respiratory infections, and neurological and other symptoms. As a result, it is plausible that microbes are an important mediator between ventilation and health (Sundell et al., 2011). In cases in which hazardous microbial agents have exclusive indoor sources (e.g., influenza virus and the bacterium *Mycobacterium tuberculosis*), the importance of outdoor air ventilation is predictable. In one study, for example, installation and operation of heat recovery ventilators (ventilation systems that bring outdoor air into buildings and transfer heat from the outgoing airstream to the incoming airstream) in Inuit homes was found to result in reductions in reported wheeze and rhinitis (Kovesi et al., 2009).

More broadly, studies in naturally and mechanically ventilated commercial buildings (Meadow et al., 2014) have demonstrated that the rate and method of ventilation may influence the composition of bacterial communities in indoor air. These relationships are affected by such factors as occupancy level, ventilation design and operation, and outdoor microbe concentrations and ecology. In a hospital study, bacteria closely related to human pathogens were relatively more abundant in rooms with lower ventilation rates, while rooms with natural ventilation had a more diverse bacterial community compared with mechanically ventilated rooms (Kembel et al., 2012). In a recent school study aimed at disentangling the impacts of ventilation and occupancy, indoor air concentrations of bacteria and fungi were dominated by indoor sources associated with human occupancy (resuspension and shedding) rather than outdoor air ventilation. This was true for all but the extreme cases of high ventilation rate and high outdoor air microbial concentration (Hospodsky et al., 2015).

Moisture Control

Decreasing Relative Humidity

Evidence indicates that decreasing relative humidity (RH), as modified by ventilation, can impact occupant microbial exposures. Increased ventilation in Swedish homes, for example, is indirectly associated with lower concentrations of airborne dust mite allergens because dust mite activity and growth are associated with high RH. In the cold, dry Scandinavian climates, increased outdoor air ventilation results in lower indoor RH levels (Sundell et al., 1995). Increased outdoor ventilation to reduce RH, especially in cold climates, is a commonly recommended approach for controlling mold growth in buildings. Sustained building RH levels greater than 70 percent are associated with fungal growth on building materials (Arundel et al., 1986), while building RH levels below 50 percent are expected to contribute to transmission of influenza (Yang et al., 2012). Although evidence suggests that greater biodiversity surrounding people's homes influences the classes of bacteria on their skin and is associated with reduced incidence of allergic disposition (atopy) (Hanski et al., 2012), there is little evidence that increased outdoor air ventilation has a benefit for immune system development.

The literature indicates general consensus that ventilation, either direct or through the control of RH, influences the concentration and ecology of microbes in buildings. In most cases, and especially for potentially beneficial microbes, this information is not sufficiently developed to support design decisions or well-founded recommendations on how to design, control, and operate ventilation systems specifically for control of microbial communities and exposures.

Remediating in Damp Buildings

Dampness and visible mold have been reported in approximately 50 percent of U.S. homes (Mudarri and Fisk, 2007; Spengler et al., 1994). Water damage and visible mold inside buildings are consistently associated with respiratory and allergic health effects in infants, children, and adults (Mendell et al., 2011). Prior research and practice have led to the development of some operative approaches for reducing associated health symptoms. Intense remediation and environmental intervention for homes with moisture sources and visible mold have been shown to reduce some microbial exposures and can reduce allergy symptoms and asthma morbidity. In research previously sponsored by the National Institutes of Health (NIH), the U.S. Department of Housing and Urban Development, and the U.S. Environmental Protection Agency (EPA), remediation that included removal of water-damaged building materials and alteration of HVAC systems resulted in a large decrease in asthma symptom days and asthma exacerbation versus a no-remediation control group (Kercsmar et al., 2006). A 2004 report of the Institute of Medicine reviews building dampness, associated mold growth, and health impacts and recommends the development of national guidelines for preventing indoor dampness, as well as economic or other incentives to spur adherence to moisture prevention practice by those that construct and manage buildings (IOM, 2004).

An additional concern regarding interventions for damp buildings is the lack of critical guidance on when to initiate costly remediation and whether remediation efforts are successful. The Institute for Inspection, Cleaning and Restoration Certification (IICRC) has established widely used guidance documents in the mold and building restoration industry, such as American National Standards Institute (ANSI)/IICRC S520 2015, *Standard and Reference Guide for Professional Mold Remediation*. The purpose of remediation is to restore the property to an acceptable state similar to that prior to the occurrence of the indoor mold contamination, designated by S520 as "Condition 1." Condition 1 is defined as "(normal fungal ecology): an indoor environment that may have settled spores, fungal fragments or traces of actual growth whose identity, location, and quantity are reflective of a normal fungal ecology for a similar indoor environment" (ANSI/IICRC, 2015, p. 16).

The major uncertainty in the S520 document and all other mold remediation guidance is that there are no accepted methods for defining and quantifying the "normal fungal ecology" reference point. The types, abundances, and concentrations of microbial taxa that constitute the normal ecology of a building are not clearly defined, and they likely differ based on such environmental variables as climate and land use (Amend et al., 2010; Kembel et al., 2014; Qian et al., 2012; Reponen et al., 2011). Such variability could potentially be addressed by the detailed ecological information leveraged by modern DNA sequencing technologies and bioinformatics analyses now used in building microbiome studies. Gaps in knowledge about what represents "normal" microbial ecology and how to interpret particular microbial findings are one reason that building microbial sampling often is not recommended. The issue of microbial sampling for building microbial assessment and remediation strategies is beyond the scope of this report.[2]

---

[2]The American Industrial Hygiene Association, for example, has a position statement on Mold and Dampness in the Built Environment (AIHA, 2013) that provides guidance on remediation efforts.

Particle Filtration

It is common practice to treat the air in building HVAC systems using particle filtration, which could be an effective intervention for removing airborne microbes. Particle filtration has been used in buildings for many decades, primarily to reduce fouling of heat transfer surfaces by particulate matter in outdoor air and airborne organic substances in the outdoor airstream. More recently, the benefits of reducing indoor fine and ultrafine particle concentrations for occupant health have been considered, leading to requirements for and use of higher levels of filter efficiency in buildings. Conventional air filter performance is a strong function of particle diameter. Smaller particles (on the order of 0.1 μm in diameter and lower) are removed primarily via diffusion, in which random motion of the particle leads to contact with a fiber in the filter and subsequent capture. Larger particles (diameters of about 1 μm and larger) are carried along by the airstream and collide with fibers as a result of impaction. Both diffusion and impaction are less effective at particle removal between roughly 0.1 μm and 1 μm. Thus, removal efficiencies are lower in this size range, which encompasses the known aerodynamic diameter ranges of some viruses, such as influenza (Lindsley et al., 2010), but this range is smaller than most indoor bacteria and fungi (Hospodsky et al., 2015).

Particle filtration efficiencies are rated based on testing using American Society of Heating, Refrigerating and Air-Conditioning Engineers (ASHRAE) Standard 52.2 (ASHRAE, 2017), which leads to filters being classified using the so-called Minimum Efficiency Reporting Value (MERV) scale.[3] MERV ratings range from 1 to 20, with higher values corresponding to more effective particle removal. ASHRAE Standard 62.1 requires MERV 8 filters upstream of wetted surfaces in HVAC systems to reduce the accumulation of organic matter on these surfaces and the subsequent likelihood of microbial growth (ASHRAE, 2010). The standard also requires MERV 6 filters in the outdoor air intake when outdoor levels of particulate matter $(PM)_{10}$ exceed ambient air quality standards and MERV 11 filters when $PM_{2.5}$ exceeds ambient standards. It is important to note that filtration will treat only air passing through the filters, making filter installation and sealing important to achieving the intended level of particle removal. Filtration is an additional loss mechanism that could be incorporated into Equation 5.1, the rate of which will be affected by removal efficiency and amount of airflow through the filter.

Low-rise residential buildings that are ventilated by infiltration and open windows do not generally provide for systemic filtration of outdoor

---

[3]MERV is a scale used to indicate the effectiveness of air filters. Fewer particles will pass through filters with a higher MERV rating.

air, although some loss of particles is expected to result from deposition as outdoor air penetrates the building envelope, with particles below 0.1 and greater than 1 µm in diameter penetrating less efficiently (Nazaroff, 2004). Thus, filtration is achieved by the circulation of indoor air through a residential HVAC system. Among U.S. residences with HVAC systems, surveys indicate that 25 percent use HVAC filters of MERV 5 or less, 60 percent use MERV 6–8, 10 percent use MERV 12, and 5 percent use MERV 16 (Stephens and Siegel, 2012). Portable air filtration systems—that is, stand-alone units typically intended to provide filtration in a single room—are employed in some applications. These units are rated using the so-called clean air delivery rate (CADR) based on an Association of Home Appliance Manufacturers (AHAM) test method that converts the particle removal rate to an equivalent volumetric airflow (dilution) rate.

The particle removal efficiency of MERV filters can be estimated as a function of particle aerodynamic diameter (Azimi et al., 2014; EPA, 2008), and the particle size distributions of several relevant viruses, bacteria, and fungi have been described (Hospodsky et al., 2015; Lindsley et al., 2010; Pastuszka et al., 2000; Yamamoto et al., 2014). Studies on human health improvements due to filtration have been reviewed for residential and commercial buildings. Results suggest that filtration does result in reductions in exposure to biologic particles, including cat, dog, and dust mite allergens, and offers modest improvements in allergy or asthma severity (Fisk, 2013). Although the fundamental information exists for testing and predicting the effects of filtration on reducing microbial exposure and associated disease, this research is not well developed.

A suite of less common air treatment technologies for particle removal, including electrostatically enhanced filtration and electrostatic precipitation, has previously been reviewed (EPA, 2008).

UV Germicidal Irradiation

Microbes can be transported to and from indoor surfaces via aerosol routes, and it is recognized by the engineering and medical communities that airborne microbes deposit onto fomites and vice versa in response to human activity and common environmental perturbations (Prussin and Marr, 2015; see also Chapter 3). Thus, disinfection practice needs to include both air and surfaces. UV germicidal irradiation (UVGI) is one common physical approach for disinfection of air and surfaces.

Exposure to light between wavelengths of 100 and 400 nanometers (nm) (UV-A, -B, -C) can damage the DNA of living organisms and result in an inability to replicate, thus rendering a cell noninfectious. UVGI, typically from low-pressure mercury vapor lamps with spectral power distribution focused at 254 nm, has been applied to disinfect indoor air and surfaces for more than 50 years (Reed, 2010). UV systems that emit wavelengths below 242 nm can generate ozone and are not appropriate for air disinfection in occupied buildings. While UVGI can be effective at microbial inactivation, its effect against the allergenic or toxigenic properties of microbes is not well described. UV disinfection has commonly been adapted for use within the upper levels of rooms for effective air disinfection, placed on cooling coils to reduce microbial growth on and fouling of coils, placed in air-supply ducts, and more recently applied in surface disinfection. Guidelines are available for use by practitioners in applying this relatively mature technology to a variety of indoor settings (ASHRAE, 2016). Although UV equipment for residential systems is readily available from HVAC vendors in the domestic consumer market, UV air treatment generally has been dominated by health care applications and specialty commercial sectors (ASHRAE, 2016).

UV Air Disinfection

Upper-level UVGI typically is applied in health care settings for the purpose of interrupting transmission of *Mycobacterium tuberculosis*. The technology entails irradiation of the upper ~20 percent of a room with UV-C lamps (reviewed by Brickner et al., 2003). This technology leverages rapid vertical mixing of air in a room rather than removal of air from a room and subsequent treatment in a ventilation duct. (It should be noted that UV sources present acute radiation exposure risks to occupants, and UV lamps must be specially louvered for in-room applications [Sliney, 2013].) Full-room experiments have demonstrated the utility of upper-level UVGI for reducing infectious bacterial exposures to occupants (Xu et al., 2003), and the effects of environmental conditions, especially RH, on UV inactivation efficiency have been documented for bacterial pathogens and surrogates (Peccia et al., 2001). There also exists epidemiologic evidence of the effectiveness of upper-level UVGI. A significantly lower incidence of influenza in a veterans hospital tuberculosis (TB) ward equipped with upper-level UVGI suggests the efficacy of this approach in decreasing airborne transmission of influenza virus (McLean, 1961). Similarly, a clinical trial in an HIV–TB ward in Lima, Peru, demonstrated a more than double reduction in TB infection rate in wards equipped with UVGI versus those without (Escombe et al., 2009).

UV Surface Disinfection

Surface disinfection by UVGI has recently focused on the use of mobile UV units in hospitals to disinfect surfaces, as well as irradiation of cooling coils in large building HVAC systems. The mobile technology is new, and the operation and workflow of these units have not been optimized. However, recent randomized trials have found a significantly reduced relative risk of infection for methicillin-resistant *Staphylococcus aureus* (MRSA) and vancomycin-resistant enterococci (VRE) when terminal UV disinfection was added to the standard cleaning regimen in hospital rooms (Anderson et al., 2017; Weber et al., 2016). In that study, large and significant reductions in the culturable concentrations of MRSA, VRE, and multidrug-resistant *Acinetobacter* were also observed on hospital room surfaces. And multiple studies suggest reductions in *Clostridium difficile* hospital infections when UV-C technology is used to treat hospital room surfaces (Levin et al., 2013; Miller et al., 2015; Nagaraja et al., 2015).

The application of UV on HVAC cooling coils has been recognized for its potential to inhibit microbial fouling of heat transfer equipment, thereby reducing pressure drops across the system and improving heat transfer (Luongo et al., 2017; Wang et al., 2016). An additional benefit of irradiating coils is the previously observed reductions in respiratory, mucosal, and musculoskeletal symptoms in workers in a building when UVGI was installed in the HVAC system. Reductions of symptoms in that study were coincident with UV operation and a large reduction in microbial and endotoxin concentration on irradiated surfaces within the ventilation system (Menzies et al., 2003).

CHEMICAL INTERVENTIONS TO REDUCE EXPOSURE TO HAZARDOUS MICROBES

Chemical interventions in the built environment focus on the inactivation of surface-bound microbes through the use of chemical disinfectants and, to a lesser extent, on the introduction of antimicrobial materials. These types of interventions include the treatment of air and surfaces through chemical disinfection and the design and use of antimicrobial materials or coatings.

Chemical Disinfection

Surface-associated microorganisms are a central component of the indoor microbiome. Important sources of microbes on surfaces include tracked-in dust, microbes shed from humans, settled airborne microbial aerosols and droplets, and microbial growth (Adams et al., 2013; Grant et al., 1989; Roberts et al., 1999; see also Chapter 3). In addition to contact- and fomite-based exposure, resuspension of microbes from flooring and elevated surfaces often is a significant source of airborne bacteria and inhalation exposure to bacteria and fungi (Bhangar et al., 2014; Hospodsky et al., 2014; Qian et al., 2012). Thus, chemical disinfection of surfaces is an important type of intervention for modifying exposure to infectious microorganisms and viruses. A significant literature describes studies testing the efficacy of disinfection in schools, health care facilities, and industrial facilities (e.g., food preparation) where elevated surfaces and flooring are considered reservoirs of infectious bacteria and viruses (Donskey, 2013).

Regarding the building microbiome, uncertainty exists in chemical and other surface disinfection practices in two important areas. First, the rate at which the redevelopment of surface microbial communities occurs is poorly understood, resulting in uncertainty in determining optimal cleaning practices. This rate is affected by building parameters that include occupancy, ventilation, building materials, and moisture. Second, growing evidence suggests that early-life exposure to house dust containing increased fungal and bacterial diversity (Dannemiller et al., 2014; Ege et al., 2011) and elevated content of some specific bacteria (Lynch et al., 2014) may be protective concerning the development of asthma and recurrent wheeze in children (see Chapter 2). The impact of chemical cleaning interventions on the exposure of children to the chemicals and chemical by-products, as well as to these microorganisms, and on health outcomes is not known.

It is also important to note that tremendous differences exist among indoor building materials, all of which have different porosities, as well as chemical compositions that affect their ability to host dirt, microbes, skin cells, hair, and other human effluents in and on which microbes survive. Specialty practices exist for introducing aerosols containing oxidants and surfactants inside built environments for the express purpose of indoor disinfection. These practices most often include, but are not limited to, the introduction of ozone, vaporized hydrogen peroxide, chlorine derivatives (vaporized hypochlorous acid and chlorine dioxide), and peracetic acid microdroplets (Boyce, 2016). Like UV irradiation, however, aerosolized chemical disinfection practices present acute exposure risks to indoor occupants that need to be carefully managed prior to utilization. Oxidizing aerosol applications in the residential sector remain limited to the remediation of large-scale water damage and use in some health care settings. These strong chemical oxidants are known to react with indoor building materials (Hubbard et al., 2009), and in the case of ozone, carbonyls are released as ozonation by-products (Poppendieck et al., 2007). Because of costs and liabilities, the health care and government building sectors will likely remain the largest users of oxidizing aerosols.

Antimicrobial Materials

The indoor built environment has a plethora of textile and nontextile surfaces constructed of a wide range of natural and synthetic materials. A significant fraction of indoor surfaces, including textiles, have incorporated antibiotics and metal nanoparticles—often silver—for the express purpose of imparting antimicrobial properties to their facade or other structural base (Chen and Schluesener, 2008). Recognizing that the association of microbes with furnishings and structural surfaces can negatively affect those in close contact, as well as the surface itself, the manufacturing of specialty surfaces incorporating broad-spectrum biocides has received increased attention for built environment design, especially in the health care sector but also in homes. Comprehensive reviews address antimicrobial finishing practices, qualitative and quantitative evaluations of antimicrobial efficacy, and methods for applying antimicrobial agents. Some of the most recent developments in antimicrobial treatment of surfaces and textiles include using various active agents, such as metal nanoparticles, quaternary ammonium salts, polyhexamethylene biguanide, triclosan, chitosan, dyes, and regenerable halamine compounds (Gao and Cranston, 2008; Hasan et al., 2013). Activating antimicrobial surfaces with such metals as copper or silver and applying liquid compounds, including biocidal paints, that confer on surfaces persistent antimicrobial activity are additional strategies that require validation and further investigation for built environment application.[4]

INTERVENTIONS TO ENCOURAGE EXPOSURE TO BENEFICIAL MICROBES

Although the focus of a variety of microbially motivated building interventions has been on reducing exposure to harmful microbes, a more recent emphasis is on encouraging exposure to potentially beneficial microbes (an idea also known as "environmental probiotics"). The nascent arena concerning beneficial microbe exposures within buildings is emerging into the following major perspectives. First, environmental probiotics may protect against colonization by and expansion of opportunistic pathogens in the environment and thus reduce human exposure to an infectious or otherwise harmful agent. This concept has initially been explored in the context of pathogen control in plumbing systems (Wang et al., 2013) and the proposed

---

[4]An antimicrobial pesticide must be registered by the U.S. Environmental Protection Agency under the Federal Insecticide, Fungicide, and Rodenticide Act (FIFRA), 7 U.S.C. § 136 et seq., prior to sale or distribution within the United States. An antimicrobial pesticide is defined as any product intended to disinfect, sanitize, reduce, or mitigate the growth or development of microbiological organisms or protect inanimate objects, industrial processes or systems, surfaces, water, or other chemical substances from contamination, fouling, or deterioration caused by bacteria, viruses, fungi, protozoa, algae, or slime. Wood preservatives and antifoulants are also classified as antimicrobial pesticides if the products have antimicrobial claims. These products are handled by EPA's Office of Pesticide Programs Antimicrobials Division. See https://www.epa.gov/pesticide-registration/antimicrobial-pesticide-registration (accessed July 27, 2017) (Communication, EPA, July 27, 2017).

use of environmental probiotic cleaning agents in the hospital environment (Caselli et al., 2016). Second, the presence of beneficial microbes in buildings may act as a source and modulator of the human microbiome, and the changes in the human microbiome thus introduced may result in the prevention or reduction of disease, although these connections remain to be fully explored (see Chapter 2).[5] The concept of environmental probiotics has gained consumer market attention, and several products are available for residential use that emit microbes claimed (by the vendor) to be beneficial. Yet, despite these promising hypotheses, the efficacy and potential drawbacks of any specific built environment probiotic have not been rigorously investigated in the peer-reviewed literature.[6] Below, two of the most compelling links between microbial exposures in buildings and protection against disease are discussed. Early work in this area suggests there may be benefits to designing, operating, and maintaining buildings to encourage exposure to microbiota from other humans, animals, plants, and biodiverse natural environments.

Indirectly Adding Microbes to Buildings: Animals in or Near Homes

A significant body of scientific literature supports the hypothesis that some microbial exposures are protective against the development of asthma (von Mutius, 2016; see also Chapter 2). In nonfarm environments, opportunities for beneficial microbial exposures may come through dog ownership. Evidence from cross-sectional epidemiology studies demonstrates that ownership of pets (especially dogs) is associated with reduced risk of allergic sensitization (Ownby et al., 2002). Fujimura and colleagues (2014) exposed mice to dust collected from homes with and without dogs. They found that exposure to dog-associated house dust resulted in a gut microbiome that was enriched in *Lactobacillus*, which in turn protected the mice against airway allergen challenge and virus infection. The presence of dogs is known to exert a strong influence on the microbiome of homes and to result in increased bacterial and fungal diversity (Dannemiller et al., 2016b; Dunn et al., 2013).

---

[5]As discussed in Chapter 2, many open questions remain. For example, accumulating evidence suggests that perinatal exposures to microorganisms are important in establishing the human microbiome in early life, but the effects of microbial exposures in adulthood are much less well understood.

[6]Environmental probiotics would require registration under FIFRA if claims of pesticidal effect were made; however, if there were no claims of pesticidal effect, the product might not need to be registered. Probiotic pesticide products would be handled by EPA's Office of Pesticide Programs Biopesticide and Pollution Prevention Division (Communication, EPA, July 27, 2017).

Increasing Outdoor Biodiversity Around Occupied Buildings

It has been estimated that 66 percent of humanity will live in cities by 2050 (UN, 2015), and current trends of increasing urban density are resulting in the loss and increased fragmentation of green spaces. Evidence suggests that human well-being in urban areas is linked to neighborhood greenness, and recent research, discussed below, suggests that this link may be driven in part by indoor exposure to the diverse microbial communities associated with plants. Questions remain, however, as to how outdoor landscape features and the presence of indoor plants influence indoor microbiome quality.

The increase in some illnesses in higher-income, urbanized societies may be associated with a trend of failing immunoregulation and poorly regulated inflammatory responses in humans. It has been hypothesized that these immune system failures are due to a lack of exposure to organisms (also sometimes called "old friends") from humankind's evolutionary past that needed to be tolerated and therefore evolved roles in driving immunoregulatory mechanisms (Rook, 2013).

Through modern living and reduced exposure to outdoor spaces, plants, and animals, humans may lose contact with the commensal microorganisms transmitted by their mother, other people, animals, and the environment (Rook et al., 2014). Hanski and colleagues (2012) explored the connections among land use, the human microbiome, and allergy risk. They used DNA sequencing technology to compare the skin microbiome and allergic disposition of adolescents living with more or less forest and agricultural land within a 3 km radius from their homes. The authors observed that in healthy individuals, greater green (vegetative) space around the home was associated with higher concentrations of skin Proteobacteria. They also found that healthy individuals had a greater diversity of Gammaproteobacteria on their skin relative to individuals with allergic sensitization. Finally, they showed that individuals' greater amounts of bacteria from the genus *Acinetobacter* (which belongs to the class Gammaproteobacteria) on the skin produced greater amounts of interleukin-10 (an anti-inflammatory cytokine that is known to increase immune tolerance). Together, these findings suggest that land use and the environmental context surrounding a building influence the human microbiome and health (see Ruokolainen et al., 2015, for further analysis of this hypothesis and dataset). Hanski (2014) interprets his research in a way that builds on the "old friends" concept, as framed by the "biodiversity hypothesis." According to that hypothesis, reduced contact of people with nature (and in particular with plants and their associated microbial communities) may adversely affect the human microbiome and immune function. (See Stamper et al., 2016, for a recent review of the topic.)

Potential interventions to increase indoor exposure to vegetation have been suggested—for example, the use of plant "biowalls" in buildings, which harbor microbial communities and have been claimed to play potential roles in filtering volatile organic gases. Such walls have aesthetic value, and investigations continue into their roles in affecting air quality (Darlington et al., 2000; Russell et al., 2014). However, the health effects of human exposure to vegetation-associated microorganisms in such walls remain unknown, and the walls also require careful design to manage the moisture they introduce into the indoor environment.

A FRAMEWORK FOR ASSESSING BUILT ENVIRONMENT INTERVENTIONS

Approaches to dealing with the myriad factors involved in the control of microbial exposures in buildings include tools for design and assessment and the development of new building technologies. There is growing evidence of benefits from interventions that reduce exposures to hazardous microbes or encourage exposures to certain beneficial ones. However, real or proposed interventions may conflict with economic, indoor air quality, energy, and other human health–related building goals or constraints. Recognized and potential negative health, energy, and economic trade-offs associated with ventilation- and cleaning-based interventions are presented below, along with holistic strategies for designing interventions that can reduce or eliminate these trade-offs.

Material Balance Modeling

Central to assessment of built environment interventions is understanding the fate and transport of and human exposure to microbes in indoor air. All buildings are unique, but the dynamics of microbes and microbial communities within the built environment are controlled by a narrower regime of mostly physical processes (reviewed by Nazaroff [2016]). With the general exception of moisture damage, which allows for bacterial and fungal growth on building materials and in floor dust (Dannemiller et al., 2016a; Mudarri and Fisk, 2007; Spengler et al., 1994), microbes in air and on surfaces of buildings likely contain relatively low levels of metabolic activity, if at all. Thus, the dynamics and assemblages of microbial communities are largely a response to physical rather than chemical gradients. These physical processes link air with surfaces and sources, and an understanding of these processes is essential for making quantitative assessments to track the effectiveness of interventions and the liabilities of the trade-offs involved. Models of these processes often result in the ability to predict indoor concentrations and human exposures. If dose-response information is available, microbial risk analysis can be added to reflect health as an endpoint more directly (Fabian et al., 2014).

While not commonly applied to microbes in buildings, dispersion modeling approaches have significant potential for use as an assessment tool. The basis of a dispersion model is a mass (or material) balance. The role of a mass balance in understanding systems was introduced in Chapter 1 and represented in Equation 5.1. It is reintroduced here as an aid to modeling microbial concentrations and movements in a built environment and providing a priori exposure estimates for impacts of different interventions. Figure 5-1 depicts a number of physical variables that are important to incorporate into dispersion models. This simplified aerosol balance at steady state shows that the amount of biomass entering from outside through ventilation plus the amount entering from outside through infiltration and the contributions from indoor "generation" sources (growth, resuspension, and shedding) is equivalent to the amount of indoor biomass that is flowing or leaking out of the building plus other indoor "losses" (deposition, decay, and removal by filtration).

Dispersion models are commonly applied to assess indoor air concentrations and exposures for particles or chemicals. Computer programs for simulating air movement and transport of airborne particle and chemical contaminants in buildings have existed since the 1980s, and they continue to be developed and applied to a range of building performance issues (e.g., CONTAM [Dols and Polidoro, 2015; see also Chapter 4]). These programs fall into two broad categories—micro and macro. Micro models can be used to predict detailed airflow patterns and airborne contaminant concentrations in a building space at scales of centimeters or less using computational fluid dynamics. Macro models simulate whole buildings, typically representing each space (e.g., room, hallway, vertical shaft) as a single node at a uniform concentration and a single pressure. Both types of models can be used to model many processes relevant to microbes, including filtration and deposition on surfaces, and they could be modified to include a broader suite of processes relevant to microbial exposure.

The application of models specific to microbes is limited largely by the lack of microbial information with which to run dispersion models for assessing exposure, limited dose-response information for microbial agents, and poor understanding of the variety of responses to microbes within a human population. Because particle size drives many of the important physical processes detailed in Figure 5-1, more information on the size distributions of indoor air microbes of importance is necessary. These limitations are most extreme for potentially beneficial microbes, as large uncertainty exists with respect to the beneficial agents, the necessary doses, and the relevant sectors of the human population.

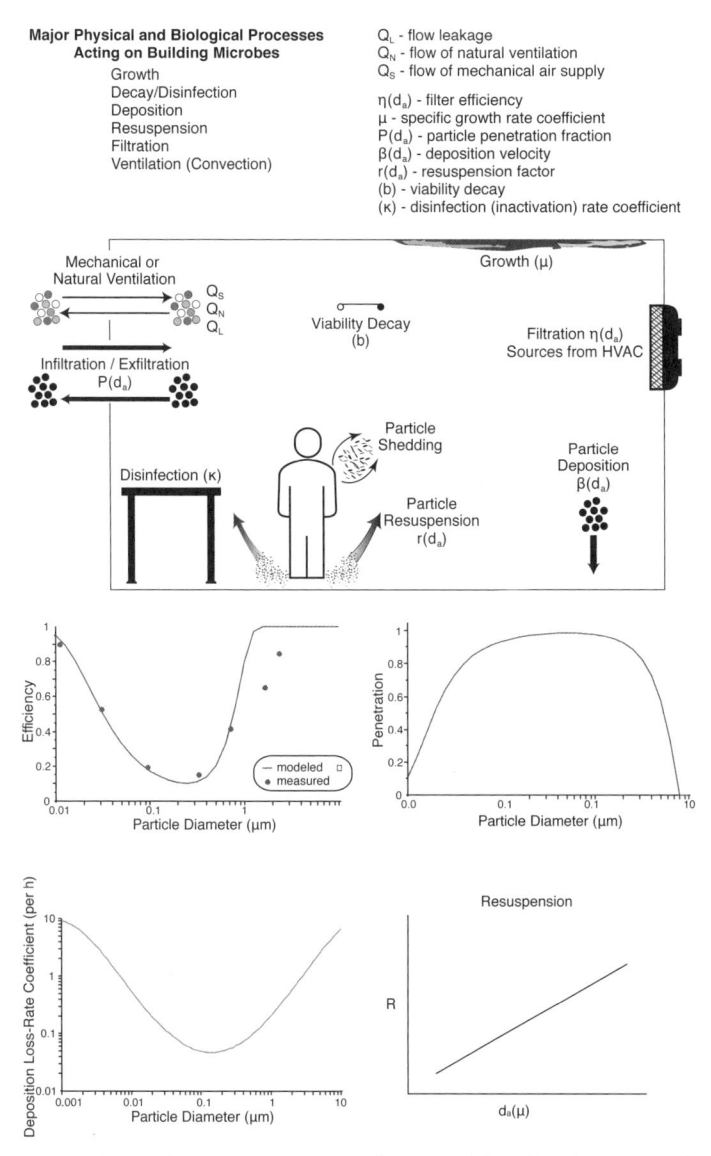

FIGURE 5-1 Physical processes govern the assembly of indoor microbial communities, with values and coefficients representing ventilation sources and losses. The term d_a is the aerodynamic diameter of a particle (or cell), and parameters that include (d_a), such as $P(d_a)$ or $\beta(d_a)$, indicate that the value of this parameter is influenced by the aerodynamic diameter of a particle.
SOURCES: Filtration efficiency, penetration efficiency, and deposition plots modified from Nazaroff (2004); data for resuspension rates and house figure from Thatcher and Layton (1995).

Balancing Ventilation and Energy Usage

Through dilution and humidity control, increased outdoor air ventilation with filtration is a common approach for removing hazardous microbes, allergens, and toxins from indoor air. A potential negative trade-off for this increased flow of outdoor air is an increase in energy consumption. When outdoor and indoor temperature and humidity conditions differ, energy in proportion to the rate at which outdoor air enters a building is required to raise or lower the temperature and adjust the water content necessary to condition the air (Persily, 2016), in accordance with

$$q_s = Q\rho C_p \Delta T + Q\rho h_w \Delta W \quad \text{(Equation 5.2)}$$

where q_s (energy/time) is the energy consumption associated with changing the outdoor ventilation air temperature and water content to the desired indoor conditions. The first term on the right-hand side of this equation is the sensible energy consumption, where Q (volume/time) is the rate at which outdoor air enters the building, ρ (mass/volume) is the outdoor air density, C_p (energy/mass per temperature difference) is the specific heat of air, and ΔT is the indoor–outdoor temperature difference. The second term on the right-hand side is the latent energy consumption, where h_w (energy/mass) is the latent heat of vaporization of water, and ΔW is the indoor–outdoor humidity ratio difference (ASHRAE, 2013). In addition, fan energy is required to provide ventilation air in mechanically ventilated buildings and to overcome pressure drops across filters. In mechanically ventilated buildings, the relationship between ventilation and energy consumption is complex and depends on the type of HVAC system and the associated control and operating strategies, and it can be evaluated only through detailed simulation or measurement efforts. In U.S. Mid-Atlantic states, for example, elevated energy consumption for ventilation is associated with heating requirements in the winter and dehumidification and cooling in the summer, while spring and fall operations can make use of economizers to introduce outdoor air that requires less conditioning during these seasons. Depending on outdoor climate and conditions and the use of strategies that include economizers, increased building envelope tightness, heat recovery ventilation, and demand-controlled ventilation, increasing outdoor ventilation rates to a level that improves indoor air quality in general and reduces exposure to indoor-generated microbial contaminants does not necessarily come at significantly increased energy or capital costs (Persily and Emmerich, 2012). The economic and energy costs of ventilation and the balance between these costs and benefits have not been rigorously studied.

Balancing Ventilation and Outdoor Air Quality

Growing evidence from the biodiversity hypothesis indicates a potential health benefit of human exposure to a diverse cohort of microorganisms in outdoor air. Such exposures may be facilitated through increased unfiltered outdoor air ventilation. An unintentional consequence of the increased use of mechanical ventilation with particle filtration is a shift in the indoor air microbiome away from the outdoor microbiome (Kembel et al., 2012; Meadow et al., 2014). Some limitations may be inherent in ventilation as a result of its nonselective nature. Depending on the locations and sensitivities of occupants, outdoor air may introduce into occupied spaces ambient or localized air quality hazards in both particulate matter and gas phase, as well as allergenic fungi and pollen. In areas with poor outdoor air quality, increased ventilation introduces biologic particles (associated with crustal materials), chemicals, and particulate matter into the built environment. Filtration of incoming outdoor air may reduce the influence of hazardous outdoor particles, but it will not remove gases and may also eliminate potentially beneficial outdoor microbes associated with an outdoor biodiverse environment. Enhanced filtration of outdoor air, especially to reduce $PM_{2.5}$ exposure, requires the use of filters with higher MERV ratings and can also incur significant capital and operational costs. However, lower-pressure-drop, high-efficiency filtration technology is being developed that reduces the associated operating costs. Recognizing the inherent link between outdoor and indoor air, further efforts to produce good indoor air quality will be influenced by efforts and legislation aimed at improving outdoor ambient air quality.

Balancing Microbial Removal with Exposure to UV Radiation and Cleaning Products

As noted earlier, several trade-offs are associated with efforts to kill or remove hazardous microbes in the built environment. A primary concern has been exposure to cleaning chemicals, which are common sources of VOCs in indoor air. Both so-called green and conventional cleaning products emit primary chemical aerosols and may also result in secondary aerosol formation (Nazaroff and Weschler, 2004). A second concern is the presence of antibiotics or antimicrobial chemicals in many consumer cleaning products and the risk of producing resistant bacterial strains (Aiello and Larson, 2003). Moreover, as noted previously in the discussion of UV germicidal air and surface treatments, these technologies present radiation exposure or ozone generation hazards, respectively, for occupants, and systems must be designed and operated to manage these risks.

An emerging concern with respect to cleaning and disinfection practices is the unintentional removal of beneficial bacteria and other microbes. The tools used for these purposes are imprecise and provide no opportunity to remove pathogens selectively without also removing microbial "old friends" that may be present or reducing diversity (Rook et al., 2014).

Expanding Models to Include Health Outcomes, Economics, and Energy

The output of the dispersion models described earlier in this chapter includes indoor air concentrations and human exposures. However, approaches to designing sustainable built environment interventions will not be derived through standard engineering practices focused on the optimization of a single variable. To truly capture the complex nature of interventions and associated trade-offs, these models for microbes need to be expanded to consider health, energy, and economics. The strength of this broader approach is the potential to link building characteristics quantitatively to health outcomes and compare costs of health care and interventions. Recently, for example, indoor airflow and contaminant dispersion models (CONTAM) have been coupled with energy evaluation models (EnergyPlus) to capture interdependencies between airflow and heat transfer and thereby directly link indoor air quality and energy analyses (Dols et al., 2016).

Important considerations include the socioeconomic status of home/building occupants and unique environments. Buildings are commonly private, and there are no regulations concerning biological exposures to drive research and practice. While prevention of infection from drinking water and food is regulated and expected by the general public, few expectations are focused on transmission of infectious diseases or exacerbation of allergenic disease in buildings. The enthusiasm and organization of efforts to develop rational interventions are thereby limited. In addition, residents who rent or are of low socioeconomic status will have a limited ability to pay for interventions or innovative technologies. Such environments as schools, health care facilities, and heavily water-damaged buildings may require specialized approaches and trade-offs, which likely will be different for different building types. For example, ventilation with efficient filtration in hospitals is necessary to reduce the incidence of hospital-associated infection. In hospitals, but perhaps not homes with young children, these needs may override concerns about encouraging biodiverse microbial exposures.

SUMMARY OBSERVATIONS AND KNOWLEDGE GAPS

Summary Observations

A number of potential physical and chemical interventions that can significantly affect indoor environmental quality can be undertaken in the built environment. The discussion in this chapter has focused particularly on examples of interventions that affect indoor air.

Buildings with tight envelopes and well-designed mechanical or engineered natural ventilation provide more potential to control and modulate microbial exposures relative to buildings ventilated by unintentional air leakage. The design of effective ventilation-based interventions will require a greater understanding of the potential (and limitations) of these ventilation schemes in modulating indoor microbial exposures and risks.

Understanding and improving the environment outside of buildings should be considered part of intervention design. The onus for improved indoor environmental quality cannot reside solely with building operations, particularly because air and water from outside sources are important inputs to indoor environments. Instead of reliance on the building envelope to control all exposures, researchers will need to consider the broader perspectives for meeting the diverse goals of improving microbial indoor air quality, recognizing which goals cannot be achieved absent good outdoor air quality.

Critical guidance is lacking on when to initiate interventions for damp buildings and on how to gauge the success of these interventions. Built environment interventions are most commonly proposed or undertaken with a goal of reducing exposure to microorganisms that may have negative health effects. As discussed in prior chapters, the presence of dampness, water damage, and visible mold in buildings is associated with negative respiratory health effects, and as a result, remediation aimed at drying and removing building materials affected by these conditions from indoor environments is often considered. Significant questions remain about what constitutes normal microbial ecology in different building types and under different conditions. This information will be needed to delineate impacted versus normal microbial ecologies to enable understanding and assessing the impact of interventions.

Interventions for promoting human exposure to beneficial microbes are in a nascent stage. Research continues to explore whether microbes with beneficial effects can be identified; how or whether exposures to these microorganisms in the built environment are associated with various types of health impacts; and the building conditions that promote, hinder, or alter these exposures.

Quantitative frameworks provide valuable insights to support the design of interventions and to understand the trade-offs among potentially competing priorities. Such frameworks can be used to better understand the anticipated effects of interventions and to aid in the design of intervention approaches. To design and deploy interventions that promote human and environmental health, such frameworks as building airflow and contaminant transport models, risk analyses, and building energy models will need to be linked systematically and holistically to infrastructure design and occupant health data.

Knowledge Gaps

On the basis of the above summary observations and the information developed in this chapter on interventions in the built environment, the committee identified the following goals for research to address knowledge gaps and advance the field:

1. **Improve understanding of "normal" microbial ecology in buildings of different types and under different conditions.** This information provides important input into assessment of and decision making about potential interventions. In the case of damp building remediation to alter mold exposures, for example, if remediation goals are to fit within a "normal fungal ecology" standard, research will be required to define the normal fungal (and other microbial) ecologies for buildings within specific geographic areas and climates. DNA sequencing and the associated bioinformatics may be well suited to assessing the natural microbial ecology of buildings.
2. **Further explore the concept of interventions that promote exposure to beneficial microorganisms, and whether and under what circumstances these might promote good health.** Provided that microbes with beneficial health effects are identified, additional research will be needed both to determine whether these microbes, while in the built environment, can be transmitted to humans and impact the human microbiome and to design and test interventions that encourage and control exposure to these microbes in buildings.
3. **Obtain additional data necessary to support the use of a variety of quantitative frameworks for understanding and assessing built environment interventions.** A variety of information will need to be incorporated into models that link building and microbial information systematically and holistically to additional design, energy, environmental, and health data. To apply dispersion and risk models effectively for microbes, more research will be required to understand the size distributions of health-relevant microbiota; more de-

tailed information will be needed on the emission rates of microbes in buildings; and clear dose-response information will be needed for health-relevant microorganisms. To understand the economic implications of interventions, more research will be required to determine such parameters as health care costs of microbe-associated disease (or prevention of disease via beneficial microbes) and the energy, greenhouse gas, and other implications of interventions, such as the energy trade-offs associated with increasing outdoor air ventilation to control infectious disease.

REFERENCES

Adams, R. I., M. Miletto, J. W. Taylor, and T. D. Bruns. 2013. Dispersal in microbes: Fungi in indoor air are dominated by outdoor air and show dispersal limitation at short distances. *The ISME Journal* 7(7):1751-7362.

Aiello, A. E., and E. Larson. 2003. Antibacterial cleaning and hygiene products as an emerging risk factor for antibiotic resistance in the community. *The Lancet Infectious Diseases* 3(8):501-506.

AIHA (American Industrial Hygiene Association). 2013. *Position statement on mold and dampness in the built environment.* https://www.aiha.org/government-affairs/PositionStatements/P-Mold-03-26-13.pdf (accessed July 18, 2017).

Amend, A. S., K. A. Seifert, R. Samson, and T. D. Bruns. 2010. Indoor fungal composition is geographically patterned and more diverse in temperate zones than in the tropics. *Proceedings of the National Academy of Sciences of the United States of America* 107(31):13748-13753.

Anderson, D. J., L. F. Chen, D. J. Weber, R. W. Moehring, S. S. Lewis, P. F. Triplett, M. Blocker, P. Becherer, J. C. Schwab, L. P. Knelson, Y. Lokhnygina, W. A. Rutala, H. Kanamori, M. F. Gergen, and D. J. Sexton. 2017. Enhanced terminal room disinfection and acquisition and infection caused by multidrug-resistant organisms and Clostridium difficile (the Benefits of Enhanced Terminal Room Disinfection study): A cluster-randomised, multicentre, crossover study. *The Lancet* 389(10071):805-814.

ANSI/IICRC (American National Standards Institute/Institute of Inspection, Cleaning and Restoration Certification). 2015. *S520: 2015 Reference guide for professional mold remediation.* Las Vegas, NV: IICRC.

Arundel, A. V., E. M. Sterling, J. H. Biggin, and T. D. Sterling. 1986. Indirect health effects of relative humidity in indoor environments. *Environmental Health Perspectives* 65:351-361.

ASHRAE (American Society of Heating, Refrigerating and Air-Conditioning Engineers). 2010. *ANSI/ASHRAE Standard 62.1-2010. Ventilation for acceptable indoor air quality.* Atlanta, GA: ASHRAE.

ASHRAE. 2013. *Fundamentals handbook.* Atlanta, GA: ASHRAE.

ASHRAE. 2016. Ultraviolet air and surface treatment proceedings. In *TC 2.9.* https://tc0209.ashraetcs.org (accessed May 1, 2017).

ASHRAE. 2017. Standard 52.2: Method of Testing General Ventilation Air-Cleaning Devices for Removal Efficiency by Particle Size (ANSI Approved). Atlanta, GA: ASHRAE.

Azimi, P., D. Zhao, and B. Stephens. 2014. Estimates of HVAC filtration efficiency for fine and ultrafine particles of outdoor origin. *Atmospheric Environment* 98:337-346.

Bhangar, S., J. A. Huffman, and W. W. Nazaroff. 2014. Size-resolved fluorescent biological aerosol particle concentrations and occupant emissions in a university classroom. *Indoor Air* 24(6):604-617.

Bornehag, C. G., J. Sundell, L. Hägerhed-Engman, and T. Sigsgaard. 2005. Association between ventilation rates in 390 Swedish homes and allergic symptoms in children. *Indoor Air* 15(4):275-280.

Boyce, J. M. 2016. Modern technologies for improving cleaning and disinfection of environmental surfaces in hospitals. *Antimicrobial Resistance & Infection Control* 5(1):10.

Brickner, P. W., R. L. Vincent, M. First, E. Nardell, M. Murray, and W. Kaufman. 2003. The application of ultraviolet germicidal irradiation to control transmission of airborne disease: Bioterrorism countermeasure. *Public Health Reports* 118(2):99-114.

Caselli, E., M. D'Accolti, A. Vandini, L. Lanzoni, M. T. Camerada, M. Coccagna, A. Branchini, P. Antonioli, P. G. Balboni, D. Di Luca, and S. Mazzacane. 2016. Impact of a probiotic-based cleaning intervention on the microbiota ecosystem of the hospital surfaces: Focus on the resistome remodulation. *PLOS ONE* 11(2):e0148857.

Chen, X., and H. J. Schluesener. 2008. Nanosilver: A nanoproduct in medical application. *Toxicology Letters* 176(1):1-12.

CIBSE (Chartered Institution of Building Services Engineers). 2005. *Natural ventilation in non-domestic buildings*. CIBSE application manual AM10. London, UK: CIBSE.

Dannemiller, K. C., M. J. Mendell, J. M. Macher, K. Kumagai, A. Bradman, N. Holland, K. Harley, B. Eskenazi, and J. Peccia. 2014. Next-generation DNA sequencing reveals that low fungal diversity in house dust is associated with childhood asthma development. *Indoor Air* 24(3):236-247.

Dannemiller, K. C., C. J. Weschler, and J. Peccia. 2016a. Fungal and bacterial growth in floor dust at elevated relative humidity levels. *Indoor Air* 27(2):354-363.

Dannemiller, K. C., J. F. Gent, B. P. Leaderer, and J. Peccia. 2016b. Influence of housing characteristics on bacterial and fungal communities in homes of asthmatic children. *Indoor Air* 26(2):179-192.

Darlington A., M. Chan, D. Malloch, C. Pilger, and M. A. Dixon. 2000. The biofiltration of indoor air: Implications for air quality. *Indoor Air* 10(1):39-46.

Dols, W. S., and B. J. Polidoro. 2015. *CONTAM user guide and program documentation*. NIST Technical Note 1887. Gaithersburg, MD: National Institute of Standards and Technology.

Dols, W. S., S. J. Emmerich, and B. J. Polidoro. 2016. Coupling the multizone airflow and contaminant transport software CONTAM with EnergyPlus using co-simulation. *Building Simulation* 9(4):469-479.

Donskey, C. J. 2013. Does improving surface cleaning and disinfection reduce health care-associated infections? *American Journal of Infection Control* 41(5, Suppl.):S12-S19.

Dunn, R. R., N. Fierer, J. B. Henley, J. W. Leff, and H. L. Menninger. 2013. Home life: Factors structuring the bacterial diversity found within and between homes. *PLOS ONE* 8(5):e64133.

Ege, M. J., M. Mayer, A.-C. Normand, J. Genuneit, W. O. C. M. Cookson, C. Braun-Fahrländer, D. Heederik, R. Piarroux, and E. von Mutius. 2011. Exposure to environmental microorganisms and childhood asthma. *New England Journal of Medicine* 364(8):701-709.

EPA (U.S. Environmental Protection Agency). 2008. *Critical assessment of building air cleaner technologies*. EPA/600/R-08/053. Washington, DC: EPA.

Escombe, A. Roderick, D. A. J. Moore, R. H. Gilman, M. Navincopa, E. Ticona, B. Mitchell, C. Noakes, C. Martínez, P. Sheen, R. Ramirez, W. Quino, A. Gonzalez, J. S. Friedland, and C. A. Evans. 2009. Upper-room ultraviolet light and negative air ionization to prevent tuberculosis transmission. *PLOS Medicine* 6(3):e1000043.

Fabian, M. P., G. Adamkiewicz, N. K. Stout, M. Sandel, and J. I. Levy. 2014. A simulation model of building intervention impacts on indoor environmental quality, pediatric asthma, and costs. *Journal of Allergy and Clinical Immunology* 133(1):77-84.

Fisk, W. J. 2013. Health benefits of particle filtration. *Indoor Air* 23(5):357-368.

Fujimura, K. E., T. Demoor, M. Rauch, A. A. Faruqi, S. Jang, C. C. Johnson, H. A. Boushey, E. Zoratti, D. Ownby, N. W. Lukacs, and S. V. Lynch. 2014. House dust exposure mediates gut microbiome Lactobacillus enrichment and airway immune defense against allergens and virus infection. *Proceedings of the National Academy of Sciences of the United States of America* 111(2):805-810.

Gao, Y., and R. Cranston. 2008. Recent advances in antimicrobial treatments of textiles. *Textile Research Journal* 78(1):60-72.

Grant, C., C. A. Hunter, B. Flannigan, and A. F. Bravery. 1989. The moisture requirements of moulds isolated from domestic dwellings. *International Biodeterioration* 2(4):259-284.

Hanski, I. 2014. Biodiversity, microbes and human well-being. *Ethics in Science and Environmental Politics* 14(1):19-25.

Hanski, I., L. von Hertzen, N. Fyhrquist, K. Koskinen, K. Torppa, T. Laatikainen, P. Karisola, P. Auvinen, L. Paulin, M. J. Mäkelä, E. Vartiainen, T. U. Kosunen, H. Alenius, and T. Haahtela. 2012. Environmental biodiversity, human microbiota, and allergy are interrelated. *Proceedings of the National Academy of Sciences of the United States of America* 109(21):8334-8339.

Hasan, J., R. J. Crawford, and E. P. Ivanova. 2013. Antibacterial surfaces: The quest for a new generation of biomaterials. *Trends in Biotechnology* 31(5):295-304.

Hospodsky, D., H. Yamamoto, W. W. Nazaroff, and J. Peccia. 2014 (unpublished). *Size-resolved, baseline characterization data for indoor emission rates of bacteria, fungi, and particulate matter.*

Hospodsky, D., N. Yamamoto, W. W. Nazaroff, D. Miller, S. Gorthala, and J. Peccia. 2015. Characterizing airborne fungal and bacterial concentrations and emission rates in six occupied children's classrooms. *Indoor Air* 2(6):641-652.

Hubbard, H., D. Poppendieck, and R. L. Corsi. 2009. Chlorine dioxide reactions with indoor materials during building disinfection: Surface uptake. *Environmental Science & Technology* 43(5):1329-1335.

IOM (Institute of Medicine). 2004. *Damp indoor spaces and health.* Washington, DC: The National Academies Press.

Kembel, S. W., E. Jones, J. Kline, D. Northcutt, J. Stenson, A. M. Womack, B. J. M. Bohannan, G. Z. Brown, and J. L Green. 2012. Architectural design influences the diversity and structure of the build environment. *The ISME Journal* 6(8):1469-1479.

Kembel, S. W., J. F. Meadow, T. K. O'Connor, G. Mhuireach, D. Northcutt, J. Kline, M. Moriyama, G. Z. Brown, B. J. M. Bohannan, and J. L. Green. 2014. Architectural design drives the biogeography of indoor bacterial communities. *PLOS ONE* 9(1):e87093.

Kercsmar, C. M., D. G. Dearborn, M. Schluchter, L. Xue, H. L. Kirchner, J. Sobolewski, S. J. Greenberg, S. J. Vesper, and T. Allan. 2006. Reduction in asthma morbidity in children as a result of home remediation aimed at moisture sources. *Environmental Health Perspectives* 114(10):1574-1580.

Kovesi, T., C. Zaloum, C. Stocco, D. Fugler, R. E. Dales, A. Ni, N. Barrowman, N. L. Gilbert, and J. D. Miller. 2009. Heat recovery ventilators prevent respiratory disorders in Inuit children. *Indoor Air* 19(6):489-499.

Levin, J., L. S. Riley, C. Parrish, D. English, and S. Ahn. 2013. The effect of portable pulsed xenon ultraviolet light after terminal cleaning on hospital-associated Clostridium difficile infection in a community hospital. *American Journal of Infection Control* 41(8):746-748.

Lindsley, W. G., F. M. Blachere, R. E. Thewlis, A. Vishnu, K. A. Davis, G. Cao, J. E. Palmer, K. E. Clark, M. A. Fisher, R. Khakoo, and D. H. Beezhold. 2010. Measurements of airborne influenza virus in aerosol particles from human coughs. *PLOS ONE* 5(11):e15100.

Luongo, J. C., J. Brownstein, and S. L. Miller. 2017. Ultraviolet germicidal coil cleaning: Impact on heat transfer effectiveness and static pressure drop. *Building and Environment* 112:159-165.

Lynch, S. V., R. A. Wood, H. Boushey, L. B. Bacharier, G. R. Bloomberg, M. Kattan, G. T. O'Connor, M. T. Sandel, A. Calatroni, E. Matsui, C. C. Johnson, H. Lynn, C. M. Visness, K. F. Jaffee, P. J. Gergen, D. R. Gold, R. J. Wright, K. Fujimura, M. Rauch, W. W. Busse, and J. E. Gern. 2014. Effects of early-life exposure to allergens and bacteria on recurrent wheeze and atopy in urban children. *Journal of Allergy and Clinical Immunology* 134(3):593-601. doi:10.1016/j.jaci.2014.04.018.

McLean, R. L. 1961. The mechanism of spread of Asian influenza: General discussion. *American Review of Respiratory Disease* 83(2P2):36-38.

Meadow, J. F., A. E. Altrichter, S. W. Kembel, J. Kline, G. Mhuireach, M. Moriyama, D. Northcutt, T. K. O'Connor, A. M. Womack, G. Z. Brown, J. L Green, and B. J. M. Bohannan. 2014. Indoor airborne bacterial communities are influenced by ventilation, occupancy, and outdoor air source. *Indoor Air* 24(1):41-48.

Mendell, M. J., A. G. Mirer, K. Cheung, M. Tong, and J. Douwes. 2011. Respiratory and allergic health effects of dampness, mold, and dampness-related agents: A review of the epidemiologic evidence. *Environmental Health Perspectives* 119(6):748-756.

Menzies, D., and J. Bourbeau. 1997. Building-related illnesses. *New England Journal of Medicine* 337(21):1524-1531.

Menzies, D., J. Popa, J. A. Hanley, T. Rand, and D. K. Milton. 2003. Effect of ultraviolet germicidal lights installed in office ventilation systems on workers' health and wellbeing: Double-blind multiple crossover trial. *The Lancet* 362(9398):1785-1791.

Miller, R., S. Simmons, C. Dale, J. Stachowiak, and M. Stibich. 2015. Utilization and impact of a pulsed-xenon ultraviolet room disinfection system and multidisciplinary care team on Clostridium difficile in a long-term acute care facility. *American Journal of Infection Control* 43(12):1350-1353.

Mudarri, D., and W. J. Fisk. 2007. Public health and economic impact of dampness and mold. *Indoor Air* 17(3):226-235.

Nagaraja, A., P. Visintainer, J. P. Haas, and J. Menz, G. P. Wormser, and M. A. Montecalvo. 2015. Clostridium difficile infections before and during use of ultraviolet disinfection. *American Journal of Infection Control* 43(9):940-945.

Nazaroff, W. W. 2004. Indoor particle dynamics. *Indoor Air* 14(S7):175-183.

Nazaroff, W. W. 2016. Indoor bioaerosol dynamics. *Indoor Air* 26(1):61-78.

Nazaroff, W. W., and C. J. Weschler. 2004. Cleaning products and air fresheners: Exposure to primary and secondary air pollutants. *Atmospheric Environment* 38(18):2841-2865.

Ownby, D. R., C. Johnson, and E. L. Peterson. 2002. Exposure to dogs and cats in the first year of life and risk of allergic sensitization at 6 to 7 years of age. *Journal of the American Medical Association* 288(8):963-972.

Pastuszka, J. S., U. Kyaw Tha Paw, D. O. Lis, A. Wlazło, and K. Ulfig. 2000. Bacterial and fungal aerosol in indoor environment in Upper Silesia, Poland. *Atmospheric Environment* 34(22):3833-3842.

Peccia, J., H. M. Werth, S. Miller, and M. Hernandez. 2001. Effects of relative humidity on the ultraviolet induced inactivation of airborne bacteria. *Aerosol Science and Technology* 35(3):728-740.

Persily, A. K. 2016. Field measurement of ventilation rates. *Indoor Air* 26(1):97-111.

Persily, A. K., and S. J. Emmerich. 2012. Indoor air quality in sustainable, energy efficient buildings. *HVAC&R Research* 18(1-2):4-20.

Poppendieck, D. G., H. F. Hubbard, C. J. Weschler, and R. L. Corsi. 2007. Formation and emissions of carbonyls during and following gas-phase ozonation of indoor materials. *Atmospheric Environment* 41(35):7614-7626.
Prussin, A. J., and L. C. Marr. 2015. Sources of airborne microorganisms in the built environment. *Microbiome* 3(1):78.
Qian, J., D. Hospodsky, N. Yamamoto, W. W. Nazaroff, and J. Peccia. 2012. Size-resolved emission rates of airborne bacteria and fungi in an occupied classroom. *Indoor Air* 22(4):339-351.
Reed, N. G. 2010. The history of ultraviolet germicidal irradiation for air disinfection. *Public Health Reports* 125(1):15-27.
Reponen, T., S. Vesper, L. Levin, E. Johansson, P. Ryan, J. Burkle, S. A. Grinshpun, S. Zheng, D. I. Bernstein, J. Lockey, M. Villareal, G. K. Hershey, and G. LeMasters. 2011. High environmental relative moldiness index during infancy as a predictor of asthma at 7 years of age. *Annals of Allergy, Asthma & Immunology* 107(2):120-126.
Roberts, J. W., W. S. Clifford, G. Glass, and P. G. Hummer. 1999. Reducing dust, lead, dust mites, bacteria, and fungi in carpets by vacuuming. *Archives of Environmental Contamination and Toxicology* 36(4):477-484.
Rook, G. A. W. 2013. Regulation of the immune system by biodiversity from the natural environment: An ecosystem service essential to health. *Proceedings of the National Academy of Sciences of the United States of America* 110(46):18360-18367.
Rook, G. A. W., C. L. Raison, and C. A. Lowry. 2014. Microbial "old friends," immunoregulation and socioeconomic status. *Clinical & Experimental Immunology* 177(1):1-12.
Ruokolainen, L., L. von Hertzen, N. Fyhrquist, T. Laatikainen, J. Lehtomäki, P. Auvinen, A. M. Karvonen, A. Hyvärinen, V. Tillmann, O. Niemelä, M. Knip, T. Haahtela, J. Pekkanen, and I. Hanski. 2015. Green areas around homes reduce atopic sensitization in children. *Allergy* 70(2):195-202.
Russell, J. A., Y. Hu, L. Chau, M. Pauliushchyk, I. Anastopoulos, S. Anandan, and M. S. Waring. 2014. Indoor-biofilter growth and exposure to airborne chemicals drive similar changes in plant root bacterial communities. *Applied and Environmental Microbiology* 80(16):4805-4813.
Schulze, T., and U. Eicker. 2013. Controlled natural ventilation for energy efficient buildings. *Energy and Buildings* 56:221-232.
Seppänen, O., and W. J. Fisk. 2002. Association of ventilation system type with SBS symptoms in office workers. *Indoor Air* 12(2):98-112.
Sliney, D. 2013. Balancing the risk of eye irritation from UV-C with infection from bioaerosols. *Photochemistry and Photobiology* 89(4):770-776.
Spengler, J., L. Neas, S. Nakai, D. Dockery, F. Speizer, J. Ware, and M. Raizenne. 1994. Respiratory symptoms and housing characteristics. *Indoor Air* 4(2):72-82.
Stamper, C. E., A. J. Hoisington, O. M. Gomez, A. L. Halweg-Edwards, D. G. Smith, K. L. Bates, K. A. Kinney, T. T. Postolache, L. A. Brenner, G. A. W. Rook, and C. A. Lowry. 2016. Chapter 14. The microbiome of the built environment and human behavior: Implications for emotional health and well-being in postmodern western societies. In *International review of neurobiology*, edited by J. F. Cryan and G. Clarke. Cambridge, MA: Academic Press. Pp. 289-323.
Stephens, B., and J. A. Siegel. 2012. Penetration of ambient submicron particles into single-family residences and associations with building characteristics. *Indoor Air* 22(6):501-513.
Sundell, J., M. Wickman, G. Pershagen, and S. L. Nordvall. 1995. Ventilation in homes infested by house-dust mites. *Allergy* 50(2):106-112.

Sundell, J., H. Levin, W. W. Nazaroff, W. S. Cain, W. J. Fisk, D. T. Grimsrud, F. Gyntelberg, Y. Li, A. K. Persily, A. C. Pickering, J. M. Samet, J. D. Spengler, S. T. Taylor, and C. J. Weschler. 2011. Ventilation rates and health: Multidisciplinary review of the scientific literature. *Indoor Air* 21(3):191-204.

Thatcher, T. L., and D. W. Layton. 1995. Deposition, resuspension, and penetration of particles within a residence. *Atmospheric Environment* 29(13):1487-1497.

UN (United Nations). 2015. *World urbanization prospects: The 2014 revision.* ST/ESA/SER.A/366. New York: UN, Department of Economic and Social Affairs, Population Division. https://esa.un.org/unpd/wup/Publications/Files/WUP2014-Report.pdf (accessed July 20, 2017).

von Mutius, E. 2016. The microbial environment and its influence on asthma prevention in early life. *Journal of Allergy and Clinical Immunology* 137(3):680-689.

Wang, H., M. A. Edwards, J. O. Falkinham, and A. Pruden. 2013. Probiotic approach to pathogen control in premise plumbing systems? A review. *Environmental Science & Technology* 47(18):10117-10128.

Wang, Y.., C. Sekhar, W. P. Bahnfleth, K. W. Cheong, and J. Firrantello. 2016. Effectiveness of an ultraviolet germicidal irradiation system in enhancing cooling coil energy performance in a hot and humid climate. *Energy and Buildings* 130:321-329.

Weber, D. J., H. Kanamori, and W. A. Rutala. 2016. "No touch" technologies for environmental decontamination: Focus on ultraviolet devices and hydrogen peroxide systems. *Current Opinion in Infectious Diseases* 29(4):424-431.

Xu, P., J. Peccia, P. Fabian, J. W. Martyny, K. P. Fennelly, M. Hernandez, and S. L. Miller. 2003. Efficacy of ultraviolet germicidal irradiation of upper-room air in inactivating airborne bacterial spores and mycobacteria in full-scale studies. *Atmospheric Environment* 37(3):405-419.

Yamamoto, N., W. W. Nazaroff, and J. Peccia. 2014. Assessing the aerodynamic diameters of taxon-specific fungal bioaerosols by quantitative PCR and next-generation DNA sequencing. *Journal of Aerosol Science* 78:1-10.

Yang, W., S. Elankumaran, and L. C. Marr. 2011. Concentrations and size distributions of airborne influenza A viruses measured indoors at a health centre, a day-care centre and on aeroplanes. *Journal of The Royal Society Interface* 8(61):1176-1184.

Yang, W., S. Elankumaran, and L. C. Marr. 2012. Relationship between humidity and influenza A viability in droplets and implications for influenza's seasonality. *PLOS ONE* 7(10):e46789.

6

Moving Forward: A Vision for the Future and Research Agenda

Research on the communities of bacteria, viruses, fungi, and other microbes found in built environments illustrates the multiplicity of interactions among indoor microbiomes, their influences on human occupants, and how choices about building design and operation affect microbial communities. Prior chapters in this report have summarized lessons learned from research efforts in this field that extend back decades. Yet, knowledge in many of these areas remains incomplete. How could indoor microbiomes be changed and their impact on occupants be enhanced or reduced if advancing microbial knowledge could be translated into practical application? Since knowledge about the interactions among indoor microbiomes, human occupants, and built environments is not yet at an actionable level, this chapter lays out a vision for the future of buildings informed by microbial understanding and provides a research agenda for making progress toward achieving this vision.

A VISION FOR THE FUTURE OF THE FIELD: MICROBIOME-INFORMED BUILT ENVIRONMENTS

In the committee's vision for the future, greater understanding of indoor environments will result in buildings that support a more productive, healthier population at lower cost and with reduced impacts on the outdoor environment. This reality could be achieved by harnessing current and future knowledge about the relationships among the built environments, microbial communities, and human occupants and applying it through improved practice.

This vision, which is still far from being a reality, is not to be interpreted as direct recommendations or conclusions. But it is driven by a desire to design, construct, and operate buildings that support occupant health and well-being while promoting sustainability and resilience. The vision also takes into account several trends that will continue to impact building design and operation, including climate change; aging building stock; increasing urbanization; the adaptive reuse of existing buildings; and the increasing use of chemicals indoors, including antibiotics and antimicrobials. Buildings of the future that emphasize sustainability and overall population health will be well positioned in a world that addresses the effects of climate at regionally distinct levels, incorporating a variety of technological changes. These trends affect many of the trade-offs discussed in Chapter 5—for example, the need for buildings that can adapt to changing outdoor conditions; requirements to improve the energy efficiency of buildings; and the advantages and disadvantages of natural, mechanical, and hybrid ventilation systems. As a result, future buildings will reflect attempts to optimize occupant health, energy use, and other features in concert, based on a thorough consideration of public health implications. They will need to accommodate occupant comfort and preferences for individual environmental control (Boerstra, 2016), as well as draw on new technologies that support predictive and adaptive management of building conditions, including reactions to changing outdoor conditions. Future buildings also will need to reflect the economic realities that underpin decision making about building design and operation.

To design these future buildings, a number of challenges will need to be overcome. These are multidimensional challenges that will not be simple to address and that will likely require significant investment and buy-in from both public and private entities. The committee's vision for the future includes the components detailed below.

Researchers will have a much deeper understanding of the effects of indoor microbial communities on human health, and the connections among exposure, response, and health outcomes will be established. Scientists and practitioners will know how environmental microbial exposures result in physiologic responses linked to significant health impacts and will be able to quantify how these exposures are connected, in turn, to particular physical, chemical, and biological conditions of the built environment. They also will have gained greater understanding of which types of people, in which types of indoor settings, experience adverse or beneficial health effects as a result of particular exposures. Refinements of this understanding will be important for shaping guidance that can have broad application while accounting for the existence of significant individual variability.

The growth, establishment, and evolution of indoor microbiomes will be better understood. Understanding of the behavior and functions of indoor microbial communities, as well as the factors that affect their proliferation and activities, will improve. Scientists will have a concrete understanding of which building design and operation choices impact indoor microbiome dynamics and how these choices impact microbiomes positively or negatively from a building health and human health perspective.

The sources of microorganisms in buildings that impact human health and well-being positively or negatively will have been identified and understood. Not only will potentially beneficial and harmful microbes present in indoor environments have been identified, but their differential effects on humans of various ages, sexes, and health status will also be understood to inform building engineers and facilities managers about how to operate and maintain buildings to promote the health of their occupants. Once this knowledge base has been achieved, it will be possible to design, construct, operate, and maintain buildings in a manner that will reduce harmful microbes of concern while supporting building sustainability and health goals. For example, design features will be implemented to manage water, airflow, and light so as to allow for an abundance of beneficial microbes and a reduction in microbes with negative impacts. To prevent detrimental microbial effects associated with dampness, future buildings will include features to minimize or mitigate water damage. Many of these features—such as providing accessibility to concealed spaces that may be sources of microbes or employing enclosure assemblies that minimize condensation—are available with today's technology but are not consistently incorporated into buildings. Similarly, heating, ventilation, and air conditioning (HVAC) systems will be designed with features that reduce condensation and water accumulation, as well as with improved air filtration and reliable outside air ventilation for controlling humidity and indoor pollutants.

Advanced technologies that facilitate indoor environmental quality and energy efficiency will have been developed, installed, and embraced in building operation. Buildings of the future will incorporate improved technologies to support building operation, including sensing and self-actuated maintenance. To this end, detection and response strategies will be required in both unconditioned and occupied spaces, because the former spaces can have very different environmental conditions and microbial communities but still be well connected to occupied spaces. Specific examples include the incorporation of sensors to detect water penetration, coupled with new or modified materials capable of self-sealing to minimize leaks. Sensors for and responses to the performance of air and water filtration, occupant density, and indoor and outdoor air and water quality, or sensors to identify situations in which outdoor air is more healthful than indoor air, coupled with automatic controls to change airflow to optimize indoor environmental quality, will also be important. Finally, sensors that sample air, water, dust, and other media to determine concentrations of potentially beneficial and potentially harmful microbes will allow improved control of the environment. Many of these sensors may already exist but are not being deployed routinely or effectively, and many others can be developed through collaborative teams of engineers, biochemists, and materials scientists.

People will be informed about and engaged in maintaining healthy indoor environments. Occupants and facility managers of both residential and nonresidential buildings will be knowledgeable about the conditions that create problematic indoor microbial environments and how to avoid these conditions through such activities as system maintenance and cleaning. The development of personal sensors and monitors, along with guidance, education, and training, can enable occupants to understand which practices in their indoor spaces are associated with microbial proliferation and diversity, either beneficial or harmful. These efforts will be useful for identifying when a person is releasing or being exposed to undesirable microbes, such as infectious agents and allergens. Sensor data will inform intervention recommendations—for example, how occupant behavior and building operations can be altered to reduce occupant exposures. In addition, manufacturers and members of the building trades will be trained in how to construct and maintain buildings to promote and support healthy indoor environments. Best practices to reduce negative or promote positive microbial conditions will ultimately need to be embedded in building code requirements and professional guidance documents, along with the implementation and adoption of systems to support improved building maintenance.

The benefits of connections to the outdoors will be better understood and, where useful, incorporated into the design and operation of buildings. Research will improve understanding of the impact of physical and visual connections to the outdoors, as well as the variability in temperature, light, airflow, and humidity provided by a connection with the outdoor environment. Building features and environmental connections that embrace the outdoors may contribute positively or negatively to healthful indoor environments and to the optimal management of indoor microbial communities. As transportation energy shifts from combustion to electricity, for example, outdoor air quality will improve, but overlaid on such changes are the effects of climate change and urban densification; each of these changing elements may affect the overall trade-offs between the benefits and costs of outdoor air ventilation. Where beneficial, indoor–outdoor connections will be strengthened in buildings and building systems to support occupant health, energy efficiency, and resiliency. The quality of the water and air coming into buildings by design, as well as entering via unintended routes, has effects on the health of building occupants, on the building microbiome, and on building materials. Improving the quality of outdoor air and delivered water can impact the health status of building occupants and affect the indoor microbiome.[1]

A RESEARCH AGENDA FOR ACHIEVING THE VISION

To realize the vision just presented, significant gaps in the knowledge needed to translate building, microbiome, and clinical research findings into practical application that have been identified in previous chapters need to be addressed. To this end, partnerships are needed across scientific disciplines, bridging U.S. and international research expertise and with communities of practice in clinical medicine and in the design, reinvention, and operation of buildings. The parameters that constitute a beneficial indoor microbiome have not yet been defined, much less the specific building designs, construction, materials, sensors, and operating approaches that will establish and sustain such a microbiome. Also necessary is to go beyond current identification and characterization of microbial taxa in indoor environmental samples to provide greater understanding of microbial functional activities and to clarify whether and how built environment microbiomes impact human health. Agreement has not yet been achieved on standardized microbial and building data to collect; on sampling and analytical protocols; and on data-sharing practices, which would facilitate cross-comparison of results. And providing solid evidence of health effects connected to indoor microbial exposures will require additional studies that contribute more quantitative and reproducible exposure and response data.

Built environment microbiomes include not only viable bacteria, viruses, and microbial eukaryotes but also dormant and dead microorganisms, microbial components, microbially produced chemicals such as volatile organic compounds (VOCs), and other metabolites. As complex as these microbiomes are, however, they are only one dimension of the still more complex exposures humans encounter indoors, which include many other types of chemicals present in buildings, as well as inorganic particulate matter, that can serve as contributing or confounding factors. Ultimately, links to human health are likely to depend on exposures to mixtures of airborne, waterborne, and surface-residing contaminants, which remain poorly characterized and understood. In addition, research will need to be conducted with occupants in diverse socioeconomic circumstances, housing conditions, and ecologic situations so potential variables can be considered and benefits that may ultimately result from the application of new knowledge can be shared.

---

[1] Improvements to municipal water sources, outdoor pollution, and other such dimensions will influence built environment microbiomes; detailed discussion of these features is beyond the scope of this study.

Gaining understanding of the interacting microbial, physical, chemical, and human systems that make up the built environments in which people live and work and translating this understanding into improved building design is a long-term goal that will not be achieved immediately. Progress can be made, however, in advancing this field. The multidisciplinary research agenda presented below includes 12 research areas that are priorities for making progress in achieving 5 major objectives:

1. Characterize interrelationships among microbial communities and built environment systems of air, water, surfaces, and occupants.
2. Assess the influences of the built environment and indoor microbial exposures on the composition and function of the human microbiome, on human functional responses, and on human health outcomes.
3. Explore nonhealth impacts of interventions to manipulate microbial communities.
4. Advance the tools and research infrastructure for addressing microbiome–built environment questions.
5. Translate research into practice.

The research priorities for making progress toward each of these objectives are detailed in the following sections. Together, they constitute a research program that builds on the current state of research, identified knowledge gaps, and future directions presented in prior chapters.

Characterize Interrelationships Among Microbial Communities and Built Environment Systems of Air, Water, Surfaces, and Occupants

Buildings impact microbial colonization and transport, and further research is needed to identify the factors associated with building environments that are permissive or restrictive for bacterial, viral, and eukaryotic microbial growth and distribution. Despite examples of the known occurrence of pathogens and harmful microbes in the built environment, it is likely that most microorganisms present in a well-maintained and dry building will have no impact and that some may even have a beneficial impact on human occupants. And similarly, some microorgansims found even in damp or poorly maintained environments are likely to have no impact on human occupants. Continued work is needed to study what constitutes both harmful and healthful indoor microbiomes and to identify the aspects of building design and operation that affect microbial communities. A scientific understanding of the interrelated contributions of water, air, and surfaces to microbial distribution and transport will be critical, as will an improved understanding of the influences of the interactions and behaviors of human occupants.

Priority Research Areas

1. Improve understanding of the relationships among building site selection, design, construction, commissioning, operation, and maintenance; building occupants; and the microbial communities found in built environments. Areas for further inquiry include fuller characterization of interactions among indoor microbial communities and materials and chemicals in built environment air, water, and surfaces, along with further studies to elucidate microbial sources, reservoirs, and transport processes.
2. Incorporate the social and behavioral sciences to analyze the roles of the people who occupy and operate buildings, including their critical roles in building and system maintenance.

A better understanding of the important building attributes and a clearer identification of microbial sources associated with potential harm or benefit can drive the generation of new hypotheses and the testing of interventions aimed at control of the sources and distributions of microbial communities. Sufficient understanding of these relationships is needed as a foundation for translating knowledge into advances in building design, commissioning, and maintenance practices; corresponding professional standards and building regulations; and future investments directed at monitoring and mitigating problems or promoting benefits.

Studying human occupants and their behaviors and activities would provide further knowledge of the effects of human management of built environments. Humans will always influence the indoor environment through their presence as important sources of indoor microorganisms and through their behaviors. Occupant comfort and perception also will continue to be critical factors in building design and operation. Effective studies on built environment microbiomes likely will need to incorporate additional measures of occupant perception, behavior, and motivation. As knowledge is obtained, occupants and facilities managers will need to be educated on how their modifications of built environments affect the indoor microbiome. Risk communicators and behavioral scientists will need to be involved in the creation and implementation of this training to ensure that it is communicated effectively and does not cause misunderstanding, or even fear, among facilities managers or occupants.

Several issues can be addressed by research that meets the goals of the two priority research areas detailed above. Examples include identifying key building attributes that are critical to the survival, activity, or death of bacterial, viral, and eukaryotic microbial communities, and discovering how variations in indoor environmental conditions, such as air temperature, humidity, and the condition of water in premise plumbing and other indoor water systems, affect these communities. The level of detail needed to capture and analyze these relationships will be substantial given the variations in these attributes in current and future built environments, compounded by occupant behaviors and facility management practices. Other research questions include understanding how ecological and evolutionary processes affect the composition, diversity, succession, stability, and activities of indoor microbiomes; understanding the modes of transport of bacterial, viral, and eukaryotic microbes in the built environment and the relevant air, water, and surface transfer mechanisms and interrelationships; and ensuring that studies include the impact of concealed spaces (e.g., inside walls and floor and ceiling cavities). For example, genomic analysis of building materials of different types and when exposed to moisture at damaging levels can further elucidate the bacterial and fungal ecology connected with these conditions and provide useful information on potential sources of indoor microbial exposures.

Assess the Influences of the Built Environment and Indoor Microbial Exposures on the Composition and Function of the Human Microbiome, on Human Functional Responses, and on Human Health Outcomes

Future research to explore the composition and behavior of microbiomes of the built environment will be critical to identifying qualities or states of built environment microbial communities that lead to healthful indoor environments for building occupants. To make substantial progress toward this goal, researchers will need to determine the nature and scale of microbial impacts on human health. Although a variety of studies in this area have been or are being conducted, much remains unknown. Important objectives include both characterizing the negative impacts of microbial communities and their constituents that have adverse effects and capturing the positive impacts of beneficial microbial communities. It will be critical to understand at least four aspects of the impacts on human health of built environment microbiomes: how individual microbes affect human health; how the community of microorganisms and mixed exposures affect human health; how changes in the community of microorganisms affect humans, and vice versa; and what the mechanisms are through which exposures result in health outcomes.

Interactions between building microbiomes and humans are inherently bidirectional. Human-associated and environmental microbiomes may affect and be affected by humans. Humans contribute to indoor microbiomes by shedding microorganisms, but the numbers and types of microorganisms they disseminate change with their health status, behavior, and other factors. Occupant density is likely to influence indoor microbial abundance and composition, while such human actions as use of chemicals and cleaning practices will disturb and alter microbial communities. In turn, human factors such as age, genetics, and health status will likely affect human responses to microbial exposures.

Drawing on prior culture-based studies and the wealth of new data obtained through "omics" techniques, researchers have made strides in characterizing microbial taxa and their ecological dynamics in different types of built environments. But these characterizations need to be pursued further and applied to understand how built environment microbial exposures affect human health, including quantification of exposure (which microorganisms and how many) and clearer causal connections to immune and metabolic responses. Answering these and other related questions will require studies designed to provide evidence that connects environmental microbial exposures to health effects. A range of study types will be necessary to interrogate the many relationships and connections between human health and microbial exposures, including, but not limited to, controlled human exposure studies, studies using animal models, longitudinal cohort studies, intervention studies, coupled modeling and risk assessment studies, and studies designed to understand dose-response associations.

Priority Research Areas

3. Use complementary study designs—human epidemiologic observational studies (with an emphasis on collection of longitudinal data), animal model studies (for hypothesis generation and validation of human observational findings), and intervention studies—to test health-specific hypotheses.
4. Clarify how timing (stage of life), dose, and differences in human sensitivity, including genetics, affect the relationships among microbial exposures and health. These relationships may be associated with protection or risk and are likely to have different strengths of effect, parameters that are important to understand further.
5. Recognize that human exposures in built environments are complex and encompass microbial agents, chemicals, and physical materials. Develop exposure assessment approaches to address how combinations of exposures influence functional responses in different human compartments (e.g., the lungs, the brain, the peripheral nervous system, and the gut) and downstream health outcomes at different stages of life.

There are many open questions and areas of investigation that can be addressed through research to meet the goals of these research priorities. Example themes include deepening the emerging knowledge on the effects of early-life exposures to bacterial, viral, or fungal microbes in the built environment; exploring how these exposures affect the human microbiome and human physiologic responses in ways that are beneficial to health; and elucidating the specific components required for such beneficial effects. Much less evidence currently exists for health benefits associated with later-in-life microbial exposures relative to early-life exposures, a topic that needs to be better understood. Understanding the mechanisms of action linking exposures to health effects—for example, mediated by human immune or metabolic responses—also will be critical, as will understanding the effects on health of the mixed exposures that naturally occur in the built environment, including exposures to microorganisms and microbial molecules, other built environment chemicals, and inorganic particulate matter. A number of underexplored dimensions are also associated with occupants of lower socioeconomic status, including a higher frequency of lower-quality housing and environmental conditions; a potential lack of routine building maintenance or repairs; and a lack of resources and support systems in circumstances known to affect microbial growth, such as following flooding. Research directed at understanding the effects of low socioeconomic status on the nexus of built environments, microbial communities, and occupant health will therefore be valuable.

Explore Nonhealth Impacts of Interventions to Manipulate Microbial Communities

Microorganisms and microbial communities in the built environment also have effects other than those on occupant health, such as enhanced or reduced degradation and corrosion of building materials and water systems. Although the ability to design and operate buildings that are healthful is one of the most compelling goals for research in this field, it will also be important to understand such nonhealth impacts and how they affect sustainability, costs, and other parameters important to assessing the impacts and trade-offs of building design, operation, and maintenance choices. This knowledge can inform the assessment of interventions, the development of practical guidance, and decision making.

Priority Research Area

6. Improve understanding of energy, environmental, and economic impacts of interventions that modify microbial exposures in built environments, and integrate the relevant data into existing built environment–microbial frameworks for assessing the effects of potential interventions.

Research to address this goal will need to consider diverse building types and locations, as these factors affect the materials and systems used in a building and how it is operated. Pursuing a fuller understanding of the interactions between microorganisms and building and construction materials will be useful. Smart materials or smart coatings may hold potential in this area.

The impacts and trade-offs associated with interventions in the built environment intended to affect microbial communities can have economic, energy, and sustainability dimensions beyond occupant comfort and health. Research exploring these impacts and trade-offs might consider interventions in building design and operation that include changes to building ventilation and filtration, temperature and humidity control, air and surface sterilization, and maintenance practices. Other types of interventions aimed at affecting microbial communities include the use of antimicrobial surfaces and exploration of the concept of environmental probiotics. How construction materials can be designed or tailored to address specific microbial activity is one research avenue that could be encouraged. Because the outdoor environment, including outdoor air quality and the quality of water entering a building, influence the indoor environment through building systems, studies also could usefully deepen understanding of how outdoor environmental interventions affect the assessment of indoor microbial environmental quality and occupant health. Potential interventions in this area might include, for example, landscaping to improve local biodiversity. Information gained through research addressing this priority area can also provide information about impacts on trade-offs beyond health to contribute to the development of new models, as well as evidence to support the development of future monitoring sensors and response innovations.

Advance the Tools and Research Infrastructure for Addressing Microbiome–Built Environment Questions

Identification, characterization, and quantification tools from multiple fields can be brought to bear in studying microbiome–built environment–human interactions. Advances in genomic sequencing and analysis already have improved researchers' assessments of microbial diversity in indoor environments beyond what was previously possible. However, an important role remains for techniques that can provide complementary information on microbial functions and on relevant physical and chemical conditions within the built environment. Improvements to tools and methods will be important for analyzing the suites of bacterial, viral, fungal, and other eukaryotic microbes (such as algae and protozoa) in the built environment. Advances in modeling and in development of an agreed-upon data commons also will provide crucial foundations for further knowledge and experimentation in the microbiomes of the built environment field.

Priority Research Areas

> 7. Refine molecular tools and methodologies for elucidating the identity, abundance, activity, and functions of the microbial communities present in built environments, with a focus on enabling more quantitative, sensitive, and reproducible experimental designs.

Although progress has been made in characterizing diverse communities of built environment microorganisms, a sufficiently detailed understanding of the functional activities of microbial communities and their associations with built environment and occupant factors is lacking. Understanding these associations is required to provide the basis for assessing rational interventions to promote health and sustainability. Measurement of the functional activities of microbial communities draws on information that extends beyond genomics, including information on viability, metabolism, and interactions with other microbes in the built environment community and with human physiologic systems. Molecular tool development will support improved analysis of collected samples. In addition, more studies are currently available on bacterial and fungal organisms in indoor microbiomes compared with viruses and nonfungal eukaryotes, resulting in greater knowledge of some built environment microbial populations relative to others. These gaps will need to be filled to increase the meaningful interpretation of microbial community findings.

Questions that could be addressed to help meet the goals of this research priority include further development of tools to provide improved quantitative information, such as absolute abundance data, to support modeling of community dynamics and dose-response relationships, and the development of more sensitive and reproducible tools for measuring microbial functional activities. Also useful will be the development of nondestructive measurement methods for sampling microbial characteristics in and on building materials, which will help ensure that realistic sample measurements are being taken from these materials. As noted above, collecting, measuring, and categorizing nonbacterial and nonfungal components of built environment microbiomes, such as viruses, archaea, and protists, will be valuable as well.

> 8. Refine building and microbiome sensing and monitoring tools, including those that enable researchers to develop building-specific hypotheses related to microbiomes and that assist in conducting intervention studies.

Both research and practical tools are needed to inform an understanding of the intersection between built environments and microbiomes. These tools can include real-time sensors and monitors in buildings to measure environmental conditions, as well as microbial properties. For example, there is a need to develop better, more rapid methods for detecting moisture levels on surfaces (water activity) and relating these levels to air humidity, material moisture content, temperatures, and dew points. Improvements in the methods used to measure indoor environmental conditions that impact the microbiome are needed as part of efforts to better characterize these interacting systems. These tools are also critical in obtaining building metadata to complement microbial data for studies aimed at understanding exposures, as well as in supporting further hypothesis generation.

Improvement in sensor technology is continually driven by technological developments in computation, materials science, engineering, and other fields. The integration of expertise from these fields with expertise from the microbiologic and clinical sciences can provide new tools to support future investigations. A variety of research can be envisioned to help meet this goal. For example, new building sensors could be developed to enhance the collection and analysis of data on a building's physical and chemical environment, to measure human and animal occupancy, and to detect microbial growth or microbial VOCs in conjunction with the development of guidance on where to place building sensors and arrays. Sensors that can measure cumulative exposures will also be useful. The output of these sensors will support the ongoing construction and use of data-driven models.

9. Develop guidance on sampling methods and exposure assessment approaches that are suitable for testing microbiome–built environment hypotheses.
10. Develop a data commons with data description standards and provisions for data storage, sharing, and knowledge retrieval. Creating and sustaining the microbiome–built environment research infrastructure would promote transparent and reproducible research in the field, increase access to experimental data and knowledge, support the development of new analytic and modeling tools, build on current benchmarking efforts, and facilitate improved cross-study comparison.

Capturing the attributes of the built environment and its management in a consistent manner will be critical to extrapolating results from microbial research and applying them to practice and to advancing building system design and management. Defining the sampling approaches most relevant to answering particular questions will be important (e.g., when studying associations between human exposures and health outcomes).

Sample collection and handling influence the microbial results obtained from built environment studies; the establishment of common understandings and guidance to inform sampling and analysis can be valuable. In addition, standardizing the descriptions of building attributes collected in studies and using such approaches as conducting round-robin studies of standardized samples to understand laboratory-to-laboratory variability will enable comparison of results obtained by independent research groups using diverse methods.

These features would all usefully be part of a data commons for the microbiome–built environment research community. This commons will need to include an agreed-upon set of metadata to be collected in experiments, including agreed-upon criteria for recording building conditions, as well as criteria and systems, such as databases, for sharing the methods, tools, and results of microbial research. One significant challenge will be determining a balance among collecting as much detailed building, microbial, and human information as possible; experimental practicality; and cost. Efforts in these directions have already been undertaken, and potential partners for further development of a data commons may exist in multiple federal agencies and professional societies. Achieving such a data commons will require engaging researchers, microbial ecologists, building scientists, informatics experts, and others (such as health researchers) working in the field of microbiomes of the built environment to develop practices and standards that meet experimental needs and are acceptable to the community undertaking such studies. Efforts to achieve community agreement around data collection, data standards, and data sharing will need to continue, as these areas represent foundational components of future research in the field.

11. Develop new empirical, computational, and mechanistic modeling tools to improve understanding, prediction, and management of microbial dynamics and activities in built environments.

A variety of modeling tools can be implemented in studying microbiome–built environment–human interactions. These tools include models of air and water flows throughout buildings, transport pathways of air- and waterborne microorganisms, and occupant behavior. New modeling tools that improve predictions of microbial persistence, transmission, and health outcomes and incorporate data on intervention costs and tradeoffs are likely to have significant influence on the field. Mining of detailed microbiome data requires intensive computational approaches; it is not a simple task. However, it is essential to capture and account for the physical, chemical, and biologic dynamics of indoor environments so that these constraints can be statistically integrated with microbial dynamics and microbial outcomes resulting from changes in built environment conditions can be predicted. A variety of building airflow and contaminant transport models exist, but microbial data need to be tied to these models to enable modeling of human exposures to indoor microorganisms and subsequent health impacts. Introducing microbial, indoor air quality, and health variables into the computational tools used by the building design and engineering community to monitor and predict heat, air, moisture, and contaminant transport would be one useful development. For example, incorporating experimental data into models to link building and HVAC system design with interventions aimed at promoting human and environmental health by changing microbial exposures may provide insights to support future research agendas.

Translate Research into Practice

Ongoing research efforts, along with the development of tools and methods, ultimately lead to the question of which interventions can and should be undertaken in built environments to alter buildings and their operation, built environment microbial communities, and occupant behaviors, as well as how those interventions with positive health and sustainability impacts can be promoted. Although significant fundamental research to characterize and manage microbiome–built environment interactions remains to be carried out, studies already undertaken provide a basis for further exploring these important issues. With long-term interest and investment, the field will be poised to design and test interventions in built environments that affect microbiomes in predictable ways and to develop strategies for integrating health, economic, energy, and other data to support informed decision making on which interventions to implement and at what point.

As knowledge is gained, it will be important to translate these advances into practice and to communicate and engage effectively with the diverse stakeholders that design, operate, maintain, live, and work in the built environment. These stakeholders will need guidance tailored to their goals and needs, whether it be professional practice guidelines or guidance for occupants in a range of building types. Guidance targeting occupants also will need to take account of differences in age, health status, and economic resources to inform such practices as cleaning and maintenance.

Priority Research Areas

12. Support the development of effective communication and engagement materials to convey microbiome–built environment information to diverse audiences, including guidance for professional

building design, operation, and maintenance communities; guidance for clinical practitioners; and information for building occupants and homeowners. Social and behavioral scientists should be involved in creating and communicating these materials.

To continue moving toward practical application will require exploring microbiomes, built environments, and human health as an integrated system. Insights from studies in multiple areas—such as microbiology, human and built environment microbiomes, human health, indoor exposures to chemicals and particulate matter, and building system design and performance—will need to be combined. The research agenda detailed above can make progress in answering the question of what is gained by exploring built environments, as has been done in other types of ecosystems, and looking at microorganisms, buildings, and occupants in communities rather than in isolation. Answering these and other questions will require the involvement of experts from multiple fields working in concert, as well as the engagement of practitioners from the building community who are responsible for building design and operation and from the clinical community who focus on human health.

Integration across multiple disciplines to address scientific and societal challenges is a broad priority for many agencies and organizations (NAE and NASEM, 2017; NRC, 2014). Yet, achieving deep and sustained engagement that combines disciplines is difficult. A 2015 National Research Council report addresses approaches to fostering collaboration and cooperation in research teams and may contain lessons that can be applied to the built environment field (NRC, 2015). For example, the establishment of centers that bring together researchers with diverse scientific and professional backgrounds is a common approach to tackling scientific and institutional challenges associated with integrated research, and several such centers have emerged to study microbiome–built environment interactions. As with many challenging and multidisciplinary research topics, however, no single agency or organization covers the intersection of building design and operation, environmental microbiomes, and human health. Agencies or foundations interested in pursuing this research integration could consider such options as establishing collaborative funding or incorporating requirements for disciplinary integration into research solicitations. Despite these many challenges, a future informed by knowledge about indoor microbiomes holds promise for improving both human health and built environments, and it will depend on effective collaboration and on the sharing of knowledge and expertise.

REFERENCES

Boerstra, A. C. 2016. *Personal control over indoor climate in offices: Impact on comfort, health and productivity*. Eindhoven, Netherlands: Technische Universiteit Eindhoven.

NAE (National Academy of Engineering) and NASEM (National Academies of Sciences, Engineering, and Medicine). 2017. *A new vision for center-based engineering research*. Washington, DC: The National Academies Press.

NRC (National Research Council). 2014. *Convergence: Facilitating transdisciplinary integrations of life sciences, physical sciences, engineering, and beyond*. Washington, DC: The National Academies Press.

NRC. 2015. *Enhancing the effectiveness of team science*. Washington, DC: The National Academies Press.

A

An Assessment of Molecular Characterization Tools[1]

Information extracted from microbiome molecular measurement tools is intended to reconstruct and ultimately predict critical microbiome features that include relative and absolute abundances of microbial taxa, persistence, impact on human health, and transmissibility. Application of the measurement tools is driven by access to observational data and intentional hypothesis-driven experiments. Given its central role in elucidating the microbiome world, tools for observational experimentation are the primary focus of this appendix (Gilbert et al., 2016; Wang and Jia, 2016). The key objective of these efforts is to determine site-specific microbial content across space and time, along with the corresponding building and environment conditions. Critical to these efforts is the collection of relevant metadata and data sharing to support interpretation of measurement results and accurate reproducible microbiome models. The development of mathematical and computational models for microbiome dynamics, the refinement of statistical techniques that guide sampling design, and the linkage of models to data are important parts of the toolkit of microbiome researchers. But in this appendix, the focus is on the basic data that are needed to parameterize and test such models in the first place.

Tools for controlled experiments are more commonly applied at the macro rather than the molecular level and are not the focus of this appendix. Nonetheless, molecular manipulation could emerge in the future as an

---

[1] Mention of commercial products or organizations does not imply endorsement by the National Academies of Sciences, Engineering, and Medicine or by members of the Committee on Microbiomes of the Built Environment: From Research to Application.

237

important tool (Biteen et al., 2016). Intentional molecular interventions could include introduction of a specific organism, gene, community composition, or chemical product into the environment to test a hypothesis or to achieve a desired impact. As a result of the vast diversity and largely uncultivated status of built environment microorganisms, analyzing the structure, functions, activities, and dynamics of microbial communities, especially under natural settings, remains an enormous challenge and focus of existing research. Over the last few decades, to meet this challenge, a variety of open-format (e.g., high-throughput sequencing, mass spectrometry–based proteomic and metabolomic approaches) and closed-format (e.g., functional gene arrays, protein and metabolite arrays) detection technologies have been developed and used to address questions about microbial ecology at the frontier of knowledge (Roh et al., 2010; Vieites et al., 2009; Zhou et al., 2015). Substantial biological insights have been obtained from a variety of ecosystems important to human health (e.g., Alivisatos et al., 2015), as well as from analyses of foodstuffs, systems subject to climate changes (Xue et al., 2016), industrial settings, and agriculture, and more broadly across the environmental sciences (Long et al., 2016).

The open- and closed-format technologies are fundamentally different in sample preparation, quality control, data processing and analysis, performance, and applications (Zhou et al., 2015). Recently, numerous studies have examined the performance of various types of technologies, but most of these studies are related to high-throughput sequencing and microarrays. High-throughput sequencing includes primarily polymerase chain reaction (PCR) amplification-based target gene sequencing (TGS) with phylogenetic (e.g., 16S rRNA) or functional (e.g., *amoA* and *nifH*) targets, and shotgun metagenome sequencing. The discussion that follows is focused mainly on the performance of these types of high-throughput technologies in terms of key performance issues, such as specificity, resolution, sensitivity, biologic activities, quantification, and reproducibility, within the context of microbial communities, particularly the microbiomes of the built environment.

EVALUATING THE CAPABILITIES OF CURRENT TOOLS

Table A-1 presents a summary of characterization tools, organized by input type, tool type, and detection format (open or closed). This summary shows the diverse set of molecular measurement tools that are available and distills their strengths and weaknesses with respect to analytical certainty and interpretive power. In many cases a tool may include multiple measurement types. For example, there are both targeted and untargeted metabolite measurement tools.

It is important to consider the ability of molecular measurement tools to recover information relevant to the following features: relative and absolute abundances of microbial taxa, persistence, potential health impacts, and transmissibility. To assess these capabilities, the following criteria are considered in the sections that follow: specificity, taxonomic resolution, sensitivity and organism coverage, organism viability, biologic activity, functional coverage, toxicologic potential, quantification, and reproducibility.

Table A-1 is meant to reflect the positive attributes of each tool; the text identifies limitations and caveats. The table highlights the fact that there are a number of new and emerging molecular measurement tools designed for different types of measurements, with each tool having varying capabilities. Each criterion is discussed in greater detail in the following sections, along with the rationale for the qualitative assessments shown in the table. These qualitative metrics are meant to provide a consensus perspective on microbiome measurement tools through a profile of their strengths and gaps in meeting the goals of recovering microbiome information from built environments. It should be noted that it is difficult, if not impossible, to make straightforward, point-by-point direct comparisons among different technologies because of their broad diversity and distinct characteristics. Therefore, an attempt has been made to highlight the major differences among the various technologies at very coarse levels. The following discussion focuses on the issues important to microbial ecology rather than reviewing the various technologies comprehensively.

Open- and Closed-Format Tools

"Open-format" refers to "technologies whose potential experimental results cannot be anticipated prior to performing the analysis, and thus, the experimental outcome is considered open" (Zhou et al., 2015, p. 2). In contrast, "closed-format" refers to "detection technologies whose range of potential experimental results is defined prior to performing the analysis, and thus, the experimental outcome is considered closed" (Zhou et al., 2015, p. 2).

Closed-format nucleic acid–based tools are more adept at identifying known organisms, but they provide limited information on biologic identity. In general, fundamental questions of reproducibility remain for all of the high-throughput measurement tools. This limitation is due, in part, to their exquisite sensitivity and thus potential for high reporting variability, but also to a lack of common benchmarks and the reliance on an evolving collection of bioinformatics tools and databases. In metagenomics, "open" formats can also be called "untargeted," and closed formats can be called "targeted." Open-based formats such as shotgun sequencing show considerable potential to elucidate specific functions and capture unknown microbial material; however, the sampling efforts and costs to use open-based

TABLE A-1 Overview of Molecular Characterization Tools

Tools	Input	Format	Specificity	Taxonomic Resolution	Sensitivity	Organism Coverage
Amplicon sequencing of phylogenetic markers (e.g., 16S)	DNA/RNA	Open	Targeted	Genus/family	Rare detection	Conserved primers
Amplicon sequencing of functional markers	DNA/RNA	Open	Targeted	Species/strains	Rare detection	Conserved primers
Whole-community shotgun sequencing	DNA/RNA	Open	Off targets	Species/strains	Variable	Untargeted
Phylogenetic gene arrays	DNA/RNA	Closed	Targeted	Genus/family	Rare detection	High multiplex
Functional gene arrays	DNA/RNA	Closed	Targeted	Species/strains	Rare detection	High multiplex
Culturing	Cellular	Open	Targeted	Strain	Variable	Culturable
Direct microscopy	Cellular	Open	Targeted	Family	Variable	Morphology
Mass spectrometry–based proteomics	Proteins	Open	Off targets	Genus	No	Untargeted
Array-based proteomics	Proteins	Closed	Targeted	Genus	Rare detection	High multiplex
Mass spectrometry–based metabolomics	Small molecules	Open	Off targets	None	Variable	Untargeted
Array-based metabolomics	Small molecules	Closed	Targeted	None	Variable	High multiplex

NOTES: Because various technologies have different characteristics, it is difficult to make straightforward, point-by-point direct comparisons. Therefore, this table attempts to highlight the major differences among various technologies at a coarse level for general comparison; accuracy may be lost in some cases in this simple table. A hyphen (-) indicates tools that exhibit less variability and may not include replicates in reporting. "High multiplex" indicates a tool that can track thousands of distinct targets but requires prior knowledge of the target.

APPENDIX A

Organism Viability	Biologic Activity	Functional Coverage	Toxicologic Potential	Quantification	Reproducibility
No	No	Inferred	No	Relative	Middle
No	No	Single gene	Yes	Relative	Middle
No	No	Multi-genes	Yes	Relative	Low
No	No	Single gene	No	Absolute	High
No	No	Multi-genes	Yes	Absolute	High
Yes	No	Single organism	Yes	Absolute	-
Yes	Yes	No	Yes	Absolute	-
No	No	Peptide fragment	Yes	Relative	Low
No	No	Peptide fragment	Yes	Absolute	High
No	Yes	Metabolites	No	Relative	Low
No	Yes	Metabolites	No	Absolute	High

tools in indoor environments continue to prevent their use in cases where large numbers of samples are needed.

Specificity

Specificity refers to the fraction of recovered biologic material belonging to a target microbial community of interest. For example, if shotgun sequencing is undertaken on an air filter intended to monitor indoor microbial communities with pollen particles, a large portion of the recovered genomic material may yield plant genomes (Be et al., 2015). The specificity metric is particularly important for analyzing environmental samples because there could be numerous homologous sequences for each gene present in a sample. Various technologies, such as target gene sequencing, shotgun metagenome sequencing, and gene arrays, are commonly used for detecting specific organisms of interest. The specificity of target gene sequencing is determined primarily by means of the primers used for PCR amplification of the target genes. After amplification, single nucleotide differences can be resolved by subsequent high-throughput sequencing. Thus, theoretically, highly specific detection of different phylogenetic groups, species, strains, ecotypes, populations, genes, and/or single nucleotide polymorphisms can be achieved, depending on the target genes and sequencing depths, as well as the complexity of communities examined.

To detect broader groups of organisms, highly conserved degenerate primers generally are designed, such as those used for amplifying 16S rRNA genes for bacteria and archaea, the internal transcribed spacer (ITS) for fungi, and 18S and 28S for eukaryotes in general (Cole et al., 2007; Fischer et al., 2016; Tedersoo et al., 2015). On the one hand, the higher the degree of conservation of the primers among different organisms, the broader is the phylogenetic scope of the organisms that can be detected. On the other hand, the acquired data are potentially less specific to the target genes/organisms of interest. Also, various primer sets can be designed for adopting next-generation sequencing (NGS) for phylogenetic marker genes, but their specificity and hence detection broadness vary greatly among different primer sets (Cole et al., 2007). Appropriate selection of primer sets for amplification also depends on a variety of factors such as research questions and objectives, sequencing platforms, and community composition. For instance, the primers for amplifying V3–V4 regions of 16S rRNA genes (280 bp fragment) have been used for Illumina sequencing platforms. Computationally, these primer sets should be able to amplify both bacteria and archaea, but in practice, archaea generally are poorly recovered. Thus, further development is needed for the detection of various types of archaea across different environments.

Although some earlier studies showed that detection specific to individual microbial species or strains could be achieved based on 16S rRNA genes (Loy et al., 2002; Rudi et al., 2000; Urakawa et al., 2002), detection specificity is still problematic (Zhou and Thompson, 2002). Analytic technologies based on functional genes and other noncoding sequences have advantages in specifically detecting individual species or strains (Zhou, 2003). For example, both target sequencing of functional genes and shotgun sequencing of whole communities are capable of providing highly specific information at the level of nucleotide differences on both known and novel genes and pathways (Hess et al., 2011; Mackelprang et al., 2011; Qin et al., 2012; Tringe et al., 2005; Venter et al., 2004). However, PCR amplification biases (Engelbrektson et al., 2010; Kunin et al., 2010; Lemos et al., 2012; Schloss et al., 2011), sequencing errors, and chimeric sequences (Edgar, 2013; Pinto and Raskin, 2012; Schloss et al., 2011) are inherent in sequencing technology, and they will have considerable impacts on detection specificity. The existence of nontarget contaminant DNAs in sequencing libraries will greatly affect detection specificity for the shotgun sequencing approach, which is a particular problem for host-associated microbiome studies in which sequence data may be predominantly from the host (Gevers et al., 2012; Kuczynski et al., 2012; Zhou et al., 2015). In addition, the uncertainty involved in selecting various bioinformatic tools for data processing could have significant impacts on detection specificity (Nayfach and Pollard, 2016; Sinclair et al., 2015). Contamination continues to present informatics challenges in metagenomics, with results showing errors in distinguishing a live organism of interest from nucleic acids isolated from laboratory contaminants (Merchant et al., 2014; Salter et al., 2014).

High specificity can be achieved under stringent hybridization conditions (e.g., 45°C plus 40 percent formamide for GeoChip 4.0, and 67°C plus 10 percent formamide for GeoChip 5.0) with GeoChip-based functional gene arrays for detecting specific taxa in analyses of environmental samples (Tu et al., 2014; Wang et al., 2014). Various controlled studies have demonstrated that such hybridization stringencies could differentiate sequences with <90–92 percent identity (Zhou, 2009). Also, unlike sequencing technologies, contaminated nontarget DNAs should have less impact on detection specificity (Zhou et al., 2015). However, low-level cross-hybridization to nontarget genes/strains always occurs. The challenge is to resolve true hybridization signals from nonspecific noises without ambiguity (Zhou et al., 2015). In addition, on the one hand, array hybridization with probes is quite specific. On the other, since the detection is defined by probe sets on arrays, depending on the hybridization stringencies used, novel genes and highly divergent genes are not detected by array hybridization (Zhou et al., 2015). Consequently, array hybridization–based detection is not suitable for discovery of novel organisms (Zhou et al., 2015).

Taxonomic Resolution

Taxonomic resolution defines the amount of taxonomic information that is recoverable from genetic variation for each microbe in the target community (Hanson et al., 2012), with maximal resolution being an individual clone or strain and coarser resolution at higher phylogenetic levels such as genus and family. Taxonomic resolution is a critical issue for achieving appropriate detection specificity, sensitivity, and quantification to address questions related to microbial distributions, biogeography, activities, functions, and dynamic succession in response to treatments and environmental changes (Zhou et al., 2015). Although the use of phylogenetically accurate markers (e.g., 16S, 18S, 28S rRNA genes, ITS) in surveys radically changed the view of microbial diversity, distribution, and evolution, most studies are still based on information from short segments of these phylogenetic markers (Uyaguari-Diaz et al., 2016), typically 200–300 bp. Because of low rates of molecular evolution, it can be difficult to obtain fine-scale resolution at the desired species/strain level with phylogenetic markers such as 16S rRNA genes. In addition, because of the existence of sequencing errors and chimera and the lack of sufficiently accurate reference sequences, resolving classification at fine-scale taxonomic resolution (e.g., species) is even more difficult using limited short segments of phylogenetic marker genes (Jovel et al., 2016). Consequently, most bioinformatics tools provide only annotated information to the level of genera or above (Ritari et al., 2015), and hence the majority of microbial ecology studies are restricted to analyses at coarse taxonomic levels, such as differences by phylum or class (Jovel et al., 2016; Singer et al., 2016). Recently, various approaches have been developed based on single molecule sequencing using third-generation sequencing technologies such as PacBio (Singer et al., 2016) and Nanopore technology (Benitez-Paez et al., 2016) to obtain full length of sequencing of phylogenetic markers, which potentially provide more accurate and finer taxonomic resolution of microbial communities and better predict metabolic potentials. However, the experimental cost associated with such technologies remains quite high, although rapidly decreasing.

Recent informatics work shows that increased resolution of operational taxonomic units (OTUs) is possible with similarity metric alternatives to traditional sequence identity cutoffs applied to read clustering (Eren et al., 2016; Nguyen et al., 2016). New algorithms are being developed to use longer gene regions from new long-read sequencing technology (Singer et al., 2016), which improves differentiation of closely related species within a single microbial community profile. In addition, differing use of reference-based and reference-free read clustering can impact the reported community structure (He et al., 2015).

Compared with phylogenetic gene markers, functional gene markers such as *nifH* genes and other coding sequences have higher taxonomic resolution (Scholz et al., 2016; Zhou et al., 2003, 2016). As a result, technologies based on functional genes, such as shotgun sequencing and GeoChip-based functional gene arrays (He et al., 2010a; Tu et al., 2014), could resolve organismal differences at the species/strain level. High taxonomic resolution is important for gauging such treatment effects as experimental warming (Xue et al., 2016), examining fine-scale biogeographic patterns (Liang et al., 2015; Zhou et al., 2008), and understanding microbial evolution (Kashtan et al., 2014; Nayfach and Pollard, 2016; Shapiro et al., 2012). For instance, DNA-based microarrays have demonstrated species/strain-level resolution in a wide range of environmental conditions (Be et al., 2013; Devault et al., 2014; Liebich et al., 2006; Tiquia et al., 2004; Zhou et al., 2010). It is important to note that, despite the advantages in providing greater taxonomic resolution relative to rRNA genes, functional gene biomarkers are more vulnerable to the effects of horizontal gene transfer (HGT); this is particularly the case for those that are frequently plasmidborne, such as those involved in metal resistance and organic contaminant degradation (Zhou et al., 2008).

Analytic tools for shotgun sequencing continue to be actively developed. Genomic data can theoretically come from any part of each microbial community member's genome. Thus, the information, which extends beyond an isolated marker gene, can increase taxonomic resolution, moving from differentiating genera and species to tracking individual genetically differentiated populations (e.g., bacteria within a strain), and documenting evolutionary changes at even short time scales (Greenblum et al., 2015). The two principal informatics research tracks are read binning and metagenomic assembly. Metagenomic assembly holds the greatest promise for maximizing accurate community characterization since it attempts to reconstruct each organism's genome for maximal taxonomic resolution. It will also be important to develop common reference materials to support analysis. However, there are fundamental challenges to assembly that currently preclude its use as the sole technical approach (Ghurye et al., 2016). One barrier to assembly is lack of sequencing depth for each organism, a roadblock that is driven by limits in sequencing fidelity and throughput. Two open analysis challenges remain—disentangling individual genomes within a population of closely related organisms and resolving genome complexity, particularly in the case of microbial eukaryotes (Sangwan et al., 2016). These two problems could become easier to solve as read length increases and error rates are reduced with emerging sequencing technologies (Beitel et al., 2014; White et al., 2016). For microbial communities of the built environment, and with the use of current sequencing technology, the two practical features for determining informatics success in complete metagenomic assembly are ensuring sufficient biomass to retrieve representative sampling of DNA and limited community complexity to ensure adequate sequencing depth of each organism. Progress recently has been made in overcoming read depth limitations by developing new assembly approaches that bin and assemble reads that exhibit covariation patterns across multiple related microbiome samples (Imelfort et al., 2014; Nielsen et al., 2014). An open challenge remains in developing robust informatics tools to resolve mobile genetic elements, which are important for determining whether genes or pathways proliferate to multiple kinds of organisms (Jørgensen et al., 2014). Similarly, phage is a potential source of altering community composition (Koskella and Brockhurst, 2014) and stimulating lateral gene transfer, yet phage diversity and abundance are rarely addressed in existing microbiome studies (Rosario and Breitbart, 2011).

Sensitivity and Organism Coverage

Sensitivity measures the fraction of known taxa (for convenience, "species") present in the microbiome that are detected even when the organisms are present at low abundance. Organism coverage defines the fraction of organisms detected regardless of whether they are known a priori or not. Thus, a targeted sequencing method can be of high sensitivity by amplifying and sequencing a single gene copy, even if occurring at ultra-low abundance. Yet, the same method would exhibit poor organism coverage since it would be designed to target a gene that likely can occur only in a limited number of organisms. The open-detection formats are best suited to maximizing organism coverage; however, even these tools are typically limited by access to reference databases needed to make a species assignment.

Sensitivity is a critical parameter for detection, particularly for complex environmental samples in which many populations exist in low abundances (Rhee et al., 2004; Wu et al., 2001, 2006). Sensitivity generally can be assessed based on absolute amounts of template materials (e.g., DNA and RNA) needed for analyses, and the lowest percentage of populations within a community can be detected. Although the former has been reported in various studies, the information on the latter is sparse because sophisticated experiments need to be designed with special efforts on implementation.

Because target gene sequencing involves PCR amplification, typically with 25–35 cycles, its detection sensitivity is generally expected to be as high as that of PCR amplification. Highly sensitive detection can be achieved with NGS (de Boer et al., 2015; Leonard et al., 2015). In general, 1–10 ng of DNA is used for library preparation for target sequencing with the Illumina platform. Shotgun sequencing typically requires 1 µg DNA for library preparation with sonication for DNA fragmentation without PCR amplification. Various low numbers of PCR amplification cycles (typically 6) could also be used in library preparation for shotgun sequencing with the Illumina platform, depending on the amount of starting material. If biomass is extremely low (e.g., 1 ng), high PCR amplification numbers (e.g., 18 cycles) can be used in library preparation for shotgun sequencing. However, this approach is sensitive to dominant populations in the sample, which can be oversampled. Consequently, it might be difficult to detect rare taxa (Zhou et al., 2015). Further studies might be necessary to provide explicit evidence for the lowest abundance of a population in a complex community that can be detected using these technologies.

Sensitive detection can also be obtained with functional gene arrays. With the current version of GeoChip fabricated by Agilent printing technology and updated protocols, 0.2–1.0 µg of community genomic DNA is needed for direct labeling and hybridization, a sample size that can suffice for analyzing environmental samples from many habitats, such as soils, marine sediments, bioreactors, and wastewater treatment plants (Zhou, 2009; Zhou et al., 2015). Nucleic acids (1–500 ng) can also be representatively amplified using whole-community genome DNA (1 ng) amplification (WCGA) (Wu et al., 2006) or whole-community RNA amplification (WCRA) (500 ng) (Gao et al., 2007) if biomass is extremely low. Very low concentrations of DNA (~10 fg, ~2 bacterial cells) can be detected using a modified amplification method (Wu et al., 2006). Furthermore, recent studies showed that the Agilent-based functional gene arrays are highly sensitive, with a detection limit as low as 5×10^{-4} to 5×10^{-5} proportion of populations within a complex soil community in terms of DNA concentration (µg). The phylogenetic arrays (e.g., PhyloChip) exhibit a detection limit of 10^7 copies or 0.01 percent of nucleotides hybridized to the array (Brodie et al., 2007; DeAngelis et al., 2011). Such detection sensitivity is comparable to that of PCR amplification-based target gene and shotgun sequencing.

Microbial biomass is generally very low for built environment samples, and hence sensitivity is a critical issue for built environment microbiome studies. At this stage, it is not clear whether the available sequencing- and array-based detection technologies are sensitive enough for built environment microbiome studies. Rigorous systematic examinations of this issue within the context of the built environment are needed.

Despite progress in increasing the accuracy of metagenomic assembly, read binning software tools remain an important complement to improve sensitive organism detection in metagenomic shotgun sequencing since there are fewer restrictions on minimum read depth. This feature enables identification of low-abundance species, which can be important for accurate community profiling (Segata et al., 2013). Rare taxa, if present, can potentially explode in numbers if environmental conditions change. Unsupervised binning uses short sequence frequency profiles to group sequencer reads with similar profiles and can be useful both for characterizing novel organisms (Liao et al., 2014) and for identifying novel organisms common across multiple samples (Alneberg et al., 2014). Supervised approaches are widely used and match reads against a reference database for taxonomic identification. The use of multilevel hash tables, suffix arrays, de Bruiin graphs, and other related searchable data structures offers the potential to explore large metagenomic datasets against a comprehensive genome database (Ames et al., 2015). A clear challenge is the recognition that reference databases are dynamic and must be updated regularly to reflect the increasing number of available sequenced organisms.

Organism Viability

Viability refers to the measurable ability of a microorganism to replicate under artificial (engineered) or natural conditions. Because culture-based analyses can assess viability, such analyses historically have been preferred over non-culture-based methods, which with many classical protocols cannot assess viability. Culture-based approaches are valid for assessing air and surface samples for some infectious pathogens, for example. In the context of the built environment, however, adverse health effects include more than infectious diseases that depend on viability for disease transmission. Evidence suggests that adverse health effects can be caused not only by inhalation of viable airborne pathogens but also by inactive microorganisms and their fragments and component parts (Miller, 1992). There is evidence suggesting that nonviable fungi, their spores, and their fragments all can cause respiratory illness and chronic systemic illness (Burge, 1990; Flannigan et al., 1991; Sorenson et al., 1987; Su et al., 1992). The same is true for some species of airborne bacteria (Flannigan, 1992). Genera included in indoor bioaerosols and on surfaces are those that are pathogenic (e.g., *Aspergillus niger*), those that are toxigenic (e.g., *Stachybotrys atra*), and those that are allergenic (e.g., *Aspergillus versicolor*).

Inhalation exposure to microbial toxins and allergens contained in/on spores or inactive microbial fragments is not detected by viability-based culturing methods. To better understand the ecology of bioaerosols and to study the effects of engineering controls on the indoor environment, methods are needed to differentiate between metabolically competent and nonviable airborne microorganisms as they exist in situ. In short, a dead microbe is not an irrelevant dust particle, but may have serious health consequences.

Conventional culture-based approaches, of course, have intrinsic limitations for characterizing airborne microorganisms. Standard plate counts underestimate the true quantity and diversity of airborne microbes, as this method is incapable of investigating the fate of slow-growing, unculturable, or inactive microbes and their fragmented parts, and culturing techniques do not span the range of environments that prompt microbial metabolic activity and reproduction. PCR has recently been adapted to characterize airborne microorganisms; however, PCR also has intrinsic limitations, particularly in gauging activity. While promising, genetic amplification methods have been reported to estimate microbial biomass and diversity inaccurately. PCR is as yet incapable of assessing in situ activity; provides no measure of microbial fractions; and is labile to some ubiquitous environmental interferences, particularly trace concentrations of heavy metals (Alvarez et al., 1995; MacNeil et al., 1995).

Fluorochrome enumeration has been adapted to measure airborne bacterial concentrations in indoor environments using the DNA intercalating agents acridine orange or ethidium bromide (Griffiths et al., 1996; Moschandreas et al., 1996; Palmgren et al., 1986; Terzieva et al., 1996). In a bench-scale chamber study (550 cm^3), Terzieva and colleagues (1996) directly enumerated *Psuedomonas fluorescens* captured in impingers using proprietary membrane integrity dyes to assess viability. In many ecological and environmental studies, fluorochromes have been coupled with various heterocyclic tetrazolium dyes and nalidixic acid to determine not only total microorganism numbers but also the fraction of metabolically active microorganisms (Hernandez et al., 1999; Maki and Remsen, 1981; McFeters, 1995; Rodriguez et al., 1992; Tabor and Neihof, 1982; Trevors, 1985). However, in most indoor aerosol and surface studies, the detection and quantification of metabolically active microorganisms have, until recently, been limited to agar plate count methods, where sampling methods, nutritional requirements, and culturability bias the results (Burge, 1990; Buttner et al., 1993; Flannigan, 1993; Hinds, 1982; Jensen et al., 1992; Marchand et al., 1995; Pillai et al., 1996; Teltsch and Katzenelson, 1978).

Biologic Activity

"Biologic activity" refers to any enzymatically mediated metabolic function, or fraction thereof, that contributes to the persistence of a microorganism as a living entity, regardless of its ability to propagate. DNA-based metagenomics provides a "snapshot" of the diversity and functional potential of various microbial taxa with potentially functional populations in a community, but it is not clear whether the populations of these taxa are actively engaged in metabolic activity and reproduction (which is required for a population to persist). To address such questions, other "omics" approaches, including metatranscriptomics for assessing mRNAs, metaproteomics for measuring proteins, and metametabolomics for monitoring metabolites, as well as stable isotope probing, are more appropriate (Zhou et al., 2015).

Metatranscriptomics typically involves random sequencing of microbial community mRNA (DeLong, 2009; Moran, 2009; Moran et al., 2012; Shi et al., 2009; Sorek and Cossart, 2010). Total RNA is first extracted from a microbial community, generally with rRNA removal. Then, mRNA is amplified, converted into cDNA, and sequenced. Because of the low relative abundance of mRNA in total cellular RNA (e.g., 1–5 percent) and the lack of poly(A) tails, prokaryotic rRNA is often removed, and mRNA is amplified before sequencing to improve detection sensitivity (He et al., 2010b; Sorek and Cossart, 2010; Stewart et al., 2010). Effectively removing rRNA can be labor-intensive, time-consuming, and challenging.

Metatranscriptomics has been widely used for characterizing microbial communities from different habitats, such as soil (Urich et al., 2008), seawater (Frias-Lopez et al., 2008; Poretsky et al., 2009; Stewart et al., 2012), human/animal microbiomes (Giannoukos et al., 2012), and activated sludge (Yu and Zhang, 2012). Results from such studies demonstrate that metatranscriptomics provides a powerful approach to functionally characterizing microbial communities. Although metatranscriptomics is attractive, the challenge is how to obtain sufficiently high quality of total community RNA and to completely remove rRNA prior to sequencing. Obtaining sufficient community RNAs for metranscriptomics could be particularly difficult for microbial communities associated with the built environment, given the challenges of low biomass in samples.

Analysis of community RNA can also be performed via functional gene arrays (Xue et al., 2016). The main advantage of the array-based approach is that total community RNAs can be used for direct hybridization without the need for removing rRNAs, so the activity of less abundant taxa and populations can be more easily discerned, and hence the results can be more quantitative. However, many key genes of known functions and novel genes can be missed if the probes on the arrays are not representative of the diversity of the community examined; thus, the activity information may be constrained to known taxa and functional populations.

The general belief is that only mRNA, proteins, and metabolite levels can be used to assess activities of individual taxa and populations (Nawy, 2013). However, in many cases, the activities of functional genes/populations can be inferred based on changes in DNA abundances, particularly if time-series data are available. That is, community members or functions that are more (or less) active in certain conditions would have increased (or decreased) abundances reflected in higher (or lower) abundance of associated DNA. Thus, DNA-based measurements can be appropriate for signifying changes in functional activities in these cases (e.g., a growing microbial population), potentially providing a good alternative for mRNA and protein measurements. Because the half-life of mRNA is generally very short, it might in any case be less suitable for comparing functional activities of genes/populations at ecological time scales (e.g., months, seasons, years). It is also more challenging to process RNA appropriately in the field, so the development of alternative approaches could broaden the situations in which microbiomes can be assessed in the built environment. Consequently, in many ecological studies, DNA-based abundance changes have been used for assessing the activities of functional genes/populations (Van Nostrand et al., 2011). However, it is important to be cautious because assessing functional activities in natural settings can be complicated. For instance, active genes could be coupled to nonactive genes within a single genome (e.g., presence of a nonexpressed *nif* operon), making the discrimination between active and nonactive genes/populations impossible. In this case, DNA-based abundance changes would be unsuitable for measuring functional activities. Thus, ideally, a combination of both DNA- and mRNA-based measurements, as well as protein- and metabolite-based measurements, would be used to assess the presence and activity of genes/populations in a complementary and mutually reinforcing fashion.

To optically determine microbial activity in situ, redox dyes that serve as nonspecific substrates for microbial respiration have been employed to stain actively respiring microorganisms within aerosol samples. A tetrazolium analog that intracellularly reduces to a fluorescing formazan is now commercially available in high purity (Rodriguez et al., 1992). These compounds provide for concurrent fluorescence determinations of bacterial numbers and metabolic activity regardless of propagation potential. Fungal conidial and hyphal viability have been quantitatively assessed using fluorochromes that mark fungal esterase activity. Fluorescein diacetate is a colorless substrate that is intracellularly cleaved to brightly fluorescing free fluorescein by active fungal esterase enzymes. It has been found to assess the viability of fungal hyphae and conidia accurately and precisely compared with standard germination tests (Firstencel et al., 1990; Jensen and Lysek, 1991). This compound provides for concurrent epifluorescent determinations of fungal mass and activity.

Functional Coverage

Functional coverage is the amount of gene function potential recovered from the microbiome sample. This remains a particular challenge, because even recovery of the complete complement of proteins, genes, or metabolites does not automatically yield an accurate functional assessment since the function may remain unknown.

Metagenomic-based gene annotation gives a preliminary description of biochemical process potential in microbial communities with transcriptionally active pathways identified through RNA-based metagenomic sequencing. Functional gene annotation operates primarily by identifying genes through sequence homology search using translated amino acid query sequences. There have been a few efforts to identify compact genetic signatures unique to gene families of interest (e.g., Kaminski et al., 2015) and applied to tracking antibiotic resistance genes in indoor environments (Hartmann et al., 2016). However, most gene identification pipelines continue to benefit from longer or assembled reads to recover the correct gene families (Carr and Borenstein, 2014). Functional annotation is reliant on the completeness of reference databases and multiple reference gene databases, such as KEGG (Kanehisa et al., 2014) and eggNOG (Huerta-Cepas et al., 2016). Gene function profiling is commonly separated from the organism, and abundance quantification of functional categories is calculated by measuring the numbers of reads that map to a set of reference protein sequences in a family or profile hidden Markov model (Eddy, 1998). Gene pathway abundance can then be inferred from gene family abundance. In addition to direct gene measurements, 16S data have been used as a proxy to retrieve reference genomes thought to be closely related to the 16S OTUs (Langille et al., 2013). However, particular attention needs to be paid to the potential for OTUs to match directly with the correct sequenced genomes and the potential for intra species and strain gene gain and loss, which would lead to incorrect taxonomic identification.

Integrating molecular data from multiple sources—genome, transcript, metabolite, and protein—presents an emerging opportunity to identify more accurately the biologic processes that explain how the diverse elements of the microbiome persist in a specific environment (Jansson and Baker, 2016; Quinn et al., 2016). For metaproteomics, however, determining the completeness of the proteins recovered remains a challenge, and applying an accurate protein database search strategy, ideally informed by sample-specific metagenomic data, influences the observable microbiome protein content (Armengaud, 2016; Herbst et al., 2016). Complete recovery of the biochemical structures directly from mass spectrometers has yet to be fully addressed, in part because of the lack of well-annotated databases, with only an estimated 1.8 percent of spectra annotated from untargeted mass spectrometry experiments (da Silva et al., 2015; Wang et al., 2016). Considerable efforts continue on developing databases of efficiently encoded spectra data to support a more complete recognition of unlabeled measured chemical compounds (Quinn et al., 2017). Recent efforts have demonstrated the utility of topic modeling, which illustrates the potential to improve chemical substructure feature encoding and also improve sensitivity in automated biochemical annotation (van der Hooft et al., 2016). Such approaches will also benefit from being integrated with molecular networking (Watrous et al., 2012) and genome mining strategies (Cimermancic et al., 2014; Donia et al., 2014). Genomic data, however, continue to serve as a convenient information source, and previously annotated metabolic pathways have been used to infer metabolic profiles using the relative abundance of annotated enzymes with some comparisons with direct measurements of metabolites (Noecker et al., 2016). These methods have the potential to predict metabolic processes that originate from individual genes and organisms and to enter into mechanistic models for community dynamics (Mendes-Soares et al., 2016). The models have so far been restricted to limited curated metabolic pathway data and rarely take multiorganism interactions such as competition or facilitation into account (Henry et al., 2016).

Of fundamental importance are tools that use multiple microbiome samples to build models that predict and explain the persistence of microbial communities and their functional characteristics. Co-association networks are a common approach used to predict pairwise organism interactions, including cooperation and competition, based on co-occurrence patterns in multiple samples (Faust et al., 2015). Inferring organism interaction networks remains challenging for several reasons, in part because compositional data present underlying dependencies between organisms, with normalization procedures that can confound some correlation methods (McMurdie and Holmes, 2014), and large-dimensional feature space can lead to sparse feature representation and model overfitting (Cardona et al., 2016). Multi-organism dependencies (beyond pairwise) may be important but require additional samples to support rigorous statistical confidence. Recent work has employed the use of synthetic community data to show how the choice of method can impact the inferences about any organism relationships detected (Weiss et al., 2016). General extensions from organism interaction networks to gene interaction networks have not been as well developed but could yield new predictive models of important molecular processes required for successful microbial communities (Boon et al., 2014).

Toxicologic Potential

Toxicologic potential reflects the ability of an individual chemical agent, or combination of chemical agents, to alter the normal metabolic functions or replication of a cell such that its life cycle is shortened, or its functioning is compromised relative to the condition where such exposure did not occur. Toxicity is classically realized in a dose-response scenario beyond a distinct threshold dose in biologic systems.

Indoor pollutants include airborne microscopic particulate matter comprised in whole or in part of biogenic materials, which are often termed "bioaerosol." By this definition, bioaerosols include all airborne microorganisms regardless of viability or ability to be recovered by culture; additionally, the term encompasses their fractions, other biopolymers, and products from all varieties of living things (ACGIH, 1999). Bioaerosols originate from occupants (e.g., humans, pets, houseplants), but they can also drift in from external sources, as well as originate from building materials in high-moisture environments and/or experiencing water damage, even after structures have been considered refurbished by modern construction practices.

Numerous publications on indoor air quality report that airborne biologic particles can range in aerodynamic diameter from 0.01 µm to 100 µm (ACGIH, 1999); in many environments, airborne bacteria, fungi, their fragments, and other biopolymeric materials may fall into a size range that can penetrate into human lungs (<3 µm) (Górny et al., 2002; Reponen et al., 2001). While only intact microorganisms can be infectious, and culturable numbers of airborne bacteria have been positively correlated with adverse respiratory symptoms (Björnsson et al., 1995), toxic, hypersensitive, and allergic reactions can also be caused by microorganism fragments or their biochemical by-products (Burrell, 1991; WHO, 1990). The health "penumbra" of the microbiome extends well beyond living, reproducing microbes. Well-known examples of potent biogenic factions, which are collectively referred to in this context as "biomarkers," include endotoxin, a compound found in the outer membranes of Gram-negative bacteria cell walls (ACGIH, 1999); many peptides from bacterial and fungal cell walls and metabolic products (ACGIH, 1999; Miller, 1992); β-(1-3)-D-glucans, found in fungal cell walls (ACGIH, 1999); and mycotoxins, products of fungal metabolism (Robbins et al., 2000). The exocellular toxins produced by the airborne bacteria responsible for whooping cough, *Bordetella pertussis*, serve as an unfortunate example of a reemerging toxigenic disease—the agents for which have never been recovered from ambient aerosol by conventional culture techniques.

While air quality indices and recommended threshold exposure levels are well defined in terms of certain chemical compounds and particulate matter masses, they are inadequately defined regarding airborne or surface-borne contaminants of biologic origin. In contrast with the scientific grounding for wastewater and drinking water regulation, bona fide toxicologic characterization of aerosols is only beginning to emerge in the aerosol and built environment community (Brook et al., 2010; Li et al., 2003), making it difficult to devise effective building regulations.

Only the culturable portion of bacteria in the atmosphere has been studied in detail (Hernandez et al., 1999; Moschandreas et al., 1996; Tong and Lighthart, 1999), and it is clear that air quality regulations have been biased by analytic reliance on culture over the last generation (Flannigan, 1997; Heidelberg et al., 1997; Henningson et al., 1997; MacNaughton et al., 1997). It is now known that many microorganisms (>99.9 percent in some environments) are not readily cultured using routine media and growth conditions (Amann et al., 1995; Pace, 1997). The most basic genetic characterizations are only beginning to be applied to the atmospheric environment (Womack et al., 2010), and toxicology assays have not been systematically adapted to determine the relationship between the amount of ambient (bio)aerosols and the likelihood of inducing stress responses in accepted cellular exposure models (Douwes et al., 2003).

Those limited bioaerosol regulations that do exist for indoor environments are in the form of guidelines based on culturable airborne microbe concentrations from grab samples, without taking into account other assays that could better indicate the potential for impacts on human and ecosystem health (Rao et al., 1996). There is also little recognition that transmission depends upon host traits as well, which in addition to such factors as age, health condition, and genetic variation in susceptibility, could in the built environment reflect behavioral patterns of space occupancy and use. Organizations such as the North Atlantic Treaty Organization (NATO) and the World Health Organization (WHO) have concurred that there is a pressing need to develop more accurate and robust methods for characterizing the biologic contributions to total exposure (aerosol) loads (Maroni et al., 1995; WHO, 1990), yet only in the past few years have basic toxicology perspectives emerged in the indoor air quality arena beyond the characterization of occupational exposures.

Quantification

Quantification refers to the basic definable unit of biologic measurement, which can be unambiguous and identified by a referenced and accepted definition (i.e., colony-forming unit). Determination of the different microbial taxa that make up indoor bioaerosols and are surface associated is particularly important because gross abundance can serve as a meaningful indicator of exposure to airborne respiratory health risks and allergens. Fluorochromes are now available that are specific for different groups of microorganisms and thus allow for representative estimates of the range of populations that make up the microbiological fraction of indoor aerosols. Since they are used in a direct visualization technique, stains considerably reduce the potential for inaccuracies in estimating microorganism numbers in environmental samples regardless of the medium (aerosol or surface). The fluorochrome stains acridine orange (AO), 5-(4,6-dichlorotriazinyl) aminofluorescein (DTAF), 4´6-diamino-2-phenylindole (DAPI), and calcofluor M2R target different biomolecules and have been used successfully in (built) environment samples for total microorganism counts and size measurements (Bloem et al., 1995; Hernandez et al., 1999; Hobbie et al., 1977; Wagner et al., 1994). Well-tested fluorochrome stains have been used to directly enumerate (by means of microscopy) three major microbial taxa that make up airborne and surface populations, and reviews on the subject relevant to the indoor environment are available (Peccia and Hernandez, 2006).

Various quantitative parameters (e.g., absolute abundance, relative abundance, and copy numbers) are used to capture different biologic properties of a taxon or gene in a community (Nayfach and Pollard, 2016). Ideally, absolute abundances of individual taxa or genes in a community are desired for subsequent statistical analysis and model prediction. However, obtaining accurate absolute abundance data using high-throughput sequencing technologies is challenging. Because of inherently high variations in experimental protocols and bioinformatics, measurements can be quite different among different samples even under identical conditions. Therefore, most microbial ecology analyses with sequencing data are based on relative rather than absolute abundance. This is an important impediment in relating microbiome data to fundamental models in population, community, and ecosystem ecology, where absolute numbers matter. Moreover, even relative abundance calculations for universal marker genes are limited by amplification bias (Brooks et al., 2015). When unbiased, such data provide relative measures of community composition, which need not correspond with absolute abundance. Careful attention needs to be paid to using statistical methods that account for compositional bias to avoid false correlations across multiple samples with systematic differences in sample characteristics, such as sequencing depth (Kurtz et al., 2015). Computationally, metagenomic sequencing provides additional information, with the potential to measure copies of recovered genes and genomes irrespective of the organisms present. Relying on universal marker genes remains the most common approach for abundance profiling (Manor and Borenstein, 2015; Nayfach et al., 2016).

There are additional efforts to apply statistically rigorous abundance quantification in terms of genome abundance; however, more work is needed to evaluate existing datasets on real conditions of microbiomes in built environments (McLoughlin, 2016). Nevertheless, modeling abundance using information beyond a selected set of marker genes could improve flexibility, permitting one to quantify abundance in cases where marker genes are inaccessible. A promising recent development is the peak-to-trough ratio (Korem et al., 2015) and a related extension (Brown et al., 2016). These methods observe that bacterial replication begins at a distinct replication site, and a replication rate can be inferred by measuring the change in sequencing depth for each genome, starting from the origin of replication to the distal portion of the genome. The replication rates are used to infer bacterial growth rates from a single microbiome snapshot and present an important emerging informatics technique for estimating replication rates for microbial communities in the built environment.

Because traditional PCR amplification is involved in amplicon-based target sequencing, it appears that target gene sequencing is not quantitative in complex communities as previously demonstrated (Pinto and Raskin, 2012; Tremblay et al., 2015; Zhou et al., 2011). This limitation is consistent with the previous observations about pyrotag sequencing studies (Engelbrektson et al., 2010) and with the general consensus that traditional PCR amplification is not quantitative (Qiu et al., 2001; Suzuki and Giovannoni, 1996). Various strategies have been proposed for alleviating the amplification biases on quantification, including combining several amplifications (>3) and using fewer cycle numbers (e.g., 25 or no more than 30 cycles) to avoid PCR product saturation. However, some studies have shown that quantitative estimation of organismal abundance can be obtained with deep amplicon sequencing of 16S rRNA genes in a simple microbial community (Avramenko et al., 2015; de Boer et al., 2015). Nevertheless, great caution is necessary in drawing quantitative inferences about microbial community diversity, and in particular absolute abundances, in comparative studies based on amplicon sequencing data.

There are two general approaches for quantitatively assessing the abundance of organisms and functional genes in a shotgun metagenome approach. One is to classify sequencing reads according to reference databases of genes and/or genomes via alignment-based homology analyses, followed by counting the classified reads to estimate taxonomic groups and gene families (Nayfach and Pollard, 2016). The complementary approach is to perform de novo analyses by grouping shotgun sequencing and then annotating the resulting OTUs and gene families based on their homology to known genes. Theoretically, it is believed that shotgun sequencing could be quantitative (Nayfach and Pollard, 2016; Zhou et al., 2015) because shotgun sequencing of whole communities does not require amplification prior to sequencing if template DNA is sufficient, and hence it avoids many of the biases encountered in amplicon sequencing. However, quantifying organisms and functional genes in a shotgun metagenome also presents several unique challenges (Nayfach and Pollard, 2016).

First, reference databases used to date do not represent the vast majority of microbial diversity at low taxonomic levels. Consequently, it is difficult to assign taxa or genes with high confidence to estimate their abundance (Nayfach and Pollard, 2016). Also, high inherent variation in experimental protocols and uncertainty in selecting bioinformatics tools for analysis challenge the quantitative estimation of the abundances of individual taxa and genes (Clooney et al., 2016; Kerepesi and Grolmusz, 2016; Nayfach and Pollard, 2016). In addition, the massive size of short metagenome reads data makes it very difficult to compare taxa and genes across different samples. As a result, it might be impossible to obtain an absolute abundance estimation based on shotgun sequencing data alone (Nayfach and Pollard, 2016). By combining sequencing with other techniques, such as density measurements and quantitative PCR, it might be feasible to obtain estimates of absolute abundance (Nayfach and Pollard, 2016). Finally, it has recently been concluded that cellular relative abundance and average genomic copy number are the most meaningful biologic parameters that can be quantified based on shotgun sequencing reads (Nayfach and Pollard, 2016).

In contrast to sequencing-based approaches, absolute abundance of taxa and genes can be estimated based on the signal intensity from array hybridization. This is because signal intensity is derived from the extent of actual hybridization, and the hybridization signals reflect the absolute abundance for the amounts of DNAs used for hybridization. During the past decade or so, numerous studies have demonstrated strong correlations between target DNA or RNA concentrations and GeoChip hybridization signal intensities using pure cultures, mixed cultures, and environmental samples without amplification (Brodie et al., 2007; Gao et al., 2007; He et al., 2010a; Tiquia et al., 2004; Wu et al., 2006). Very good correlations were also observed between PhyloChip signal intensities and quantitative PCR copy numbers spanning five orders of magnitude (Brodie et al., 2007; Lemon et al., 2010). These results suggest that the array-based approaches are highly quantitative with environmental DNAs and could provide one avenue toward refined abundance estimation for microbial assemblages.

Reproducibility

Reproducibility needs to address both technical and biologic variations. In a deterministic world, if one could measure a sample nondestructively, then repeatedly using the same protocol applied to the same sample, one would be able to repeat one's results. More practically, if the same sample were split into multiple aliquots and sent to different labs to recover the microbiome contents, one would consider results "reproducible" if all labs returned the same answer when using the same measurement tools. A possible starting point could be to send the same raw data (rather than a starting sample) to different labs to compare the analytic results obtained. Because of inherent stochasticity, there will be sampling error, which needs to be taken into account. Moreover, the potentially dynamic nature of the microbes and the high degree of community complexity mean that natural biologic variation can prevent two labs from producing the same results even after controlling for technical variation. Nevertheless, developing an understanding of the conditions under which accurate and reproducible microbiome measurements in built environments can be made is a foundational requirement for moving investigative research toward practical building design applications that are subject to greater oversight by a broad community of stakeholders.

Reproducibility is a big concern both scientifically and ethically (AAM, 2016). Part of the irreproducibility is due to technologies themselves, because of measurement errors and biases and sampling processes. Such issues have not been appropriately recognized and addressed in microbial ecology until recently (Zhou et al., 2013). A few years ago, it was first noticed that amplicon-based sequencing approaches have very low reproducibility, with <15 percent OTU overlap among three technical replicates (Zhou et al., 2011), which is well below the theoretical expectation of 100 percent overlap among technical replicates in which the same DNA from the same samples were amplified and sequenced three times. This phenomenon has recently been well established experimentally across different laboratories (Flores et al., 2012; Ge et al., 2014; Palmer and Horn, 2012; Peng et al., 2013; Pinto and Raskin, 2012; Sinclair et al., 2015; Talley and Fodor, 2011; Vishnivetskaya et al., 2014; Wen et al., 2017; Xu et al., 2011; Zhan et al., 2014; Zhou et al., 2011), although discrepant results have been observed (Bartram et al., 2011; Kauserud et al., 2012; Mao et al., 2011; Pilloni et al., 2012). Part of the reason for such discrepancies could be the complexity of the ecosystems examined—for instance, because of microscale variation in microbial community composition (Kauserud et al., 2012; Pinto and Raskin, 2012; Zhou et al., 2011)—but part could be lab-based, such as differences in sequencing depths (Bartram et al., 2011; Lemos et al., 2012; Zhou et al., 2011) and/or variations in sequencing and sequence preprocessing approaches (Pinto and Raskin, 2012; Schloss et al., 2011). Based on random sampling theory, mathematical simulation explicitly demonstrated that low technical reproducibility for amplicon sequencing is most likely due to the artifacts associated with random sampling processes inherent in PCR amplification and sequencing (Zhou et al., 2013, 2015), which would contribute to variations because of inherent spatial and temporal heterogeneity in the microbiome itself. To achieve high technical reproducibility, several orders of magnitude more sequencing efforts are needed (Zhou et al., 2013). This makes extensive sampling that is quantitatively rich methodologically challenging. Similar challenges also exist for other "omics" technologies, such as proteomics and metabolomics.

Although shotgun sequencing avoids many of the biases encountered in amplicon sequencing, the reproducibility problem could be more severe with shotgun sequencing than with amplicon-based target sequencing. This is because sampling processes via shotgun sequencing from a community with thousands and up to hundreds of thousands of species and thousands of genes from each species are likely much more random than those via amplicon sequencing. However, no experimental evidence is available as yet to support such speculations.

In contrast, the array-based closed format has lower susceptibility relative to the sequencing-based open format for random sampling artifacts (Zhou et al., 2015). Because the number of detected taxa or genes is defined by the probe sets on the array, the overlap among technical replicates is less dependent on the level of sampling effort. Thus, high technical reproducibility would be expected, as demonstrated by mathematical simulation (Zhou et al., 2015).

Because the technical variations associated with random sampling processes could greatly overestimate microbial β-diversity, high reproducibility is critical for comparative studies to be reliable across different spatial and temporal scales and environmental gradients. Great caution is needed in quantifying and interpreting β-diversity for microbial community analysis using high-throughput metagenomics technologies, particularly next-generation sequencing. Null model approaches (such as those used in community ecology [Gotelli, 2001]) could be a valuable tool for assessing the degree of reproducibility in empirical microbiome studies. In addition, reproducibility depends on software, versions of software packages, predefined parameters of the software, and databases.

There have been several benchmarking efforts designed to demonstrate reproducibility on multiple aspects of the problem from sample collection to data analysis, yet complete, accurate, and reproducible recovery of complex communities remains challenging. For example, comparison of different DNA extraction and sequencing library preparation protocols has shown that the observed microbial community can differ dramatically (Brooks et al., 2015; Hart et al., 2015; Jones et al., 2015). In addition, there have been several published efforts to develop mock community resources. A "mock" community is a defined synthetic mixture of microbial cells or nucleic acids designed to simulate a microbial community (Highlander, 2015). To date, the vast majority of resources focus either on defining 16S-based reference material (Bokulich et al., 2016), with limited microbial diversity (Morgan et al., 2010; Tanca et al., 2013), or exclusively on the human microbiome (HMP Consortium, 2012; Sinha et al., 2015). Benchmarking studies can be expensive and challenging to do well, and it may be difficult to obtain funding for such studies.

In addition to published reports, multiple consortia have begun to form to help organize future efforts in designing reference material. Beyond physical mock communities, there are efforts to develop simulated datasets with which to benchmark computational tools, which have demonstrated that different descriptions of community structure may be found for the same input data depending on the choice of analysis tool and choice of parameters (Lindgreen et al., 2016; Randle-Boggis et al., 2016; Weiss et al., 2016; see http://www.cami-challenge.org).

Mock communities in the context of human microbiomes helped establish sequencing protocols for sequencing centers, which are working to expand their sequencing capability to support microbial community profiling (Gohl et al., 2016). Mock communities have been used to evaluate the impact of analytic parameters on functional annotation (Nayfach et al., 2015). However, considerably less attention has been given to designing mock communities and benchmarking standards so as to better reflect indoor built environments. An open question remains of how best to use existing benchmarking efforts to guide future validation work in the built environment. In addition, opportunities to build reference material that better captures living biologic material in a controlled environment would further enhance existing reference material resources (Ling et al., 2015).

Data Sharing and Metadata on Buildings and Building Systems

A key component to support reproducibility and the maturation of the microbiome modeling field is ensuring that experimental data and software are accessible to the research community. This requires publicly accessible data repositories such as the Sequence Read Archive, where raw genomic data can be housed, as well as data-sharing standards to ensure sufficiently complete and accurate descriptions of the experimental conditions used to generate new microbiome data (Leinonen et al., 2011).

Studies of indoor microbial communities and the factors that affect them necessarily need to characterize the buildings in which the studies are conducted. This information is essential for understanding how the features of buildings and building systems influence these communities, in order to design and operate buildings to reduce the likelihood of detrimental health outcomes and to increase the potential for beneficial communities in the future.

LOOKING TO THE FUTURE

Various high-throughput technologies of both open and closed formats have been developed and used for the analysis of microbial communities. Each has advantages and disadvantages in terms of specificity, resolution, sensitivity, activity measurement, quantification, and reproducibility. Generally speaking, while the closed-format technologies have advantages for hypothesis-driven comparative studies, open-format technologies are excellent for exploratory discovery studies (Zhou et al., 2015). They can be integrated in a complementary fashion to address complex biologic questions and objectives (Zhou et al., 2015). Also, careful experimental design is as important as the selection of various "omics" technologies. Because all high-throughput technologies have inherently high noise, increasing biologic replicates is critical for ameliorating the impacts of technical variations on drawing biologic conclusions. The numbers of biologic replicates needed depend on the biologic systems examined, the research questions, the objectives, and the magnitudes of biologic variations. In addition, because of various potential technical difficulties associated with specificity, sensitivity, quantification, resolution, and/or reproducibility, it is extremely helpful to apply high-throughput "omics" technologies for relative comparisons, which are the comparisons between communities (e.g., between treatment and control samples), and thus typically the ratio (treatment/control) is used. This is especially important when dealing with environmental samples of unknown composition (He et al., 2007). Because the signal ratios (based on sequencing reads or hybridization intensity) of treatment samples to control samples are used, the effects of errors, biases, sampling processes, and bioinformatics uncertainty can potentially be canceled out if the biases and errors are more or less similar between the treatment and control samples (He et al., 2007; Zhou et al., 2015). This could be the best use of "omics" data to address biologic questions of interest.

REFERENCES

AAM (American Academy of Microbiology). 2016. *Promoting responsible scientific research*. Washington, DC: AAM. https://www.asm.org/images/Colloquia-report/Promoting_Responsible_Scientific_Research.pdf (accessed July 8, 2017).

ACGIH (American Conference of Governmental Industrial Hygienists). 1999. *Bioaerosols: Assessment and control*, 1st ed., edited by J. M. Macher. Cincinnati, OH: ACGIH.

Alivisatos, A. P., M. J. Blaser, E. L. Brodie, M. Chun, J. L. Dangl, T. J. Donohue, P. C. Dorrestein, J. A. Gilbert, J. L. Green, J. K. Jansson, R. Knight, M. E. Maxon, M. J. McFall-Ngai, J. F. Miller, K. S. Pollard, E. G. Ruby, and S. A. Taha. 2015. A unified initiative to harness Earth's microbiomes. *Science* 350(6260):507-508.

Alneberg, J., B. S. Bjarnason, I. de Bruijn, M. Schirmer, J. Quick, U. Z. Ijaz, L. Lahti, N. J. Loman, A. F. Andersson, and C. Quince. 2014. Binning metagenomic contigs by coverage and composition. *Nature Methods* 11(11):1144-1146.

Alvarez, A. J., M. P. Buttner, and L. D. Stetzenbach. 1995. PCR for bioaerosol monitoring: Sensitivity and environmental interference. *Applied and Environmental Microbiology* 61:3639-3644.

Amann, R. I., W. Ludwig, and K. H. Schleifer. 1995. Phylogenetic identification and in situ detection of individual microbial cells without cultivation. *Microbiological Reviews* 59(1):143-169.

Ames, S. K., S. N. Gardner, J. M. Marti, T. R. Slezak, M. B. Gokhale, and J. E. Allen. 2015. Using populations of human and microbial genomes for organism detection in metagenomes. *Genome Research* 25:1056-1067.

Armengaud, J. 2016. Next-generation proteomics faces new challenges in environmental biotechnology. *Current Opinion in Biotechnology* 38:174-182.

Avramenko, R. W., E. M. Redman, R. Lewis, T. A. Yazwinski, J. D. Wasmuth, and J. S. Gilleard. 2015. Exploring the gastrointestinal "nemabiome": Deep amplicon sequencing to quantify the species composition of parasitic nematode communities. *PLOS ONE* 10(12):e0143559.

Bartram, A. K., M. D. J. Lynch, J. C. Stearns, G. Moreno-Hagelsieb, and J. D. Neufeld. 2011. Generation of multimillion-sequence 16S rRNA gene libraries from complex microbial communities by assembling paired-end illumina reads. *Applied and Environmental Microbiology* 77(11):3846-3852.

Be, N. A., J. B. Thissen, S. N. Gardner, K. S. McLoughlin, V. Y. Fofanov, H. Koshinsky, S. R. Ellingson, T. S. Brettin, P. J. Jackson, and C. J. Jaing. 2013. Detection of Bacillus anthracis DNA in complex soil and air samples using next-generation sequencing. *PLOS ONE* 8(9):e73455.

Be, N. A., J. B. Thissen, V. Y. Fofanov, J. E. Allen, M. Rojas, G. Golovko, Y. Fofanov, H. Koshinsky, and C. J. Jaing. 2015. Metagenomic analysis of the airborne environment in urban spaces. *Microbial Ecology* 69(2):346-355.

Beitel, C. W., L. Froenicke, J. M. Lang, I. F. Korf, R. W. Michelmore, J. A. Eisen, and A. E. Darling. 2014. Strain- and plasmid-level deconvolution of a synthetic metagenome by sequencing proximity ligation products. *PeerJ* 2:e415.

Benitez-Paez, A., K. J. Portune, and Y. Sanz. 2016. Species-level resolution of 16S rRNA gene amplicons sequenced through the MinION™ portable nanopore sequencer. *GigaScience* 5:4. doi:10.1186/s13742-016-0111-z.

Biteen, J. S., P. C. Blainey, Z. G. Cardon, M. Chun, G. M. Church, P. C. Dorrestein, S. E. Fraser, J. A. Gilbert, J. K. Jansson, R. Knight, J. F. Miller, A. Ozcan, K. A. Prather, S. R. Quake, E. G. Ruby, P. A. Silver, S. Taha, G. van den Engh, P. S. Weiss, G. C. Wong, A. T. Wright, and T. D. Young. 2016. Tools for the microbiome: Nano and beyond. *ACS Nano* 10(1):6-37.

Björnsson, E., D. Norback, C. Janson, J. Widstrom, U. Palmgren, G. Strom, and G. Boman. 1995. Asthmatic symptoms and indoor levels of micro-organisms and house dust mites. *Clinical & Experimental Allergy* 25(4):423-431.

Bloem, J., M. Veninga, and J. Shepherd. 1995. Fully automatic determination of soil bacterium numbers, cell volumes, and frequencies of dividing cells by confocal laser scanning microscopy and image analysis. *Applied and Environmental Microbiology* 61:926-936.

Bokulich, N. A., J. R. Rideout, W. G. Mercurio, A. Shiffer, B. Wolfe, C. F. Maurice, R. J. Dutton, P. J. Turnbaugh, R. Knight, and J. G. Caporaso. 2016. Mockrobiota: A public resource for microbiome bioinformatics benchmarking. *mSystems* 1(5):e00062-16. doi:10.1128/mSystems.00062-16.

Boon, E., C. J. Meehan, C. Whidden, D. H.-J. Wong, M. G. Langille, and R. G. Beiko. 2014. Interactions in the microbiome: Communities of organisms and communities of genes. *FEMS Microbiology Reviews* 38(1):90-118.

Brodie, E. L., T. Z. DeSantis, J. P. M. Parker, I. X. Zubietta, Y. M. Piceno, and G. L. Andersen. 2007. Urban aerosols harbor diverse and dynamic bacterial populations. *Proceedings of the National Academy of Sciences of the United States of America* 104(1):299-304.

Brook, R. D., S. Rajagopalan, C. A. Pope, III, J. R. Brook, A. Bhatnagar, A. V. Diez-Roux, F. Holguin, Y. Hong, R. V. Luepker, M. A. Mittleman, A. Peters, D. Siscovick, S. C. Smith, Jr., L. Whitsel, and J. D. Kaufman. 2010. Particulate matter air pollution and cardiovascular disease: An update to the scientific statement from the American Heart Association. *Circulation* 121(21):2331-2378.

Brooks, J. P., D. J. Edwards, M. D. Harwich, M. C. Rivera, J. M. Fettweis, M. G. Serrano, R. A. Reris, N. U. Sheth, B. Huang, P. Girerd, J. F. Strauss, III, K. K. Jefferson, and G. A. Buck. 2015. The truth about metagenomics: Quantifying and counteracting bias in 16S rRNA studies. *BMC Microbiology* 15:66.

Brown, C. T., M. R. Olm, B. C. Thomas, and J. F. Banfield. 2016. Measurement of bacterial replication rates in microbial communities. *Nature Biotechnology* 34:1256-1263.

Burge, H. A. 1990. Bioaerosols: Prevalence and health effects in the indoor environment. *Journal of Allergy and Clinical Immunology* 86:687-701.

Burrell, R. 1991. Microbiological agents as health risks in indoor air. *Environmental Health Perspectives* 95:29-34.

Buttner, M. P., P. V. Scarpino, and C. S. Clark. 1993. Monitoring airborne fungal spores in an experimental indoor environment to evaluate sampling methods and the effects of human activity on air sampling. *Applied and Environmental Microbiology* 59:219-226.

Cardona, C., P. Weisenhorn, C. Henry, and J. A. Gilbert. 2016. Network-based metabolic analysis and microbial community modeling. *Current Opinion in Microbiology* 31:124-131.

Carr, R., and E. Borenstein. 2014. Comparative analysis of functional metagenomic annotation and the mappability of short reads. *PLOS ONE* 9(8):e105776.

Cimermancic, P., M. H. Medema, J. Claesen, K. Kurita, L. C. Wieland Brown, K. Mavrommatis, A. Pati, P. A. Godfrey, M. Koehrsen, J. Clardy, B. W. Birren, E. Takano, A. Sali, R. G. Linington, and M. A. Fischbach. 2014. Insights into secondary metabolism from a global analysis of prokaryotic biosynthetic gene clusters. *Cell* 158(2):412-421.

Clooney, A. G., F. Fouhy, R. D. Sleator, A. O. Driscoll, C. Stanton, P. D. Cotter, and M. J. Claesson 2016. Comparing apples and oranges?: Next generation sequencing and its impact on microbiome analysis. *PLOS ONE* 11(2):e0148028.

Cole, J. R., B. Chai, R. J. Farris, Q. Wang, A. S. Kulam-Syed-Mohideen, D. M. McGarrell, A. M. Bandela, E. Cardenas, G. M. Garrity, and J. M. Tiedje. 2007. The Ribosomal Database Project (RDP-II): Introducing myRDP space and quality controlled public data. *Nucleic Acids Research* 35:D169-D172.

da Silva, R. R., P. C. Dorrestein, and R. A. Quinn. 2015. Illuminating the dark matter in metabolomics. *Proceedings of the National Academy of Sciences of the United States of America* 112(41):12549-12550.

de Boer, P., M. Caspers, J. W. Sanders, R. Kemperman, J. Wijman, G. Lommerse, G. Roeselers, R. Montijn, T. Abee, and R. Kort. 2015. Amplicon sequencing for the quantification of spoilage microbiota in complex foods including bacterial spores. *Microbiome* 3:30.

DeAngelis, K. M., C. H. Wu, H. R. Beller, E. L. Brodie, R. Chakraborty, T. Z. DeSantis, J. L. Fortney, T. C. Hazen, S. R. Osman, M. E. Singer, L. M. Tom, and G. L. Andersen. 2011. PCR amplification-independent methods for detection of microbial communities by the high-density microarray phylochip. *Applied and Environmental Microbiology* 77(18):6313-6322.

DeLong, E. F. 2009. The microbial ocean from genomes to biomes. *Nature* 459(7244):200-206.

Devault, A. M., K. McLoughlin, C. Jaing, S. Gardner, T. M. Porter, J. M. Enk, J. Thissen, J. Allen, M. Borucki, and S. N. DeWitte. 2014. Ancient pathogen DNA in archaeological samples detected with a Microbial Detection Array. *Scientific Reports* 4:4245. doi:10.1038/srep04245.

Donia, M. S., P. Cimermancic, C. J. Schulze, L. C. Wieland Brown, J. Martin, M. Mitreva, J. Clardy, R. G. Linington, and M. A. Fischbach. 2014. A systematic analysis of biosynthetic gene clusters in the human microbiome reveals a common family of antibiotics. *Cell* 158(6):1402-1414.

Douwes, J., P. Thorne, N. Pearce, and D. Heederik. 2003. Bioaerosol health effects and exposure assessment: Progress and prospects. *Annals of Occupational Hygiene* 47(3):187-200.

Eddy, S. R. 1998. Profile hidden Markov models. *Bioinformatics Review* 14(9):755-763.

Edgar, R. C. 2013. UPARSE: Highly accurate OTU sequences from microbial amplicon reads. *Nature Methods* 10(10):996-998.

Engelbrektson, A., V. Kunin, K. C. Wrighton, N. Zvenigorodsky, F. Chen, H. Ochman, and P. Hugenholtz. 2010. Experimental factors affecting PCR-based estimates of microbial species richness and evenness. *The ISME Journal* 4(5):642-647.

Eren, A. M., M. L. Sogin, and L. Maignien. 2016. Editorial: New insights into microbial ecology through subtle nucleotide variation. *Frontiers in Microbiology* 7:1318.

Faust, K., G. Lima-Mendez, J.-S. Lerat, J. F. Sathirapongsasuti, R. Knight, C. Huttenhower, T. Lenaerts, and J. Raes. 2015. Cross-biome comparison of microbial association networks. *Frontiers in Microbiology* 6:1200.

Firstencel, H., T. M. Butt, and R. I. Carruthers. 1990. A fluorescence microscopy method for determining the viability of entomophthoralean fungal spores. *Journal of Invertebrate Pathology* 55(2):258-264.

Fischer, M. A., S. Gullert, S. C. Neulinger, W. R. Streit, and R. A. Schmitz. 2016. Evaluation of 16S rRNA gene primer pairs for monitoring microbial community structures showed high reproducibility within and low comparability between datasets generated with multiple archaeal and bacterial primer pairs. *Frontiers in Microbiology* 7:1297.

Flannigan, B. 1992. Indoor microbial pollutants: Sources, species, characterisation and evaluation. In *Chemical, microbiological, health and comfort aspects of indoor air quality: State of the art in SBS*, edited by H. Knöppel and P. Wolkoff. Kluwer: Dordrecht, The Netherlands. Pp. 73-98.

Flannigan, B. 1993. Approaches to assessment of the microbial flora of buildings. In *ASHRAE IAQ 1992, Environments for healthy people*. Atlanta, GA: ASHRAE. Pp. 139-145.

Flannigan, B. 1997. Air sampling for fungi in indoor environments. *Journal of Aerosol Science* 28(3):381-392.

Flannigan, B., E. M. McCabe, and F. McCarry. 1991. Allergenic and toxigenic microorganisms in houses. *Journal of Applied Bacteriology* 70(Symposium Suppl.):61S-73S.

Flores, R., J. Shi, M. H. Gail, P. Gajer, J. Ravel, and J. J. Goedert. 2012. Assessment of the human faecal microbiota: II. Reproducibility and associations of 16S rRNA pyrosequences. *European Journal of Clinical Investigation* 42(8):855-863.

Frias-Lopez, J., Y. Shi, G. W. Tyson, M. L. Coleman, S. C. Schuster, and S. W. Chisholm. 2008. Microbial community gene expression in ocean surface waters. *Proceedings of the National Academy of Sciences of the United States of America* 105(10):3805-3810.

Gao, H., Z. K. Yang, T. J. Gentry, L. Wu, C. W. Schadt, and J. Zhou. 2007. Microarray-based analysis of microbial community RNAs by whole-community RNA amplification. *Applied and Environmental Microbiology* 73(2):563-571.

Ge, Y., J. P. Schimel, and P. A. Holden. 2014. Analysis of run-to-run variation of bar-coded pyrosequencing for evaluating bacterial community shifts and individual taxa dynamics. *PLOS ONE* 9(6):e99414.

Gevers, D., M. Pop, P. D. Schloss, and C. Huttenhower. 2012. Bioinformatics for the Human Microbiome Project. *PLOS Computational Biology* 8(11):e1002779.

Ghurye, J. S., V. Cepeda-Espinoza, and M. Pop. 2016. Metagenomic assembly: Overview, challenges and applications. *Yale Journal of Biology and Medicine* 89(3):353-362.

Giannoukos, G., D. Ciulla, K. Huang, B. Haas, J. Izard, J. Levin, J. Livny, A. Earl, D. Gevers, D. Ward, C. Nusbaum, B. Birren, and A. Gnirke. 2012. Efficient and robust RNA-seq process for cultured bacteria and complex community transcriptomes. *Genome Biology* 13(3):r23.

Gilbert, J. A., R. A. Quinn, J. Debelius, Z. Z. Xu, J. Morton, N. Garg, J. K. Jansson, P. C. Dorrestein, and R. Knight. 2016. Microbiome-wide association studies link dynamic microbial consortia to disease. *Nature* 535:94-103.

Gohl, D. M., P. Vangay, J. Garbe, A. MacLean, A. Hauge, A. Becker, T. J. Gould, J. B. Clayton, T. J. Johnson, R. Hunter, D. Knights, and K. B. Beckman. 2016. Systematic improvement of amplicon marker gene methods for increased accuracy in microbiome studies. *Nature Biotechnology* 34:942-949.

Górny, R. L., T. Reponen, K. Willeke, D. Schmechel, E. Robine, M. Boissier, and S. A. Grinshpun. 2002. Fungal fragments as indoor air biocontaminants. *Applied and Environmental Microbiology* 68(7):3522-3531.

Gotelli, N. J. 2001. Research frontiers in null model analysis. *Global Ecology and Biogeography* 10(4):337-343.

Greenblum, S., R. Carr, and E. Borenstein. 2015. Extensive strain-level copy-number variation across human gut microbiome species. *Cell* 160(4):583-594.

Griffiths, W. D., I. W. Stewart, A. R. Reading, and S. J. Futter. 1996. Effect of aerosolization, growth phase and residence time in spray and collection fluids on the culturability of cells and spores. *Journal of Aerosol Science* 27(5):803-820.

Hanson, C. A., J. A. Fuhrman, M. C. Horner-Devine, and J. B. H. Martiny. 2012. Beyond biogeographic patterns: Processes shaping the microbial landscape. *Nature Reviews Microbiology* 10(7):497-506.

Hart, M. L., A. Meyer, P. J. Johnson, and A. C. Ericsson. 2015. Comparative evaluation of DNA extraction methods from feces of multiple host species for downstream next-generation sequencing. *PLOS ONE* 10(11):e0143334.

Hartmann, E. M., R. Hickey, T. Hsu, C. M. Betancourt Román, J. Chen, R. Schwager, J. Kline, G. Z. Brown, R. U. Halden, C. Huttenhower, and J. L. Green. 2016. Antimicrobial chemicals are associated with elevated antibiotic resistance genes in the indoor dust microbiome. *Environmental Science & Technology* 50(18):9807-9815.

He, Z., T. J. Gentry, C. W. Schadt, L. Wu, J. Liebich, S. C. Chong, Z. Huang, W. Wu, B. Gu, P. Jardine, C. Criddle, and J. Zhou. 2007. GeoChip: A comprehensive microarray for investigating biogeochemical, ecological and environmental processes. *The ISME Journal* 1(1):67-77.

He, Z., Y. Deng, J. D. Van Nostrand, Q. Tu, M. Xu, C. L. Hemme, X. Li, L. Wu, T. J. Gentry, Y. Yin, J. Liebich, T. C. Hazen, and J. Zhou. 2010a. GeoChip 3.0 as a high-throughput tool for analyzing microbial community composition, structure and functional activity. *The ISME Journal* 4(9):1167-1179.

He, S., O. Wurtzel, K. Singh, J. L. Froula, S. Yilmaz, S. G. Tringe, Z. Wang, F. Chen, E. A. Lindquist, R. Sorek, and P. Hugenholtz. 2010b. Validation of two ribosomal RNA removal methods for microbial metatranscriptomics. *Nature Methods* 7(10):807-812.

He, Y., J. G. Caporaso, X. T. Jiang, H. F. Sheng, S. M. Huse, and J. R. Rideout. 2015. Stability of operational taxonomic units: An important but neglected property for analyzing microbial diversity. *Microbiome* 3:20.

Heidelberg, J. F., M. Shahamat, M. Levin, I. Rahman, G. Stelma, C. Grim, and R. R. Colwell. 1997. Effect of aerosolization on culturability and viability of gram-negative bacteria. *Applied and Environmental Microbiology* 63(9):3585-3588.

Henningson, E. W., M. Lundquist, E. Larsson, G. Sandstrom, and M. Forsman. 1997. A comparative study of different methods to determine the total number and the survival ratio of bacteria in aerobiological samples. *Journal of Aerosol Science* 28(3):459-469.

Henry, C. S., H. C. Bernstein, P. Weisenhorn, R. C. Taylor, J.-Y. Lee, J. Zucker, and H.-S. Song. 2016. Microbial community metabolic modeling: A community data-driven network reconstruction. *Journal of Cellular Physiology* 231(11):2339-2345.

Herbst, F.-A., V. Lünsmann, H. Kjeldal, N. Jehmlich, A. Tholey, M. Bergen, J. L. Nielsen, R. L. Hettich, J. Seifert, and P. H. Nielsen. 2016. Enhancing metaproteomics—The value of models and defined environmental microbial systems. *Proteomics* 16(5):783-798.

Hernandez, M., S. L. Miller, D. W. Landfear, and J. M. Macher. 1999. A combined fluorochrome method for quantitation of metabolically active and inactive airborne bacteria. *Aerosol Science and Technology* 30(2):145-160.

Hess, M., A. Sczyrba, R. Egan, T.-W. Kim, H. Chokhawala, G. Schroth, S. Luo, D. S. Clark, F. Chen, T. Zhang, R. I. Mackie, L. A. Pennacchio, S. G. Tringe, A. Visel, T. Woyke, Z. Wang, and E. M. Rubin. 2011. Metagenomic discovery of biomass-degrading genes and genomes from cow rumen. *Science* 331(6016):463-467.

Highlander, S. 2015. Mock community analysis. *Encyclopedia of Metagenomics: Genes, Genomes and Metagenomes: Basics, Methods, Databases and Tools* 497-503.

Hinds, W. C. 1982. *Aerosol Technology: Properties, Behavior, and Measurement of Airborne Particles.* New York: John Wiley and Sons.

Hobbie, J. E., R. J. Daley, and S. Jasper. 1977. Use of nucleopore filters for counting bacteria by fluorescence microscopy. *Applied and Environmental Microbiology* 33:1225-1228.

HMP (Human Microbiome Project) Consortium. 2012. A framework for human microbiome research. *Nature* 486:215-221.

Huerta-Cepas, J., D. Szklarczyk, K. Forslund, H. Cook, D. Heller, M. C. Walter, T. Rattei, D. R. Mende, S. Sunagawa, M. Kuhn, L. J. Jensen, C. von Mering, and P. Bork. 2016. eggNOG 4.5: A hierarchical orthology framework with improved functional annotations for eukaryotic, prokaryotic and viral sequences. *Nucleic Acids Research* 44(D1):D286-D293.

Imelfort, M., D. Parks, B. J. Woodcroft, P. Dennis, P. Hugenholtz, and G. W. Tyson. 2014. GroopM: An automated tool for the recovery of population genomes from related metagenomes. *PeerJ* 2:e603.

Jansson, J. K., and E. S. Baker. 2016. A multi-omic future for microbiome studies. *Nature Microbiology* 1(5):16049.

Jensen, C., and G. Lysek. 1991. Direct observation of trapping activities of nematode-destroying fungi in the soil using fluorescence microscopy. *FEMS Microbiology Letters* 85(3):207-210.

Jensen, P. A., W. F. Todd, G. N. Davis, and P. V. Scarpino. 1992. Evaluation of eight bioaerosol samplers challenged with aerosols of free bacteria. *Journal of the American Industrial Hygiene Association* 53:660-667.

Jones, M. B., S. K. Highlander, E. L. Anderson, W. Li, M. Dayrit, N. Klitgord, M. M. Fabani, V. Seguritan, J. Green, D. T. Pride, S. Yooseph, W. Biggs, K. E. Nelson, and J. C. Venter. 2015. Library preparation methodology can influence genomic and functional predictions in human microbiome research. *Proceedings of the National Academy of Sciences of the United States of America* 112(45):14024-14029.

Jørgensen, T. S., A. S. Kiil, M. A. Hansen, S. J. Sørensen, and L. H. Hansen. 2014. Current strategies for mobilome research. *Frontiers in Microbiology* 5:750.

Jovel, J., J. Patterson, W. Wang, N. Hotte, S. O'Keefe, T. Mitchel, T. Perry, D. Kao, A. L. Mason, K. L. Madsen, and G. K. S. Wong. 2016. Characterization of the gut microbiome using 16S or shotgun metagenomics. *Frontiers in Microbiology* 7:459.

Kaminski, J., M. K. Gibson, E. A. Franzosa, N. Segata, G. Dantas, and C. Huttenhower. 2015. High-specificity targeted functional profiling in microbial communities with ShortBRED. *PLOS Computational Biology* 11(12):1-22.

Kanehisa, M., S. Goto, Y. Sato, M. Kawashima, M. Furumichi, and M. Tanabe. 2014. Data, information, knowledge and principle: back to metabolism in KEGG. *Nucleic Acids Research* 42:D199-D205.

Kashtan, N., S. E. Roggensack, S. Rodrigue, J. W. Thompson, S. J. Biller, A. Coe, H. Ding, P. Marttinen, R. R. Malmstrom, R. Stocker, M. J. Follows, R. Stepanauskas, and S. W. Chisholm. 2014. Single-cell genomics reveals hundreds of coexisting subpopulations in wild prochlorococcus. *Science* 344(6182):416-420.

Kauserud, H., S. Kumar, A. K. Brysting, J. Norden, and T. Carlsen. 2012. High consistency between replicate 454 pyrosequencing analyses of ectomycorrhizal plant root samples. *Mycorrhiza* 22(4):309-315.

Kerepesi, C., and V. Grolmusz. 2016. Evaluating the quantitative capabilities of metagenomic analysis software. *Current Microbiology* 72(5):612-616.

Korem, T., D. Zeevi, J. Suez, A. Weinberger, T. Avnit-Sagi, M. Pompan-Lotan, E. Matot, G. Jona, A. Harmelin, N. Cohen, A. Sirota-Madi, C. A. Thaiss, M. Pevsner-Fischer, R. Sorek, R. J. Xavier, E. Elinav, and E. Segal. 2015. Growth dynamics of gut microbiota in health and disease inferred from single metagenomic samples. *Science* 349(6252):1101-1106.

Koskella, B., and M. A. Brockhurst. 2014. Bacteria–phage coevolution as a driver of ecological and evolutionary processes in microbial communities. *FEMS Microbiology Reviews* 38(5):916-931.

Kuczynski, J., C. L. Lauber, W. A. Walters, L. W. Parfrey, J. C. Clemente, D. Gevers, and R. Knight. 2012. Experimental and analytical tools for studying the human microbiome. *Nature Reviews Genetics* 13(1):47-58.

Kunin, V., A. Engelbrektson, H. Ochman, and P. Hugenholtz. 2010. Wrinkles in the rare biosphere: Pyrosequencing errors can lead to artificial inflation of diversity estimates. *Environmental Microbiology* 12(1):118-123.

Kurtz, Z. D., C. L. Müller, E. R. Miraldi, D. R. Littman, M. J. Blaser, and R. A. Bonneau. 2015. Sparse and compositionally robust inference of microbial ecological networks. *PLOS Computational Biology* 11(5):e1004226.

Langille, M. G. I., J. Zaneveld, J. G. Caporaso, D. McDonald, D. Knights, J. A. Reyes, J. C. Clemente, D. E. Burkepile, R. L. Vega Thurber, R. Knight, R. G. Beiko, and C. Huttenhower. 2013. Predictive functional profiling of microbial communities using 16S rRNA marker gene sequences. *Nature Biotechnology* 31:814-821.

Leinonen, R., H. Sugawara, and M. Shumway. 2011. The sequence read archive. *Nucleic Acids Research* 39:D19-D21.

Lemon, K. P., V. Klepac-Ceraj, H. K. Schiffer, E. L. Brodie, S. V. Lynch, and R. Kolter. 2010. Comparative analyses of the bacterial microbiota of the human nostril and oropharynx. *mBio* 1(3):e00129-10. doi:10.1128/mBio.00129-10.

Lemos, L. N., R. R. Fulthorpe, and L. F. Roesch. 2012. Low sequencing efforts bias analyses of shared taxa in microbial communities. *Folia Microbiologica* 57(5):409-413.

Leonard, S. R., M. K. Mammel, D. W. Lacher, and C. A. Elkins. 2015. Application of metagenomic sequencing to food safety: Detection of shiga toxin-producing escherichia coli on fresh bagged spinach. *Applied and Environmental Microbiology* 81(23):8183-8191.

Li, N., M. Hao, R. F. Phalen, W. C. Hinds, and A. E. Nel. 2003. Particulate air pollutants and asthma: A paradigm for the role of oxidative stress in PM-induced adverse health effects. *Clinical Immunology* 109(3):250-265.

Liang, Y., L. Wu, I. M. Clark, K. Xue, Y. Yang, J. D. Van Nostrand, Y. Deng, Z. He, S. McGrath, J. Storkey, P. R. Hirsch, B. Sun, and J. Zhou. 2015. Over 150 years of long-term fertilization alters spatial scaling of microbial biodiversity. *mBio* 6(2):e00240-15. doi:10.1128/mBio.00240-15.

Liao, R., R. Zhang, J. Guan, and S. Zhou. 2014. A new unsupervised binning approach for metagenomic sequences based on N-grams and automatic feature weighting. *IEEE/ACM Transactions on Computational Biology and Bioinformatics* 11(1):42-54.

Liebich, J., C. W. Schadt, S. C. Chong, Z. L. He, S. K. Rhee, and J. Z. Zhou. 2006. Improvement of oligonucleotide probe design criteria for functional gene microarrays in environmental applications. *Applied and Environmental Microbiology* 72(2):1688-1691.

Lindgreen, S., K. L. Adair, and P. P. Gardner. 2016. An evaluation of the accuracy and speed of metagenome analysis tools. *Scientific Reports* 6:19233.

Ling, L. L., T. Schneider, A. J. Peoples, A. L. Spoering, I. Engels, B. P. Conlon, A. Mueller, T. F. Schaberle, D. E. Hughes, S. Epstein, M. Jones, L. Lazarides, V. A. Steadman, D. R. Cohen, C. R. Felix, K. A. Fetterman, W. P. Millett, A. G. Nitti, A. M. Zullo, C. Chen, and K. Lewis. 2015. A new antibiotic kills pathogens without detectable resistance. *Nature* 517(7535):455-459.

Long, P. E., K. H. Williams, S. S. Hubbard, and J. F. Banfield. 2016. Microbial metagenomics reveals climate-relevant subsurface biogeochemical processes. *Trends in Microbiology* 24(8):600-610.

Loy, A., A. Lehner, N. Lee, J. Adamczyk, H. Meier, J. Ernst, K. H. Schleifer, and M. Wagner. 2002. Oligonucleotide microarray for 16S rRNA gene-based detection of all recognized lineages of sulfate-reducing prokaryotes in the environment. *Applied and Environmental Microbiology* 68(10):5064-5081.

Mackelprang, R., M. P. Waldrop, K. M. DeAngelis, M. M. David, K. L. Chavarria, S. J. Blazewicz, E. M. Rubin, and J. K. Jansson. 2011. Metagenomic analysis of a permafrost microbial community reveals a rapid response to thaw. *Nature* 480(7377):368-371.

MacNaughton, S. J., T. L. Jenkins, S. Alugupalli, and D. C. White. 1997. Quantitative sampling of indoor air biomass by signature lipid biomarker analysis: Feasibility studies in a model system. *American Industrial Hygiene Association Journal* 58(4):270-277.

MacNeil, L., T. Kauri, and W. Robertson. 1995. Molecular techniques and their potential application in monitoring the microbiological quality of indoor air. *Canadian Journal of Microbiology* 41:657-665.

Maki, J. S., and C. C. Remsen. 1981. Comparison of two direct count methods for determining metabolizing bacteria in freshwater. *Applied and Environmental Microbiology* 41:1132-1138.

Manor, O., and E. Borenstein. 2015. MUSiCC: A marker genes based framework for metagenomic normalization and accurate profiling of gene abundances in the microbiome. *Genome Biology* 16:53.

Mao, Y., A. C. Yannarell, and R. I. Mackie. 2011. Changes in N-transforming archaea and bacteria in soil during the establishment of bioenergy crops. *PLOS ONE* 6(9):e24750.

Marchand, G., J. Lavoie, and L. Lazure. 1995. Evaluation of bioaerosols in a municipal solid waste recycling and composting plant. *Journal of the Air and Waste Management Association* 45:778-781.

Maroni, M., R. Axelrad, and A. Bacaloni. 1995. NATO efforts to set indoor air quality guidelines and standards. *American Industrial Hygiene Association Journal* 56(5):499-508.

McFeters, G. A., P. Y. Feipeng, B. H. Pyle, and P. S. Stewart. 1995. Physiological assessment of bacteria using fluorochromes. *Journal of Microbiological Methods* 21:1-13.

McLoughlin, K. 2016. *Benchmarking for quasispecies abundance inference with confidence intervals from metagenomic sequence data.* Technical report. Livermore, CA: Lawrence Livermore National Laboratory.

McMurdie, P. J., and S. Holmes. 2014. Waste not, want not: Why rarefying microbiome data is inadmissible. *PLOS Computational Biology* 10:e1003531.

Mendes-Soares, H., M. Mundy, L. M. Soares, and N. Chia. 2016. MMinte: An application for predicting metabolic interactions among the microbial species in a community. *BMC Bioinformatics* 17(1):343.

Merchant, S., D. E. Wood, and S. L. Salzberg. 2014. Unexpected cross-species contamination in genome sequencing projects. *PeerJ* 2:e675.

Miller, D. J. 1992. Fungi as contaminants in indoor air. *Atmospheric Environment* 26A(12):2163-2172.

Moran, M. A. 2009. Metatranscriptomics: eavesdropping on complex microbial communities. *Microbe* 4(7):329-335.

Moran, M. A., B. Satinsky, S. M. Gifford, H. Luo, A. Rivers, L.-K. Chan, J. Meng, B. P. Durham, C. Shen, V. A. Varaljay, C. B. Smith, P. L. Yager, and B. M. Hopkinson. 2012. Sizing up metatranscriptomics. *The ISME Journal* 7(2):237-243.

Morgan, J. L., A. E. Darling, and J. A. Eisen. 2010. Metagenomic sequencing of an in vitro-simulated microbial community. *PLOS ONE* 5(4):e10209.

Moschandreas, D. J., D. K. Cha, and J. Qian. 1996. Measurement of indoor bioaerosol levels by a direct counting method. *Journal of Environmental Engineering* 122(5):374-378.
Nawy, T. 2013. Probing microbiome function. *Nature Methods* 10:35. doi:10.1038/nmeth.2293.
Nayfach, S., and K. S. Pollard. 2016. Toward accurate and quantitative comparative metagenomics. *Cell* 166(5):1103-1116.
Nayfach, S., P. H. Bradley, S. K. Wyman, T. J. Laurent, A. Williams, J. A. Eisen, K. S. Pollard, and T. J. Sharpton. 2015. Automated and accurate estimation of gene family abundance from shotgun metagenomes. *PLOS Computational Biology* 11:e1004573.
Nayfach, S., B. Rodriguez-Mueller, and K. S. Pollard. 2016. An integrated metagenomics pipeline for strain profiling reveals novel patterns of bacterial transmission and biogeography. *Genome Research* 26:1-14. doi:10.1101/gr.201863.115.
Nguyen, N.-P., T. Warnow, M. Pop, and B. White. 2016. A perspective on 16S rRNA operational taxonomic unit clustering using sequence similarity. *Biofilms and Microbiomes* 2:16004. doi:10.1038/npjbiofilms.2016.4.
Nielsen, H. B., M. Almeida, A. S. Juncker, S. Rasmussen, J. Li, S. Sunagawa, D. R. Plichta, L. Gautier, A. G. Pedersen, E. Le Chatelier, E. Pelletier, I. Bonde, T. Nielsen, C. Manichanh, M. Arumugam, J.-M. Batto, M. B. Quintanilha dos Santos, N. Blom, N. Borruel, K. S. Burgdorf, F. Boumezbeur, F. Casellas, J. Doré, P. Dworzynski, F. Guarner, T. Hansen, F. Hildebrand, R. S. Kaas, S. Kennedy, K. Kristiansen, J. Roat Kultima, P. Léonard, F. Levenez, O. Lund, B. Moumen, D. Le Paslier, N. Pons, O. Pedersen, E. Prifti, J. Qin, J. Raes, S. Sørensen, J. Tap, S. Tims, D. W. Ussery, T. Yamada, P. Renault, T. Sicheritz-Ponten, P. Bork, J. Wang, S. Brunak, and S. D. Ehrlich. 2014. Identification and assembly of genomes and genetic elements in complex metagenomic samples without using reference genomes. *Nature Biotechnology* 32(8):822-828.
Noecker, C., A. Eng, S. Srinivasan, C. M. Theriot, V. B. Young, J. K. Jansson, D. N. Fredricks, and E. Borenstein. 2016. Metabolic model-based integration of microbiome taxonomic and metabolomic profiles elucidates mechanistic links between ecological and metabolic variation. *mSystems* 1.
Pace, N. R. 1997. A molecular view of microbial diversity and the biosphere. *Science* 276(5313):734-740.
Palmer, K., and M. A. Horn. 2012. Actinobacterial nitrate reducers and proteobacterial denitrifiers are abundant in N2O-metabolizing palsa peat. *Applied and Environmental Microbiology* 78(16):5584-5596.
Palmgren, U., G. Strom, P. Malmberg, and G. Blomquist. 1986. The nucleopore filter method: A technique for enumeration of viable and nonviable airborne microorganisms. *American Journal of Industrial Medicine* 10:325-327.
Peccia, J., and M. Hernandez. 2006. Incorporating polymerase chain reaction-based identification, population characterization, and quantification of microorganisms into aerosol science: A review. *Atmospheric Environment* 40:3941-3961.
Peng, X., K.-Q. Yu, G.-H. Deng, Y.-X. Jiang, Y. Wang, G.-X. Zhang, and H.-W. Zhou. 2013. Comparison of direct boiling method with commercial kits for extracting fecal microbiome DNA by Illumina sequencing of 16S rRNA tags. *Journal of Microbiological Methods* 95(3):455-462.
Pillai, S. D., K. W. Widmer, S. E. Down, and S. C. Ricke. 1996. Occurrence of airborne bacteria and pathogen indicators during land application of sewage sludge. *Applied and Environmental Microbiology* 62(1):296-299.
Pilloni, G., M. S. Granitsiotis, M. Engel, and T. Lueders. 2012. Testing the limits of 454 pyrotag sequencing: Reproducibility, quantitative assessment and comparison to T-RFLP fingerprinting of aquifer microbes. *PLOS ONE* 7(7):e40467.

Pinto, A. J., and L. Raskin. 2012. PCR biases distort bacterial and archaeal community structure in pyrosequencing datasets. *PLOS ONE* 7(8):e43093.
Poretsky, R. S., I. Hewson, S. Sun, A. E. Allen, J. P. Zehr, and M. A. Moran. 2009. Comparative day/night metatranscriptomic analysis of microbial communities in the North Pacific subtropical gyre. *Environmental Microbiology* 11(6):1358-1375.
Qin, J., Y. Li, Z. Cai, S. Li, J. Zhu, F. Zhang, S. Liang, W. Zhang, Y. Guan, D. Shen, Y. Peng, D. Zhang, Z. Jie, W. Wu, Y. Qin, W. Xue, J. Li, L. Han, D. Lu, P. Wu, Y. Dai, X. Sun, Z. Li, A. Tang, S. Zhong, X. Li, W. Chen, R. Xu, M. Wang, Q. Feng, M. Gong, J. Yu, Y. Zhang, M. Zhang, T. Hansen, G. Sanchez, J. Raes, G. Falony, S. Okuda, M. Almeida, E. LeChatelier, P. Renault, N. Pons, J.-M. Batto, Z. Zhang, H. Chen, R. Yang, W. Zheng, S. Li, H. Yang, J. Wang, S. D. Ehrlich, R. Nielsen, O. Pedersen, K. Kristiansen, and J. Wang. 2012. A metagenome-wide association study of gut microbiota in type 2 diabetes. *Nature* 490(7418):55-60.
Qiu, X. Y., L. Y. Wu, H. S. Huang, P. E. McDonel, A. V. Palumbo, J. M. Tiedje, and J. Z. Zhou. 2001. Evaluation of PCR-generated chimeras: Mutations, and heteroduplexes with 16S rRNA gene-based cloning. *Applied and Environmental Microbiology* 67(2):880-887.
Quinn, R. A., J. A. Navas-Molina, E. R. Hyde, S. J. Song, Y. Vázquez-Baeza, G. Humphrey, J. Gaffney, J. J. Minich, A. V. Melnik, J. Herschend, J. DeReus, A. Durant, R. J. Dutton, M. Khosroheidari, C. Green, R. da Silva, P. C. Dorrestein, and R. Knight. 2016. From sample to multi-omics conclusions in under 48 hours. *mSystems* 1.
Quinn, R. A., L.-F. Nothias, O. Vining, M. Meehan, E. Esquenazi, and P. C. Dorrestein. 2017. Molecular networking as a drug discovery, drug metabolism, and precision medicine strategy. *Trends in Pharmacological Sciences* 38(2):143-154.
Randle-Boggis, R. J., T. Helgason, M. Sapp, and P. D. Ashton. 2016. Evaluating techniques for metagenome annotation using simulated sequence data. *FEMS Microbiology Ecology* 92(7):fiw095. doi:10.1093/femsec/fiw095.
Rao, C., H. A. Burge, and J. C. S. Chang. 1996. Review of quantitative standards and guidelines for fungi in indoor air. *Journal of the Air & Waste Management Association* 46(9):899-908.
Reponen, T., S. A. Grinshpun, K. L. Conwell, J. Wiest, and M. Anderson. 2001. Aerodynamic versus physical size of spores: Measurement and implication on respiratory deposition. *Grana* 40:119-125.
Rhee, S.-K., X. Liu, L. Wu, S. C. Chong, X. Wan, and J. Zhou. 2004. Detection of genes involved in biodegradation and biotransformation in microbial communities by using 50-mer oligonucleotide microarrays. *Applied and Environmental Microbiology* 70(7):4303-4317.
Ritari, J., J. Salojarvi, L. Lahti, and W. M. de Vos. 2015. Improved taxonomic assignment of human intestinal 16S rRNA sequences by a dedicated reference database. *BMC Genomics* 16:1056.
Robbins, C. A., L. J. Swenson, M. L. Nealley, R. E. Gots, and B. J. Kelman. 2000. Health effects of mycotoxin in indoor air: A critical review. *Applied Occupational and Environmental Hygiene* 15(10):773-784.
Rodriguez, G. G., D. Phipps, K. Ishiguro, and H. F. Ridgeway. 1992. Use of fluorescent redox probe for direct visualization of actively respiring bacteria. *Applied and Environmental Microbiology* 58:1801-1808.
Roh, S. W., G. C. J. Abell, K.-H. Kim, Y.-D. Nam, and J.-W. Bae. 2010. Comparing microarrays and next-generation sequencing technologies for microbial ecology research. *Trends in Biotechnology* 28(6):291-299.
Rosario, K., and M. Breitbart. 2011. Exploring the viral world through metagenomics. *Current Opinion in Virology* 1(4):289-297.

Rudi, K., O. M. Skulberg, R. Skulberg, and K. S. Jakobsen. 2000. Application of sequence-specific labeled 16S rRNA gene oligonucleotide probes for genetic profiling of cyanobacterial abundance and diversity by array hybridization. *Applied and Environmental Microbiology* 66(9):4004-4011.

Salter, S. J., M. J. Cox, E. M. Turek, S. T. Calus, W. O. Cookson, M. F. Moffatt, P. Turner, J. Parkhill, N. J. Loman, and A. W. Walker. 2014. Reagent and laboratory contamination can critically impact sequence-based microbiome analyses. *BMC Biology* 12:87.

Sangwan, N., F. Xia, and J. A. Gilbert. 2016. Recovering complete and draft population genomes from metagenome datasets. *Microbiome* 4:8.

Schloss, P. D., D. Gevers, and S. L. Westcott. 2011. Reducing the effects of PCR amplification and sequencing artifacts on 16S rRNA-based studies. *PLOS ONE* 6(12):e27310.

Scholz, M., D. V. Ward, E. Pasolli, T. Tolio, M. Zolfo, F. Asnicar, D. T. Truong, A. Tett, A. L. Morrow, and N. Segata. 2016. Strain-level microbial epidemiology and population genomics from shotgun metagenomics. *Nature Methods* 13(5):435-438.

Segata, N., D. Boernigen, T. L. Tickle, X. C. Morgan, W. S. Garrett, and C. Huttenhower. 2013. Computational meta'omics for microbial community studies. *Molecular Systems Biology* 9:666.

Shapiro, B. J., J. Friedman, O. X. Cordero, S. P. Preheim, S. C. Timberlake, G. Szabo, M. F. Polz, and E. J. Alm. 2012. Population genomics of early events in the ecological differentiation of bacteria. *Science* 336(6077):48-51.

Shi, Y., G. W. Tyson, and E. F. DeLong. 2009. Metatranscriptomics reveals unique microbial small RNAs in the ocean's water column. *Nature* 459(7244):266-269.

Sinclair, L., O. A. Osman, S. Bertilsson, and A. Eiler. 2015. Microbial community composition and diversity via 16S rRNA gene amplicons: Evaluating the illumina platform. *PLOS ONE* 10(2):e0116955.

Singer, E., B. Bushnell, D. Coleman-Derr, D. Bowman, R. M. Bowers, A. Levy, E. A. Gies, J. F. Cheng, A. Copeland, H. P. Klenk, S. J. Hallam, P. Hugenholtz, S. G. Tringe, and T. Woyke. 2016. High-resolution phylogenetic microbial community profiling. *The ISME Journal* 10(8):2020-2032.

Sinha, R., C. Abnet, O. White, R. Knight, and C. Huttenhower. 2015. The microbiome quality control project: Baseline study design and future directions. *Genome Biology* 16:276.

Sorek, R., and P. Cossart. 2010. Prokaryotic transcriptomics: A new view on regulation, physiology and pathogenicity. *Nature Reviews Genetics* 11(1):9-16.

Sorenson, W. G., D. G. Frazer, B. B. Jarvis, J. Simpson, and V. A. Robinson. 1987. Trichothecene mycotoxins in aerosolized conidia of *Stachybotrys atra*. *Applied and Environmental Microbiology* 53(6):1370-1375.

Stewart, F. J., E. A. Ottesen, and E. F. DeLong. 2010. Development and quantitative analyses of a universal rRNA-subtraction protocol for microbial metatranscriptomics. *The ISME Journal* 4(7):896-907.

Stewart, F. J., O. Ulloa, and E. F. DeLong. 2012. Microbial metatranscriptomics in a permanent marine oxygen minimum zone. *Environmental Microbiology* 14(1):23-40.

Su, H. J., A. Rotnitzky, H. A. Burge, and J. D. Spengler. 1992. Examination of fungi in domestic interiors by using factor analysis—correlations and associations with home factors. *Applied and Environmental Microbiology* 58:181-186.

Suzuki, M. T., and S. J. Giovannoni. 1996. Bias caused by template annealing in the amplification of mixtures of 16S rRNA genes by PCR. *Applied and Environmental Microbiology* 62(2):625-630.

Tabor, P. S., and R. A. Neihof. 1982. Improved method for determination of respiring individual microorganisms in natural waters. *Applied and Environmental Microbiology* 43:1249-1255.

Talley, N. J., and A. A. Fodor. 2011. Bugs, stool, and the irritable bowel syndrome: Too much is as bad as too little? *Gastroenterology* 141(5):1555-1559.

Tanca, A., A. Palomba, M. Deligios, T. Cubeddu, C. Fraumene, G. Biosa, D. Pagnozzi, M. F. Addis, and S. Uzzau. 2013. Evaluating the impact of different sequence databases on metaproteome analysis: Insights from a lab-assembled microbial mixture. *PLOS ONE* 8(12):e82981.

Tedersoo, L., S. Anslan, M. Bahram, S. Polme, T. Riit, I. Liiv, U. Koljalg, V. Kisand, R. H. Nilsson, F. Hildebrand, P. Bork, and K. Abarenkov. 2015. Shotgun metagenomes and multiple primer pair-barcode combinations of amplicons reveal biases in metabarcoding analyses of fungi. *MycoKeys* 10:1-43.

Teltsch, B., and E. Katzenelson. 1978. Airborne enteric bacteria and viruses from spray irrigation with wastewater. *Applied and Environmental Microbiology* 35(2):290-296.

Terzieva, S., J. Donnelly, V. Ulevicius, S. A. Grinshpun, D. Willeke, G. N. Stelma, and K. B. Brenner. 1996. Comparison of methods for detection and enumeration of airborne microorganisms collected by liquid impingement. *Applied and Environmental Microbiology* 62:2264-2272.

Tiquia, S. M., L. Wu, S. C. Chong, S. Passovets, D. Xu, Y. Xu, and J. Zhou. 2004. Evaluation of 50-mer oligonucleotide arrays for detecting microbial populations in environmental samples. *Biotechniques* 36(4):664-670.

Tong, Y., and B. Lighthart. 1999. Diurnal distribution of total and culturable atmospheric bacteria at a rural site. *Aerosol Science and Technology* 30(2):246-254.

Tremblay, J., K. Singh, A. Fern, E. S. Kirton, S. M. He, T. Woyke, J. Lee, F. Chen, J. L. Dangl, and S. G. Tringe. 2015. Primer and platform effects on 16S rRNA tag sequencing. *Frontiers in Microbiology* 6:771.

Trevors, J. T. 1985. Effect of temperature on selected microbial activities in aerobic and anaerobically incubated sediment. *Hydrobiologia* 126(2):189-192.

Tringe, S. G., C. von Mering, A. Kobayashi, A. A. Salamov, K. Chen, H. W. Chang, M. Podar, J. M. Short, E. J. Mathur, J. C. Detter, P. Bork, P. Hugenholtz, and E. M. Rubin. 2005. Comparative metagenomics of microbial communities. *Science* 308(5721):554-557.

Tu, Q., H. Yu, Z. He, Y. Deng, L. Wu, J. D. Van Nostrand, A. Zhou, J. Voordeckers, Y. J. Lee, Y. Qin, C. L. Hemme, Z. Shi, K. Xue, T. Yuan, A. Wang, and J. Zhou. 2014. GeoChip 4: A functional gene arrays-based high throughput environmental technology for microbial community analysis. *Molecular Ecology Resources* 14(5):914-928.

Urakawa, H., P. A. Noble, S. El Fantroussi, J. J. Kelly, and D. A. Stahl. 2002. Single-base-pair discrimination of terminal mismatches by using oligonucleotide microarrays and neural network analyses. *Applied and Environmental Microbiology* 68(1):235-244.

Urich, T., A. Lanzén, J. Qi, D. H. Huson, C. Schleper, and S. C. Schuster. 2008. Simultaneous assessment of soil microbial community structure and function through analysis of the meta-transcriptome. *PLOS ONE* 3(6):e2527.

Uyaguari-Diaz, M. I., M. Chan, B. L. Chaban, M. A. Croxen, J. F. Finke, J. E. Hill, M. A. Peabody, T. Van Rossum, C. A. Suttle, F. S. L. Brinkman, J. Isaac-Renton, N. A. Prystajecky, and P. Tang. 2016. A comprehensive method for amplicon-based and metagenomic characterization of viruses, bacteria, and eukaryotes in freshwater samples. *Microbiome* 4.

van der Hooft, J. J. J., J. Wandy, M. P. Barrett, K. E. V. Burgess, and S. Rogers. 2016. Topic modeling for untargeted substructure exploration in metabolomics. *Proceedings of the National Academy of Sciences of the United States of America* 113(48):13738-13743.

Van Nostrand, J. D., L. Wu, W.-M. Wu, Z. Huang, T. J. Gentry, Y. Deng, J. Carley, S. Carroll, Z. He, B. Gu, J. Luo, C. S. Criddle, D. B. Watson, P. M. Jardine, T. L. Marsh, J. M. Tiedje, T. C. Hazen, and J. Zhou. 2011. Dynamics of microbial community composition and function during in situ bioremediation of a uranium-contaminated aquifer. *Applied and Environmental Microbiology* 77(11):3860-3869.

Venter, J. C., K. Remington, J. F. Heidelberg, A. L. Halpern, D. Rusch, J. A. Eisen, D. Wu, I. Paulsen, K. E. Nelson, W. Nelson, D. E. Fouts, S. Levy, A. H. Knap, M. W. Lomas, K. Nealson, O. White, J. Peterson, J. Hoffman, R. Parsons, H. Baden-Tillson, C. Pfannkoch, Y.-H. Rogers, and H. O. Smith. 2004. Environmental genome shotgun sequencing of the Sargasso Sea. *Science* 304(5667):66-74.

Vieites, J. M., M.-E. Guazzaroni, A. Beloqui, P. N. Golyshin, and M. Ferrer. 2009. Metagenomics approaches in systems microbiology. *FEMS Microbiology Reviews* 33(1):236-255.

Vishnivetskaya, T. A., A. C. Layton, M. C. Y. Lau, A. Chauhan, K. R. Cheng, A. J. Meyers, J. R. Murphy, A. W. Rogers, G. S. Saarunya, D. E. Williams, S. M. Pfiffner, J. P. Biggerstaff, B. T. Stackhouse, T. J. Phelps, L. Whyte, G. S. Sayler, and T. C. Onstott. 2014. Commercial DNA extraction kits impact observed microbial community composition in permafrost samples. *FEMS Microbiology Ecology* 87(1):217-230.

Wagner, M., B. Assmus, A. Hartmann, P. Hutzler, and R. Amann. 1994. In situ analysis of microbial consortia in activated sludge using fluorescently labelled, rRNA-targeted oligonucleotide probes and confocal scanning laser microscopy. *Journal of Microscopy* 176(3):181-187.

Wang, C., X. Wang, D. Liu, H. Wu, X. Lu, Y. Fang, W. Cheng, W. Luo, P. Jiang, J. Shi, H. Yin, J. Zhou, X. Han, and E. Bai. 2014. Aridity threshold in controlling ecosystem nitrogen cycling in arid and semi-arid grasslands. *Nature Communications* 5:4799.

Wang, J., and H. Jia. 2016. Metagenome-wide association studies: Fine-mining the microbiome. *Nature Reviews Microbiology* 14(8):508-522.

Wang, M., J. J. Carver, V. V. Phelan, L. M. Sanchez, N. Garg, Y. Peng, D. D. Nguyen, J. Watrous, C. A. Kapono, T. Luzzatto-Knaan, C. Porto, A. Bouslimani, A. V. Melnik, M. J. Meehan, W. T. Liu, M. Crüsemann, P. D. Boudreau, E. Esquenazi, M. Sandoval-Calderón, R. D. Kersten, L. A. Pace, R. A. Quinn, K. R. Duncan, C. C. Hsu, D. J. Floros, R. G. Gavilan, K. Kleigrewe, T. Northen, R. J. Dutton, D. Parrot, E. E. Carlson, B. Aigle, C. F. Michelsen, L. Jelsbak, C. Sohlenkamp, P. Pevzner, A. Edlund, J. McLean, J. Piel, B. T. Murphy, L. Gerwick, C. C. Liaw, Y. L. Yang, H. U. Humpf, M. Maansson, R. A. Keyzers, A. C. Sims, A. R. Johnson, A. M. Sidebottom, B. E. Sedio, A. Klitgaard, C. B. Larson, P. C. A. Boya, D. Torres-Mendoza, D. J. Gonzalez, D. B. Silva, L. M. Marques, D. P. Demarque, E. Pociute, E. C. O'Neill, E. Briand, E. J. Helfrich, E. A. Granatosky, E. Glukhov, F. Ryffel, H. Houson, H. Mohimani, J. J. Kharbush, Y. Zeng, J. A. Vorholt, K. L. Kurita, P. Charusanti, K. L. McPhail, K. F. Nielsen, L. Vuong, M. Elfeki, M. F. Traxler, N. Engene, N. Koyama, O. B. Vining, R. Baric, R. R. Silva, S. J. Mascuch, S. Tomasi, S. Jenkins, V. Macherla, T. Hoffman, V. Agarwal, P. G. Williams, J. Dai, R. Neupane, J. Gurr, A. M. Rodríguez, A. Lamsa, C. Zhang, K. Dorrestein, B. M. Duggan, J. Almaliti, P. M. Allard, P. Phapale, L. F. Nothias, T. Alexandrov, M. Litaudon, J. L. Wolfender, J. E. Kyle, T. O. Metz, T. Peryea, D. T. Nguyen, D. VanLeer, P. Shinn, A. Jadhav, R. Müller, K. M. Waters, W. Shi, X. Liu, L. Zhang, R. Knight, P. R. Jensen, B. Ø. Palsson, K. Pogliano, R. G. Linington, M. Gutiérrez, N. P. Lopes, W. H. Gerwick, B. S. Moore, P. C. Dorrestein, and N. Bandeira. 2016. Sharing and community curation of mass spectrometry data with Global Natural Products Social Molecular Networking. *Nature Biotechnology* 34(8):828-837.

Watrous, J., P. Roach, T. Alexandrov, B. S. Heath, J. Y. Yang, R. D. Kersten, M. van der Voort, K. Pogliano, H. Gross, J. M. Raaijmakers, B. S. Moore, J. Laskin, N. Bandeira, and P. C. Dorrestein. 2012. Mass spectral molecular networking of living microbial colonies. *Proceedings of the National Academy of Sciences of the United States of America* 109(26):E1743-E1752.

Weiss, S., W. Van Treuren, C. Lozupone, K. Faust, J. Friedman, Y. Deng, L. C. Xia, Z. Z. Xu, L. Ursell, E. J. Alm, A. Birmingham, J. A. Cram, J. A. Fuhrman, J. Raes, F. Sun, J. Zhou, and R. Knight. 2016. Correlation detection strategies in microbial data sets vary widely in sensitivity and precision. *The ISME Journal* 10(7):1669-1681.

Wen, C., L. Wu, Y. Qin, J. D. Van Nostrand, B. Sun, K. Xue, F. Liu, Y. Deng, and J.-Z. Zhou. 2017. Evaluation of the reproducibility of amplicon sequencing with Illumina MiSeq platform. *PLOS ONE* 12(4):e0176716.

White, R. A., E. M. Bottos, T. Roy Chowdhury, J. D. Zucker, C. J. Brislawn, C. D. Nicora, S. J. Fansler, K. R. Glaesemann, K. Glass, and J. K. Jansson. 2016. Moleculo long-read sequencing facilitates assembly and genomic binning from complex soil metagenomes. *mSystems* 1.

WHO (World Health Organization). 1990. Indoor air quality: Biological contaminants. *WHO Regional Publications. European Series* 31:1-67.

Womack, A. M., B. J. M. Bohannan, and J. L. Green. 2010. Biodiversity and biogeography of the atmosphere. *Philosophical Transactions of the Royal Society* 365(1558):3645-3653.

Wu, L., D. K. Thompson, G. Li, R. A. Hurt, J. M. Tiedje, and J. Zhou. 2001. Development and evaluation of functional gene arrays for detection of selected genes in the environment. *Applied and Environmental Microbiology* 67(12):5780-5790.

Wu, L., X. Liu, C. W. Schadt, and J. Zhou. 2006. Microarray-based analysis of subnanogram quantities of microbial community DNAs by using whole-community genome amplification. *Applied and Environmental Microbiology* 72(7):4931-4941.

Xu, L. H., S. Ravnskov, J. Larsen, and M. Nicolaisen. 2011. Influence of DNA extraction and PCR amplification on studies of soil fungal communities based on amplicon sequencing. *Canadian Journal of Microbiology* 57(12):1062-1066.

Xue, K., M. M. Yuan, Z. J. Shi, Y. J. Qin, Y. Deng, L. Cheng, L. Y. Wu, Z. L. He, J. D. Van Nostrand, R. Bracho, S. Natali, E. A. G. Schuur, C. W. Luo, K. T. Konstantinidis, Q. Wang, J. R. Cole, J. M. Tiedje, Y. Q. Luo, and J. Z. Zhou. 2016. Tundra soil carbon is vulnerable to rapid microbial decomposition under climate warming. *Nature Climate Change* 6(6):595-600.

Yu, K., and T. Zhang. 2012. Metagenomic and metatranscriptomic analysis of microbial community structure and gene expression of activated sludge. *PLOS ONE* 7(5):e38183.

Zhan, A., W. Xiong, S. He, and H. J. MacIsaac. 2014. Influence of artifact removal on rare species recovery in natural complex communities using high-throughput sequencing. *PLOS ONE* 9(5):e96928. doi:10.1371/journal.pone.0096928.

Zhou, J. Z. 2003. Microarrays for bacterial detection and microbial community analysis. *Current Opinion in Microbiology* 6(3):288-294.

Zhou, J. Z. 2009. Predictive microbial ecology. *Microbial Biotechnology* 2(2):154-156.

Zhou, J. Z., and D. K. Thompson. 2002. Challenges in applying microarrays to environmental studies. *Current Opinion in Biotechnology* 13(3):204-207.

Zhou, J. Z., B. C. Xia, H. S. Huang, D. S. Treves, L. J. Hauser, R. J. Mural, A. V. Palumbo, and J. M. Tiedje. 2003. Bacterial phylogenetic diversity and a novel candidate division of two humid region, sandy surface soils. *Soil Biology & Biochemistry* 35(7):915-924.

Zhou, J. Z., S. Kang, C. W. Schadt, and C. T. Garten, Jr. 2008. Spatial scaling of functional gene diversity across various microbial taxa. *Proceedings of the National Academy of Sciences of the United States of America* 105(22):7768-7773.

Zhou, J. Z., Z. He, J. D. Van Nostrand, L. Wu, and Y. Deng. 2010. Applying GeoChip analysis to disparate microbial communities. *Microbe* 5(2):60-65.

Zhou, J. Z., L. Wu, Y. Deng, X. Zhi, Y.-H. Jiang, Q. Tu, J. Xie, J. D. Van Nostrand, Z. He, and Y. Yang. 2011. Reproducibility and quantitation of amplicon sequencing-based detection. *The ISME Journal* 5(8):1303-1313.

Zhou, J. Z., Y. H. Jiang, Y. Deng, Z. Shi, B. Y. Zhou, K. Xue, L. Y. Wu, Z. L. He, and Y. F. Yang. 2013. Random sampling process leads to overestimation of beta-diversity of microbial communities. *mBio* 4(3):e00324-13. doi:10.1128/mBio.00324-13.

Zhou, J. Z., Z. He, Y. Yang, Y. Deng, S. G. Tringe, and L. Alvarez-Cohen. 2015. High-throughput metagenomic technologies for complex microbial community analysis: Open and closed formats. *mBio* 6(1):e02288-14. doi:10.1128/mBio.02288-14.

Zhou, J. Z., Y. Deng, L. N. Shen, C. Q. Wen, Q. Y. Yan, D. L. Ning, Y. J. Qin, K. Xue, L. Y. Wu, Z. L. He, J. W. Voordeckers, J. D. Van Nostrand, V. Buzzard, S. T. Michaletz, B. J. Enquist, M. D. Weiser, M. Kaspari, R. Waide, Y. F. Yang, and J. H. Brown. 2016. Temperature mediates continental-scale diversity of microbes in forest soils. *Nature Communications* 7:1208.

B

Study Methods

COMMITTEE COMPOSITION

The National Academies of Sciences, Engineering, and Medicine (the National Academies) appointed a committee of 16 experts to undertake the statement of task for this study. The committee was composed of members with expertise in such areas as microbial ecology, public health, building science and engineering, architecture, materials science, bioinformatics, and molecular characterization tools. Appendix C provides biographical information for each committee member.

Meetings and Information Gathering

The committee deliberated from approximately February 2016 to May 2017. To respond to its charge, the committee gathered information and data relevant to its statement of task by conducting a review of available literature and other publicly available resources, inviting experts to share perspectives at public meetings, and soliciting public comments online and in person.

The committee held four information-gathering meetings in Washington, DC, and Irvine, California, and heard from a variety of academic and private-sector researchers, as well as federal and state government officials. These meetings focused on understanding the current research being conducted in the field of the microbiomes of the built environment, as well as on identifying research needs and roadblocks in the microbiology, engineering, and building science fields.

The first meeting, held April 11–12, 2016, in Washington, DC, provided an opportunity for the committee to discuss the study with sponsoring organizations and to hear presentations from background speakers in areas relevant to study topics, including microbiology within built environments, microbiology within the International Space Station, and current engineering standards for big-box stores.

The second meeting, held June 20–12, 2016, in Washington, DC, included speakers who discussed interactions occurring between built environment microbiomes and human occupants, as well as major building systems that affect or are affected by indoor microbiomes and their impacts.

The third meeting, held October 17–18, 2016, in Irvine, California, included speakers knowledgeable about the toolkit for studying microbiome–built environment interactions, viruses and fungi in the built environment, and other topics. The meeting also included a number of younger researchers whose travel was supported by a travel award (sponsored by the Gordon and Betty Moore Foundation) and who had the opportunity to present posters on their research.

The fourth meeting, held December 1–2, 2016, in Washington, DC, included speakers who discussed the impacts of such interventions as cleaning and the development of antimicrobial materials on indoor microorganisms, the role of dermal uptake in the indoor environment, and the current state of bioinformatics pipelines and analysis needs.

Public Communication

The committee's two largest data-gathering meetings, in June and October 2016, provided opportunities to interact with additional stakeholders, including researchers and any others interested in the study topic. These participants contributed their views during open discussions following speaker presentations and through breakout sessions. The committee also worked to make its activities as transparent and accessible as possible for those who may not have been able to attend in person. The study website[1] was updated regularly to reflect the committee's recent and planned activities. Outreach efforts included a study-specific email address for comments and questions, as well as social media feeds and tags. A subscription button also was available to provide for the receipt of email updates on the study and solicitation of comments and input to be shared with the committee.

Live video streams and subsequent links to recordings of the open session presentations were also made available during the course of the study to provide an opportunity for input from those unable to attend commit-

---

[1] See http://nas-sites.org/builtmicrobiome (accessed on July 26, 2017).

APPENDIX B

tee meetings in person. Any information provided to the committee from outside sources or through the online comment tool is available by request through the National Academies' Public Access Records Office.

Invited Speakers

The following individuals were invited speakers at the committee's meetings and data-gathering sessions:

Gary Adamkiewicz
Harvard T.H. Chan School of Public Health

Rachel Adams
University of California, Berkeley

Gary Andersen
Lawrence Berkeley National Laboratory,
University of California, Berkeley

Tina Bahadori
U.S. Environmental Protection Agency

Terry Brennan
Camroden Associates

Brandon "Bubba" Brooks
University of California, Berkeley

Lisa Chadwick
National Institute of Environmental Health Sciences

Pieter Dorrestein
University of California, San Diego

Rob Dunn
North Carolina State University

Sarah Evans
Michigan State University

M. Patricia Fabian
Boston University

Elizabeth Grice
University of Pennsylvania

Robin Guenther
Perkins+Will

Jonathan "Kirk" Harris
University of Colorado Denver

Scott Jackson
National Institute of Standards and Technology

Janet Jansson
Pacific Northwest National Laboratory

Lee Ann Kahlor
The University of Texas at Austin

Benjamin Kirkup
Naval Research Laboratory

Rob Knight
University of California, San Diego

Laura Kolb
U.S. Environmental Protection Agency

Jay Lennon
Indiana University Bloomington

Susan Lynch
University of California, San Francisco

Linsey Marr
Virginia Polytechnic Institute and State University

Jennifer Martiny
University of California, Irvine

Mark Mendell
California Department of Public Health

Shelly Miller
University of Colorado Boulder

Donald Milton
University of Maryland

Paula Olsiewski
Alfred P. Sloan Foundation

Amy Pruden
Virginia Polytechnic Institute and State University

Tiina Reponen
University of Cincinnati

Charles Robertson
University of Colorado Boulder

Richard Shaughnessy
University of Tulsa

Jeffrey Siegel
University of Toronto

Joanne Sordillo
Brigham and Women's Hospital,
Harvard T.H. Chan School of Public Health

Jelena Srebric
University of Maryland

Dennis Stanke
Trane Ingersoll Rand (retired)

Brent Stephens
Illinois Institute of Technology

Phil Stewart
Montana State University

Elizabeth Stulberg
Office of the Chief Scientist,
U.S. Department of Agriculture

John Taylor
University of California, Berkeley

David Tomko
National Aeronautics and Space Administration

Kevin van den Wymelenberg
University of Oregon

Kasthuri Venkateswaran
Jet Propulsion Laboratory,
National Aeronautics and Space Administration

Michael Waring
Drexel University

Charles Weschler
Rutgers University

C

Committee Member Biographies

Joan Wennstrom Bennett, Ph.D. (NAS) (*Chair*), has been a distinguished professor of plant biology and pathology at Rutgers University since 2006. Prior to coming to Rutgers, she was on the faculty at Tulane University, New Orleans, Louisiana, for more than 30 years. The Bennett laboratory studies the genetics and physiology of filamentous fungi. In addition to mycotoxins and other secondary metabolites, research focuses on the volatile organic compounds (VOCs) emitted by fungi. These low-molecular-weight compounds are responsible for the familiar odors associated with molds and mushrooms. Some VOCs function as semiochemicals for insects, while others serve as developmental signals for fungi. The Bennett lab has tested individual fungal VOCs in model systems and found that 1-octen-3-ol ("mushroom alcohol") is a neurotoxin in Drosophila melanogaster and causes growth retardation in *Arabidopsis thaliana*. It also inhibits growth of the fungus that causes "white nose syndrome" in bat populations. In other studies, the Bennett lab has demonstrated that VOCs from living cultures of Trichoderma, a known biocontrol fungus, can enhance plant growth. Investigations on the mechanistic aspects of fungal VOC action are under way using a yeast knockout library. Dr. Bennett was associate vice president for the Office of Promotion of Women in Science, Engineering and Mathematics ("SciWomen") at Rutgers from 2006 to 2014. She is a past editor-in-chief of *Mycologia*; a past vice president of the British Mycological Society and the International Union of Microbiological Societies; and past president of the American Society for Microbiology and the Society for Industrial Microbiology and Biotechnology. She was elected to the National Academy of Sciences in 2005.

Jonathan Allen, Ph.D., is a bioinformatics scientist at the Lawrence Livermore National Laboratory. His research focuses on the development and application of new software tools to address various genome sequence analysis problems, including prediction of genetic virulence markers in viruses, detection of genetic engineering in bacteria, and eukaryotic gene prediction. Dr. Allen is currently working with the Lawrence Livermore Microbial Detection Array, which is capable of comparing the DNA of microorganisms in a specific location or environment with a vast library of stored viral, bacterial, and fungal genetic sequences.

Jean Cox-Ganser, Ph.D., is research team leader for the Field Studies Branch, Respiratory Health Division, National Institute for Occupational Safety and Health (NIOSH). For the past 15 years she has been principal investigator for research studies on the respiratory health effects of dampness and mold in office buildings and schools, and she is author or coauthor of more than 30 peer-reviewed publications, book chapters, and reports resulting from this research. Dr. Cox-Ganser is one of the most knowledgeable and influential researchers in the world on dampness, mold, and respiratory disease. Of special interest is her many years of experience guiding and participating in detailed and technically rigorous health hazard investigations of buildings. Indoor ecology is interesting, but knowledge of building structures and their operation is equally interesting and important in understanding the indoor biome.

Jack Gilbert, Ph.D., earned his Ph.D. from Unilever and Nottingham University, United Kingdom, in 2002, and received his postdoctoral training at Queens University, Canada. He subsequently returned to the United Kingdom in 2005, coming to Plymouth Marine Laboratory as a senior scientist, until his move to Argonne National Laboratory and the University of Chicago in 2010. Currently, Dr. Gilbert is in the Department of Surgery at the University of Chicago and is group leader for microbial ecology at Argonne National Laboratory. He is also associate director of the Institute of Genomic and Systems Biology, research associate at the Field Museum of Natural History, and senior scientist at the Marine Biological Laboratory. Dr. Gilbert uses molecular analysis to test fundamental hypotheses in microbial ecology. He has authored more than 200 peer-reviewed publications and book chapters on metagenomics and approaches to ecosystem ecology. He is currently working on generating observational and mechanistic models of microbial communities in natural, urban, built, and human ecosystems. He is on the advisory board of the Genomic Standards Consortium (www.gensc.org) and is the founding editor-in-chief of *mSystems* journal. In 2014 he was recognized on Crain's Business Chicago's 40 Under 40 List, and in 2015 he was listed as 1 of the 50 most influential scientists by Business Insider and in the Brilliant Ten by *Popular Science*.

Diane Gold, M.D., is a professor of medicine at Harvard Medical School and a professor in the Department of Environmental Health at the Harvard T.H. Chan School of Public Health. She is also an associate physician at Brigham and Women's Hospital. Her research focuses on the relationships between environmental exposures and the incidence or severity of respiratory diseases, including asthma. The environmental exposures considered encompass indoor allergens, including fungi, smoking, outdoor ozone, and particles. She investigates the environmental exposures that may explain socioeconomic, cultural, and gender differences that have been observed in asthma severity. These include perinatal exposures and family stress, as well as exposure to the allergens and pollutants mentioned above. Dr. Gold is also interested in the cardiopulmonary effects of particles on the elderly.

Jessica Green, Ph.D., is an ecologist and engineer who specializes in biodiversity theory and microbial systems. She is a professor of biology at the University of Oregon, where she codirects the Biology and Built Environment Center. She is also chief technology officer of Phylagen, Inc., a microbiome company based in San Francisco, and external faculty at the Santa Fe Institute. Her research blends molecular biology, data science, and bioinformatics to understand and model complex microbial communities interacting with each other, with humans, and with the environment. Dr. Green has received numerous awards, including a Blaise Pascal International Research Chair sponsored by Île-de-France, a John Simon Guggenheim Memorial Foundation Fellowship, and a TED Senior Fellowship. She received a Ph.D. in nuclear engineering and an M.S. in civil and environmental engineering from the University of California, Berkeley, and a B.S. in civil and environmental engineering from the University of California, Los Angeles.

Charles Haas, Ph.D., is L. D. Betz professor of environmental engineering and head of the Department of Civil, Architectural, and Environmental Engineering at Drexel University, where he has been since 1991. He also holds courtesy appointments in the Department of Emergency Medicine of the Drexel University College of Medicine and in the School of Public Health. He received his B.S. (biology) and M.S. (environmental engineering) from the Illinois Institute of Technology and his Ph.D. in environmental engineering from the University of Illinois at Urbana-Champaign. He served on the faculties of Rensselaer Polytechnic Institute and the Illinois Institute of Technology prior to joining Drexel. He codirected the U.S. Environmental Protection Agency (EPA)/U.S. Department of Homeland

Security University Cooperative Center of Excellence–Center for Advancing Microbial Risk Assessment (CAMRA). He is a fellow of the International Water Association, the American Academy for the Advancement of Science, the Society for Risk Analysis, the American Society of Civil Engineers, the American Academy of Microbiology, and the Association of Environmental Engineering and Science Professors. Dr. Haas is a board-certified environmental engineering member by eminence of the American Academy of Environmental Engineers. He has received the Dr. John Leal Award of the American Water Works Association and the Clarke Water Prize. Over his career, he has specialized in the assessment of risk from and control of human exposure to pathogenic microorganisms and, in particular, the treatment of water and wastewater to minimize microbial risk to human health. Dr. Haas has served on numerous panels of the National Research Council. He is a past member of the Water Science and Technology Board of the National Academies of Sciences, Engineering, and Medicine and the EPA Board of Scientific Counselors.

Mark Hernandez, Ph.D., PE, is a professor in the Department of Civil, Environmental, and Architectural Engineering at the University of Colorado Boulder. His research interests lie at the cusp of molecular biology and civil engineering, focusing on the characterization and control of biological air pollution, both natural and anthropogenic. His recent work has focused on engineering disinfection systems for airborne bacteria and viruses and on tracking bioaerosols through natural weather patterns and catastrophic events (such as Hurricane Katrina). Dr. Hernandez is a registered professional civil engineer and an active technical consultant in the commercial waste treatment and industrial hygiene sectors. He serves as an editor of *Aerosol Science and Technology* and is the director of the Colorado Diversity Initiative. He received his Ph.D. and M.S. in environmental engineering and his B.S. in civil engineering from the University of California, Berkeley.

Robert Holt, Ph.D., is an eminent scholar and Arthur R. Marshall, Jr. chair in ecology at the University of Florida. His research focuses on theoretical and conceptual issues at the population and community levels of ecological organization and the task of linking ecology with evolutionary biology. He focuses on basic research, as well as bringing modern ecological theory to bear on significant applied problems, particularly in conservation biology. He approaches ecology by moving beyond traditional analyses of single species or interacting species pairs by focusing on an immediate level of complexity (community modules), which are small sets of interacting species, patterns of interactions found across many ecosystems. Dr. Holt is currently researching how predators influence infectious disease dynamics in host populations that are also prey.

Ronald Latanision, Ph.D. (NAE), is a senior fellow at Exponent, Inc., an engineering and scientific consulting company. Prior to joining Exponent, he was director of the H. H. Uhlig Corrosion Laboratory in the Department of Materials Science and Engineering, Massachusetts Institute of Technology (MIT), and held joint faculty appointments in the Department of Materials Science and Engineering and the Department of Nuclear Engineering. He led the School of Engineering's Materials Processing Center at MIT as its director from 1985 to 1991. He is now an emeritus professor at MIT. In April 2015, he was appointed an adjunct professor in the Key Laboratory of Nuclear Materials and Safety Assessment of the Institute of Metal Research, Chinese Academy of Sciences. In addition, he is a member of the National Academy of Engineering (NAE) and a fellow of the American Academy of Arts and Sciences, ASM International, and the National Association of Corrosion Engineers (NACE) International. From 1983 to 1988, he was the first holder of the Shell distinguished chair in materials science. He was a founder of Altran Materials Engineering Corporation, established in 1992. He served as a principal and corporate vice president before assuming his role as a senior fellow at Exponent. Dr. Latanision is a member of the International Corrosion Council and serves as co-editor-in-chief of *Corrosion Reviews* with Professor Noam Eliaz of Tel Aviv University. He is editor-in-chief of the NAE quarterly publication *The Bridge*. He has served as a science adviser to the U.S. House of Representatives Committee on Science and Technology in Washington, DC. In June 2002, he was appointed by President George W. Bush to membership on the U.S. Nuclear Waste Technical Review Board, and he was reappointed for a second 4-year term by then-President Barack Obama. Dr. Latanision received a B.S. in metallurgy from The Pennsylvania State University and a Ph.D.in metallurgical engineering from The Ohio State University. He is an honorary alumnus of MIT.

Hal Levin, B.Arch., is a research architect with Building Ecology Research Group. He has conducted research and provided consultation in the areas of building impacts on occupant health and comfort, as well as on the larger environment. For almost 40 years he has been involved in research and consulting that have included the integration of knowledge about indoor and outdoor air pollution, as well as other risk factors into the design of residential, educational, and commercial buildings and communities. His work includes many efforts to design buildings with minimal negative impacts on occupants or the larger environment, including the design of ventilation, building materials selection, energy consumption, and total environmental quality. Mr. Levin is a fellow of the American Society of Heating, Refrigerating and Air-Conditioning Engineers (ASHRAE) and the American Society for Testing and Materials (ASTM). He is a contributor to chapters in several books, including *Indoor Air Quality Handbook* (McGraw-Hill, 2001), and is a former associate editor of the journal *Indoor Air*.

Vivian Loftness, FAIA, LEED AP, is a university professor and former head of the School of Architecture at Carnegie Mellon University. She is an internationally renowned researcher, author, and educator who has spent more than 30 years focusing on environmental design and sustainability, advanced building systems integration, climate and regionalism in architecture, and design for performance in the workplace of the future. She has served on 10 National Academies of Sciences, Engineering, and Medicine panels and the Board on Infrastructure and the Constructed Environment and has given four congressional testimonies on sustainability. Ms. Loftness is the recipient of the National Educator Honor Award from the American Institute of Architecture Students and the Sacred Tree Award from the U.S. Green Building Council (USGBC). She received her B.S. and M.S. in architecture from the Massachusetts Institute of Technology and served on the national boards of the USGBC, the American Institute of Architects (AIA) Committee on the Environment, Green Building Alliance, Turner Sustainability, and the Global Assurance Group of the World Business Council for Sustainable Development. She is a registered architect and a fellow of the AIA.

Karen Nelson, Ph.D., is president of the Rockville campus of the J. Craig Venter Institute (JCVI), where she has worked for the past 15 years. She was formerly director of human microbiology and metagenomics in the Department of Human Genomic Medicine at JCVI. Dr. Nelson has extensive experience in microbial ecology, microbial genomics, microbial physiology, and metagenomics. Since joining the JCVI legacy institutes, she has led several genomic and metagenomic efforts; was involved in the analysis of the microbiota of the human stomach and gastrointestinal tract; and led the first human metagenomics study on fecal material derived from three individuals, which was published in 2006. Additional ongoing studies in her group include metagenomic approaches to studying the ecology of the gastrointestinal tract of humans and animals, reference genome sequencing and analysis, studies with nonhuman primates, and studies on the relationship between the microbiome and various human and animal disease conditions. She has authored or coauthored more than 100 peer-reviewed publications and edited three books and is currently editor-in-chief of the journals *Microbial Ecology* and *Advances in Microbial Ecology*. She also serves on the editorial boards of *BMC Genomics*, *GigaScience*, and the *Central European Journal of Biology*. Dr. Nelson was a member of the National Research Council Standing Committee on Biodefense for the U.S. Department of Defense and is a member of the National Academies of Sciences, Engineering, and Medicine's Board on Life Sciences. She is a fellow of the American Society for Microbiology. She received her undergraduate degree from the University of the West Indies and her Ph.D. from Cornell University.

Jordan Peccia, Ph.D., is a professor of chemical and environmental engineering at Yale University and director of Yale environmental engineering undergraduate studies. His research group applies classical and molecular biology to solve environmental problems. The current research thrusts in his laboratory include (1) applying molecular biology techniques to investigate the diversity, origin, and fate of airborne biological material; (2) developing functional genomic approaches for controlling microalgae growth in biodiesel production; and (3) understanding human pathogen exposure and in vitro toxicity responses associated with land applied biosolids (sewage sludge).

Andrew Persily, Ph.D., is chief of the Energy and Environment Division at the National Institute of Standards and Technology and has performed research into indoor air quality and ventilation since the late 1970s. His work has included the development and application of measurement techniques for evaluating airflows and indoor air contaminant levels in a variety of building types, including large mechanically ventilated buildings and single-family dwellings. These procedures include tracer gas techniques for measuring air change rates and air distribution effectiveness, measurements of contaminant concentrations, and envelope airtightness. He has contributed to the development and application of multizone airflow and contaminant dispersal models. Dr. Persily was a vice-president of the American Society of Heating, Refrigerating and Air-Conditioning Engineers (ASHRAE) from 2007 to 2009, and he is past chair of the ASHRAE Standing Standard Project Committee (SSPC) 62.1, responsible for the revision of the ASHRAE Ventilation Standard 62. He is currently chair of Standard 189.1, Design of High-Performance Green Buildings. He is a past chair of the American Society for Testing and Materials (ASTM) Subcommittee E6.41 on Air Leakage and Ventilation Performance and past vice chair of subcommittee D22.05 on Indoor Air Quality. Dr. Persily was named an ASTM fellow and an International Society of Indoor Air Quality and Climate (ISIAQ) fellow in 2002, and an ASHRAE fellow in 2004. He received a B.A. in physics and mathematics from Beloit College in 1976 and a Ph.D. in mechanical and aerospace engineering from Princeton University in 1982.

Jizhong Zhou, Ph.D., is George Lynn Cross research professor in the Department of Microbiology and Plant Biology and director for the Institute for Environmental Genomics, University of Oklahoma; adjunct senior scientist at Lawrence Berkeley National Laboratory; and adjunct professor at Tsinghua University, Beijing, China. His expertise is in microbial ecology and genomics, with current research focused on (1) molecular community ecology and metagenomics, particularly in terrestrial soils and groundwater ecosystems important to climate changes, bioenergy, and environmental remediation; (2) experimental evolution and functional genomics of microorganisms important to environment and bioenergy; (3) pioneering development of high-throughput metagenomic technologies, particularly functional gene arrays for biogeochemical, environmental, and ecological applications; and (4) theoretical ecology, particularly ecological theories and network ecology. Dr. Zhou has authored more than 500 publications—with more than 29,000 total citations and an H-index above 90—on microbial genomics, genomic technologies, molecular biology, molecular evolution, microbial ecology, bioremediation, bioenergy, global change, bioinformatics, systems biology, and theoretical ecology. He received the Presidential Early Career Award for Scientists and Engineers in 2001; the R&D 100 Award in 2009; and the Ernest Orlando Lawrence Award in 2014, which is the U.S. Department of Energy's highest scientific recognition. He is a senior editor for the *ISME (International Society for Microbial Ecology) Journal* and *mBio* and a former editor for *Applied and Environmental Microbiology*. He is a fellow of the American Academy of Microbiology and the American Association for the Advancement of Science.

D

Glossary

Airtightness: Resistance to inward or outward air leakage through unintentional leakage in the building envelope. This air leakage is driven by differential pressures across the building envelope due to the combined effects of indoor–outdoor temperature differences, external wind, and mechanical system operation (adapted from Guyot et al., 2010).

Bacteria: Microscopic, single-celled organisms that have some biochemical and structural features different from those of animal and plant cells (IOM, 2014).

Biofilm: A thin, normally resistant, layer of microorganisms such as bacteria that forms on and coats various surfaces.[1]

Birth cohort study: An observational study that begins at or before birth of the subjects and continues to study the same individuals at later time points, typically on more than one occasion (Wadsworth, 2005).

Building envelope: The collective name given to the physical separators between the interior and exterior of a building, comprising such components as walls, floors, roofs, windows, skylights, and doors (adapted from Sherman, 2009).

---

[1] See https://www.merriam-webster.com/dictionary/biofilm (accessed July 26, 2017).

Commensal organism: An organism that derives benefits from its association with humans without causing harm.

Commensalism: Two (or more) species coexist, one deriving benefit from the relationship without harm or obvious benefit to the other (IOM, 2014).

Commissioning: The process of ensuring that systems are designed, installed, functionally tested, and capable of being operated and maintained according to the owner's operational needs (DOE, 1999, p. 9).

DNA sequencing: Determining the order of nucleotides in DNA (IOM, 2014).

Dose-response study: In the context of this report, a study that would use controlled delivery of known doses of stressors (pathogens or microbial products) to test organisms (likely animals) in order to deduce the relationship between exposed dose and the likelihood and severity of responses.

Ecology: The scientific study of the relationship between living things and their environments (IOM, 2014).

Endotoxin: A class of lipopolysaccharide-protein complexes that are an integral part of the outer membrane of Gram-negative bacteria (NRC, 2002).

Epidemiology/epidemiologic study: The study of the distribution and determinants of health-related states or events (including disease), and the application of this study to the control of diseases and other health problems.[2]

Eukaryotic: One of the three domains of life. The two other domains, bacteria and archaea, are prokaryotes and lack several features characteristic of eukaryotes (e.g., cells containing a nucleus surrounded by a membrane and with DNA bound together by proteins [histones] into chromosomes). Animals, plants, and fungi are all eukaryotic organisms (IOM, 2014).

Fomite: A surface or other inanimate object onto which a microorganism can deposit and from which it can be transferred to a host.

Genome: The complete set of genetic information in an organism. In bacteria, this includes the chromosome(s) and plasmids (extra-chromosomal DNA molecules that can replicate autonomously within a bacterial cell) (IOM, 2014).

---

[2] See http://www.who.int/topics/epidemiology/en (accessed July 16, 2017).

Genomics: The study of genes and their associated functions (IOM, 2014).

Hybrid ventilation: A ventilation approach that employs both natural and mechanical ventilation systems, potentially using different subsystems at different times of day or seasons of the year (adapted from IEA, 2006).

Infection: The invasion of the body or a part of the body by a pathogenic agent, such as a microorganism or virus. Under favorable conditions, the agent develops or multiplies, with results that may produce injurious effects (adapted from IOM, 2014).

Infiltration: The uncontrolled entry of outdoor air through unintentional openings in the building envelope, which can be driven by indoor–outdoor air pressure differences due to weather and the operation of the building (Persily, 2016).

Mechanical ventilation: The process of moving air into and within a building using ducts and powered fans or blowers, which may include means to filter, cool, heat, humidify, dehumidify, or otherwise condition the air.

Messenger RNA (mRNA): A nucleic acid molecule that is transcribed from DNA and provides instructions to the cell's translational machinery to produce specific proteins (NASEM, 2016).

Metabolome: The census of all metabolites present in any given tissue, space, or sample (adapted from Marchesi and Ravel, 2015).

Metabolomics: Systematic global analysis of nonpeptide small molecules, such as vitamins, sugars, hormones, fatty acids, and other metabolites. It is distinct from traditional analyses that target only individual metabolites or pathways (NASEM, 2016).

Metagenome: The collection of genomes and genes from the members of a microbiota/microbial community (Marchesi and Ravel, 2015).

Metagenomics: A culture-independent method used for functional and sequence-based analysis of total environmental (community) DNA (partial from IOM, 2014).

Metaproteomics: The large-scale characterization of the entire protein complement of environmental or clinical samples at a given point in time (Marchesi and Ravel, 2015).

Metatranscriptomics: The analysis of the suite of expressed RNAs (meta-RNAs) by high-throughput sequencing of the corresponding meta-cDNAs (Marchesi and Ravel, 2015).

Microbe: A microscopic living organism, such as a bacterium, fungus, protozoan, or virus (IOM, 2014).

Microbial community/microbiota: A collection of microorganisms existing in the same place at the same time (adapted from IOM, 2014).

Microbial volatile organic compound (MVOC): A volatile organic compound produced by microorganisms.

Microbiome: Refers to the entire habitat, including the microorganisms (bacteria, archaea, lower and higher eurkaryotes, and viruses), their genomes (i.e., genes), and the surrounding environmental conditions. The microbiome is characterized by the application of one or a combination of metagenomics, metatranscriptomics, and metaproteomics together with clinical or environmental metadata (adapted from Marchesi and Ravel, 2015).

Natural ventilation: The entry of outdoor air through intentional openings in the building envelope, such as windows, doors, and vents, driven by indoor–outdoor air pressure differences due to weather and the operation of the building.

Nonpathogenic: Refers to an organism or other agent that does not cause disease (adapted from Alberts et al., 2002).

Operational taxonomic unit (OTU): The taxonomic level of sampling selected by the user to be used in a study, such as individuals, populations, species, genera, or bacterial strains (IOM, 2014).

Outdoor ventilation rate: The flow rate of outside air supplied to an indoor environment (adapted from Persily, 2016).

Outside air: Air that is brought into a building from a source outside the building.

Pathogen/pathogenic: An organism or other agent that causes disease (Alberts et al., 2002).

Penetration factor: The fraction of an outdoor, airborne contaminant that reaches the interior air volume upon passing through the building envelope.

Permissive environment: An environment having suitable conditions such that microorganisms can grow or persist.

PM$_{2.5}$: Particulate matter less than 2.5 microns in diameter (NRC, 2002).

Proteomics: Analysis of the complete complements of proteins. Proteomics includes not only the identification and quantification of proteins but also the determination of their localization, modifications, interactions, and activities (NASEM, 2016).

Relative humidity: The amount of moisture in air compared with the maximal amount the air could contain at the same temperature; expressed as a percentage (NRC, 2002).

Resilience: The rate at which a community recovers to its native structure following a perturbation (IOM, 2014).

Restrictive environment: An environment that lacks suitable conditions or that contains inhibitory substances such that microbial growth does not occur and persistence is reduced.

Semivolatile organic compound (SVOC): An organic compound with a saturation vapor pressure between 10^{-2} and 10^{-8} kPa at 25°C.[3] Such compounds are less volatile and tend to have a higher molecular weight than VOCs.

Shotgun sequencing: Sequencing of a genome that has been fragmented into small pieces (IOM, 2014).

Taxa: A term used to refer to all of the organisms that fall under a particular taxonomic criterion (such as kingdom, phyla, class, order, family, genera, species, or subspecies).

Taxonomic/taxonomy: The systematic classification, identification, and nomenclature of organisms (adapted from Baron, 1996).

Transcriptomics: The study of transcripts, including the number, type, and modification, many of which can impact phenotype (NASEM, 2016).

Transmissibility: The ease with which a microorganism(s) can spread from a source to a host.

---

[3] ASTM D1356, *Standard Terminology Relating to Sampling and Analysis of Atmospheres.*

Transmission: The transfer of a microorganism(s) from a source to a host (adapted from Baron, 1996).

Ventilation: The process of supplying air to or removing air from a space for the purpose of controlling air contaminant levels, humidity, or temperature within the space (ASHRAE Standard 62.1).[4]

Virulence: A quantitative measure of pathogenicity or disease.

Virus: A small infectious agent that can replicate only inside the cells of another organism. Viruses are too small to be seen directly with a light microscope. They infect all types of organisms, from animals and plants to bacteria and archaea (IOM, 2014).

Volatile organic compound (VOC): A compound with a low molecular weight enabling its rapid evaporation into the air, where it can be inhaled. It has been defined as an organic compound with saturation vapor pressure greater than 10^{-2} kPa at 25°C (ASTM D1356).[5]

REFERENCES

Alberts, B., A. Johnson, J. Lewis, M. Raff, K. Roberts, and P. Walter. 2002. *Molecular biology of the cell*, 4th ed. New York: Garland Science.

Baron, S., editor. 1996. *Medical microbiology*, 4th ed. Galveston, TX: University of Texas Medical Branch at Galveston.

DOE (U.S. Department of Energy). 1999. *Building commissioning. The key to quality assurance.* https://www.michigan.gov/documents/CIS_EO_commissioningguide_75698_7.pdf (accessed March 18, 2017).

Guyot, G., R. Carrié, and P. Schild. 2010. Stimulation of good building and ductwork airtightness through EPBD. In *The final recommendations of the ASIEPI project: How to make EPB-regulations more effective? Summary report.* Brussels, Belgium: European Commission. Pp. 63-82. https://ec.europa.eu/energy/intelligent/projects/sites/iee-projects/files/projects/documents/asiepi_access_the_results._en.pdf (accessed May 1, 2017).

IEA (International Energy Agency). 2006. *Technical synthesis report Annex 35. Control strategies for Hybrid ventilation in new and retrofitted office and educational buildings (HYBVENT).* http://www.iea-ebc.org/fileadmin/user_upload/docs/EBC_Annex_35_tsr.pdf (accessed April 18, 2017).

IOM (Institute of Medicine). 2014. *Microbial ecology in states of health and disease: Workshop summary.* Washington, DC: The National Academies Press.

Marchesi, J. R., and J. Ravel. 2015. The vocabulary of microbiome research: A proposal. *Microbiome* 3(1):31.

NASEM (National Academies of Sciences, Engineering, and Medicine). 2016. *Genetically engineered crops: Experiences and prospects.* Washington, DC: The National Academies Press.

---

[4] ASHRAE 62.1-2016, *Ventilation for Acceptable Indoor Air Quality.*
[5] ASTM D1356, *Standard Terminology Relating to Sampling and Analysis of Atmospheres.*

NRC (National Research Council). 2002. *The airliner cabin environment and the health of passengers and crew.* Washington, DC: National Academy Press.
Persily, A. K. 2016. Field measurement of ventilation rates. *Indoor Air* 26(1):97-111.
Sherman, M. H. 2009. Infiltration in ASHRAE's residential ventilation standard. *ASHRAE Transactions* 115:887-896.
Wadsworth, M. E. J. 2005. Birth cohort studies. In *Encyclopedia of biostatistics*, 2nd ed., edited by P. Armitage and T. Colton. Hoboken, NJ: John Wiley & Sons, Ltd.